Parallel Numerical Algorithms

Prentice Hall International Series in Computer Science

C. A. R. Hoare, Series Editor

BACKHOUSE, R. C., *Program Construction and Verification*
BACKHOUSE, R. C., *Syntax of Programming languages: Theory and practice*
DEBAKKER, J. W., *Mathematical Theory of Program Correctness*
BARR, M. and WELLS, C., *Category Theory for Computing Science*
BEN-ARI, M., *Principles of Concurrent and Distributed Programming*
BIRD, R. and WADLER, P., *Introduction to Functional Programming*
BJÖRNER, D. and JONES, C. B., *Formal Specification and Software Development*
BORNAT, R., *Programming from First Principles*
BUSTARD, D., ELDER, J. and WELSH, J., *Concurrent Program Structures*
CLARK, K. L. and McCABE, F.G., *Micro-Prolog: Programming in logic*
CROOKES, D., *Introduction to Programming in Prolog*
DROMEY, R. G., *How to Solve it by Computer*
DUNCAN, E., *Microprocessor Programming and Software Development*
ELDER, J., *Construction of Data Processing Software*
ELLIOTT, R. J. and HOARE, C. A. R. (eds.), *Scientific Applications of Multiprocessors*
GOLDSCHLAGER, L. and LISTER, A., *Computer Science: A modern introduction (2nd edn)*
GORDON, M. J. C., *Programming Language Theory and its Implementation*
HAYES, I. (ed). *Specification Case Studies*
HEHNER, E. C. R., *The Logic of Programming*
HENDERSON, P., *Functional Programming: Application and implementation*
HOARE, C. A. R., *Communicating Sequential Processes*
HOARE, C. A. R., and JONES, C. B. (eds), *Essays in Computing Science*
HOARE, C. A. R., and SHEPHERDSON, J. C. (eds), *Mathematical Logic and Programming Languages*
HUGHES, J. G., *Database Technology: a software engineering approach*
INMOS LTD, *Occam 2 Reference Manual*
JACKSON, M. A., *System Development*
JOHNSTON, H., *Learning to Program*
JONES, C. B., *Systematic Software Development using VDM (2nd edn)*
JONES, C. B. and SHAW, R. C. F. (eds), *Case Studies in Systematic Software Development*
JONES, G., *Programming in occam*
JONES, G. and GOLDSMITH, M., *Programming in occam 2*
JOSEPH, M., PRASAD, V. R. and NATARAJAN, N., *A Multiprocessor Operating System*
KALDEWAIJ, A., *Programming: The Derivation of Algorithms*
KING, P. J. B., *Computer and Communications Systems Performance Modelling*
LEW, A., *Computer Science: A mathematical introduction*
MARTIN, J. J., *Data Types and Data Structures*
MEYER, B., *Introduction to the Theory of Programming Languages*
MEYER, B., *Object-oriented Software Construction*
MILNER, R., *Communication and Concurrency*
MORGAN, C., *Programming from Specifications*
PEYTON JONES, S. L., *The Implementation of Functional Programming Languages*
POMBERGER, G., *Software Engineering and Modula-2*
POTTER, B., SINCLAIR, J. and TILL, D., *An Introduction to Formal Specification and Z*
REYNOLDS, J. C., *The Craft of Programming*
RYDEHEARD, D. E. and BURSTALL, R. M., *Computational Category Theory*
SLOMAN, M. and KRAMER, J., *Distributed Systems and Computer Networks*
SPIVEY, J. M., *The Z Notation: A reference manual*
TENNENT, R. D., *Principles of Programming Languages*
TENNENT, R. D., *Semantics of Programming Languages*
WATT, D. A., *Programming Language Concepts and Paradigms*
WATT, D. A., WICHMANN, B. A. and FINDLAY, W., *ADA: language and methodology*
WELSH, J. and ELDER, J., *Introduction to Modula 2*
WELSH, J. and ELDER, J., *Introduction to Pascal (3rd edn)*
WELSH, J., ELDER, J. and BUSTARD, D., *Sequential Program Structures*
WELSH, J. and HAY, A., *A Model Implementation of Standard Pascal*
WELSH, J. and McKEAG, M., *Structured System Programming*
WIKSTRÖM, Å., *Functional Programming using Standard ML*

Parallel Numerical Algorithms

T. L. Freeman
C. Phillips

Prentice Hall
New York London Toronto Sydney Tokyo Singapore

First published 1992 by
Prentice Hall International (UK) Ltd
Campus 400, Maylands Avenue
Hemel Hempstead
Hertfordshire, HP2 7EZ
A division of
Simon & Schuster International Group

Printed and bound in Great Britain by
Dotesios Limited, Trowbridge, Wiltshire

Library of Congress Cataloging-in-Publication Data

Freeman, Len, 1950–
    Parallel numerical algorithms / Len Freeman, Chris Phillips.
        p.   cm. — (Prentice Hall International series in computer
    science)
    Includes bibliographical references and index.
    ISBN 0-13-651597-5 (pbk.)
    1. Computer algorithms. 2. Parallel computers. I. Phillips,
    Chris, 1950– II. Title. III. Series.
QA76.9.A43F74 1992   92–9556
512′.5′0285435—dc20   CIP

British Library Cataloguing in Publication Data

A catalogue record for this book is
available from the British Library

ISBN 0-13-651597-5

---

1 2 3 4 5 96 95 94 93 92

Dedicated to our wives

Dorothy and Margaret

and our daughters

Helen; Zena, Claire, Nicole and Gael

who probably suffered more than we did

Contents

Preface

The demands of both the scientific/engineering and the commercial communities for ever increasing computing power have led to dramatic improvements in computer architecture. Initial efforts concentrated on achieving high performance on a single processor, but the more recent past has been witness to attempts to harness multiple processors, with massive power being obtained through replication.

For the most part, users of parallel computing systems tend to be those with large mathematical problems to solve, with the demand for power reflecting a desire to obtain results faster and/or more accurately. Unfortunately, the existing numerical algorithms, on which we have come to rely, were developed with a uniprocessor in mind, and the transition from a serial to a parallel environment is, therefore, not straightforward.

The migration of the applications programmer with a scientific or engineering background from a serial to a parallel system forms the focus of this book. As numerical analysts, our approach is to consider those problems which commonly arise in practice, and to investigate the potential for parallelism in their solution. This contrasts with a more abstract approach in which several programming models are developed and applied to a wider variety of problems. Our practical approach is mirrored in the classes of parallel machines we consider, and the languages and tools we exploit.

We begin, in Chapter 1, with an overview of various parallel machines and outline some of the problems inherent in programming such systems. The variety of parallel systems commercially available is diverse and to cover the full range would require a much larger volume than we have produced. Rather, we concentrate on those which are, arguably, more commonly found to be of general use, namely, shared and local memory multiple instruction multiple data (MIMD) machines.

Developments in hardware have been matched by developments in software, and, in Chapter 2, we discuss the various languages and tools which are available for programming in a parallel environment. Since we are mainly concerned with mathematical problems, it is natural that we concentrate on Fortran, but other languages which explicitly support parallelism are also considered. For illustrative purposes we consider a relatively simple, but instructive, problem, namely, forming the dot product of two vectors, and develop parallel codes in these various languages. The chapter is comprehensive in nature; many of the syntactical details of the various languages described may be omitted without serious detriment to an understanding of much of the subsequent material.

Chapter 3 is the first to cover numerical methods, and we begin by looking at a number of linear algebra primitives, possible parallel implementations, and their relevance to block algorithms. As in all subsequent chapters, we give an overview of the underlying mathematics and consider possible parallel implementations on our two chosen classes of machine. Some implementations are extensions to standard approaches on serial machines; others have been developed from the outset with a multiprocessor in mind.

Building on this, Chapter 4 contains a detailed study of what is, perhaps, the most common component of a numerical algorithm, namely the need to solve a system of linear equations. Detailed program development is given in terms of dialects of Fortran which are supported on shared and local memory multiprocessors.

Chapters 5–7 have the same basic format and give an overview of parallel algorithms in various problem areas; linear algebra problems not covered in earlier chapters (Chapter 5); numerical integration, roots of nonlinear equations and optimisation (Chapter 6); and Fourier transforms and ordinary and partial differential equations (Chapter 7). Finally, in Chapter 8, we outline some of the ways in which parallel processing is moving forward.

The material covered by this book will be of relevance to final year undergraduate and postgraduate courses on parallel numerical algorithms given in departments of computer science, mathematics and engineering. It will also be of interest to professionals who wish to exploit parallel architectures, or who wish to discover some of the potential benefits. The only prerequisites are an appreciation of computing in its widest sense and some experience with Fortran 77. In the early chapters on linear algebra the mathematics is introduced fairly gently. From Chapter 5 onwards the approach is somewhat less compromising, although it is our intention that a reader with only a grounding in numerical methods should be able to understand the material.

Each chapter concludes with exercises and a list of further reading. Ideally the reader should have access to an appropriate parallel machine in order fully to solve some of the set problems. However, we realise that this will not always be the case; the reader will have to be content to produce solutions which cannot be checked by implementation, but the experience gained will be no less valuable. No worked solutions are given, but many exercises conclude with a reference to material where a solution, or part-solution, may be found.

University of Manchester   Len Freeman
University of Newcastle upon Tyne   Chris Phillips
January 1992

Acknowledgements

The germ of which this book is the fruition was conceived in the summer of 1990, following sabbatical spells that we both spent at the Centre for Mathematical Software Research, University of Liverpool, a centre for parallel algorithms research. We wish to thank Mike Delves of the Centre for giving us the opportunity to work in such a stimulating environment.

Several colleagues kindly agreed to look at early drafts of the complete manuscript and offered constructive criticisms. The final product has benefited greatly from their efforts. We wish to extend our appreciation to Ian Gladwell of the Southern Methodist University, Dallas, Texas, and Nick Higham of the University of Manchester, along with several anonymous reviewers whom we would personally like to thank, and will do so as soon as we recognise their handwriting.

There are numerous people who, knowingly or otherwise, have contributed to the material of this book in various ways, and to a greater or lesser extent. At the risk of offending those whom we inadvertently leave out, mention of Christopher Baker, David Carlisle, Rod Cook, Sven Hammarling, Terry Hewitt, Des Higham, Gabriel Howard, Dave Kendall and Mike Stoker is in order.

Throughout the preparation of the final manuscript we received considerable help and encouragement from the staff at Prentice Hall. Through Helen Martin, Editorial Director, we wish to express our appreciation of this.

Finally, but not least, we thank our families for their support and understanding during the preparation of this book, and we take this opportunity to apologise for the time spent tapping away at keyboards when we should really have been attending to their needs. They will be the first to benefit from any royalties that may accrue.

CHAPTER 1

Fundamentals

1.1   Natural parallelism

This book is concerned with solving numerical problems on multiprocessor, or parallel, computer systems. Such systems have been developed only in relatively recent times, even taking into account the rather short history of computing. This is, perhaps, surprising given that parallelism is very much a part of our everyday existence. As individuals we exhibit parallelism. The guitar player strums the strings of his instrument with one hand whilst arranging the fingers of the other hand on the neck to select the appropriate notes and chords. The one man band does this whilst a kazoo is played with the mouth, cymbals are crashed between the knees and a drum is hit by means of a pedal operated by the movement of a leg. All of these operations must be properly coordinated if the sound produced is not to be a cacophony.

Much of modern society, with its complex organisational structure, is dependent on the successful operation of parallel processing systems. At the head of democratic governments is some form of executive, often called a cabinet. Members of the cabinet, sometimes called ministers, or secretaries of state, are responsible for individual aspects of government, such as welfare, health, education or defence. Each operates to some extent independently of the others, but regular cabinet meetings are necessary to coordinate activities since the roles of the different departments are inevitably interrelated. An increase in the level of unemployment caused by a reduction in the number of troops resulting from a peace dividend will have an effect on welfare; improvements in the health care of the elderly mean that there are likely to be more mature students entering higher education, and so on. Of course, the actions of the treasury department will affect all the other (spending) departments.

Parallelism of this and other forms is an inherent part of our lives. We meet it at the supermarket checkouts, at toll booths for a bridge or tunnel, at a bank where

1

we take out money or pay bills. A further example of parallelism is a team of bricklayers building the walls of a house. The amount of communication required between individual components varies considerably from one system to another. Most of the time supermarket cashiers are able to operate entirely independently; occasionally they may need to communicate with a neighbour to ascertain the time of day, or to obtain change for a large bank note. The bricklayers, on the other hand, need to coordinate their activities more closely so that the different walls, which they build separately, meet according to some predefined scheme. In all of these systems each component is performing essentially the same type of task. The cashier looks at the items being purchased by a customer one by one and runs the bar codes across the machine which reads them. However, no two baskets are going to contain the same goods (what we might term the data). In addition, special action may be necessary to deal with items reduced in price because they are approaching their sell-by date, or with coupons offering money off selected goods, or with damaged packaging which means that items need to be replaced.

The simple everyday examples of parallel systems which we have outlined suggest some of the difficulties which need to be overcome if the systems are to operate successfully. Usually the system is controlled by some form of manager who is in overall control: a prime minister at the head of the government, a supermarket supervisor responsible for the checkouts, a foreman in charge of the team of bricklayers. It is the responsibility of the manager to ensure that the components of the system work together as a single unit and to maintain the efficient operation of the unit by minimising the amount of time components are idle. Whilst some communication between components may take place without his knowledge, there may be other communication of which he needs to be aware.

All systems have allocated to them certain resources. In the case of the government one set of resources is the money available, referred to as the budget. Whilst this is a global resource (the treasury will decide on an overall spending target) individual departments will have their own local budgets to which they are expected strictly to adhere. They can spend up to their allocation without recourse to the other departments. If they need to spend over their limit then the treasury may require the curtailing of activities elsewhere. Departments will, therefore, wish to keep local accounts detailing their spending to which other departments will not have access unless they specifically request it. For other types of resource, say, records of the population of the country, the information may need to be available to all departments at all times. It is important that the access to such information (marriage, a mother giving birth, etc.) is properly coordinated so that any changes made take place in the correct chronological order.

The systems we have described exhibit a type of parallelism that we might term *process parallelism*. This is the model in which we are most interested throughout this book. We have categorised the system as a set of cooperating components, each of which is fairly complex in nature and performs in a manner that may well be similar to that of other members of the system, although this is not essential. However, parallelism manifests itself in other ways:

- Components of a system may all obey simple instructions synchronously as dictated by a leader. At an aerobics class physical exercises may be put to music, with the physical training instructor calling out instructions to stretch arms up, to touch toes, etc. Here the only communication is between the instructor and the members of the class, all of whom perform the same operations on their arms, legs, etc. (the data) at the same time (or, at least, in theory they do). We refer to this as *array parallelism*.

- It would be possible to operate a car assembly plant using process parallelism but if a component of the system (a human being or robot) is to be responsible for the complete assembly of (the components of) a car then it needs to be capable of performing a wide variety of tasks (spot welding, engine fitting, etc.). This is not feasible and the usual model is, therefore, one of a *pipeline*, in which each functioning unit of the system is responsible for one particular role only. The part-assembled cars (the data) arrive at a stage in the assembly line one by one. If this stage is responsible for fitting the suspension, the body work and painting will already have been completed but the wheels will be fitted later. The pipeline is kept fully operational; for each fully-assembled car leaving the assembly line there is a new body shell entering.

Some systems exhibit more than one type of parallelism, for example, a large car plant may have a number of assembly lines, each responsible for the same, or different, models.

So, if parallel systems are such an important part of our everyday lives, why is it that early computer systems followed a purely sequential mode of operation? The answer lies in the additional complications at which we have already hinted. A parallel computer system requires some form of synchronisation and/or communication if it is to behave effectively. This places additional burdens on both (hardware and software) systems designers and programmers. However, the limitations of uniprocessors are being increasingly realised and it has become clear that the additional performance that users require can be achieved at relatively low cost by taking the multiprocessor route. Such reasoning motivates the computer systems we describe and exploit in this book.

1.2 Evolution of parallel systems

We have already indicated that the emphasis of this book is on a study of the development and implementation of numerical algorithms on parallel computers. Such a study lacks motivation without some prior consideration of the development of parallel machines. It is necessary to know something about the hardware before the software can be written. The situation is a throwback to the early days of uniprocessors when all programs were written in machine code. Because the architecture of parallel systems is not the central theme of the book, our consideration of this development is at an introductory level only. For a more detailed view of parallel systems and, in particular, their evolution consult the appropriate texts listed in the further reading section at the end of this chapter.

1.2.1 Development of parallel computers

Until relatively recently the standard architecture model for most digital computers was that introduced by von Neumann (see Aspray and Burks, 1987, for a collection of papers relating to von Neumann's pioneering work). The von Neumann model assumes that the program and data are held in the store of the machine and that a central processing unit (CPU) fetches instructions from the store and executes them. The instructions result in either data in the store being manipulated or information being input or output. Machines based on this model are entirely sequential in operation – one instruction is executed in each time interval.

The first general-purpose electronic digital computer, called ENIAC, was developed at the University of Pennsylvania, USA, in 1946. Programming of the machine was achieved by special wiring, and rewiring was necessary if the operations were to be modified. The first stored-program computers, and thus the first realisations of the von Neumann model, were the EDSAC (Wilkes and Renwick, 1949), and the EDVAC (von Neumann, 1945; 1987) prototype machines, leading to commercial systems such as the UNIVAC 1. These early machines used bit-serial access to memory and consequently performed bit-serial arithmetic. By 1953 the IBM 701, the first commercial computer with bit-parallel access to memory and bit-parallel arithmetic, was available; thus from the earliest days parallelism was introduced into digital computers.

The four decades that have elapsed since the development of the first commercial systems have been witness to dramatic improvements in computer technology. Between 1950 and 1980 every five-year period has seen an approximately tenfold increase in the achievable performance of computers (Hockney and Jesshope, 1988, p. 3). These improvements were partly the result of advances in semiconductor technology, but also arose from an evolution of the original von Neumann model in which the requirement of purely serial processing was abandoned.

The development of parallel computers has been at two levels, both of which have attempted to enhance the performance of a particular class of machine. Early efforts centred on the development of high performance *supercomputers* to solve very large scientific problems such as are encountered in accurate weather forecasting. The most popular supercomputer of the mid-1970s was the Cray-1, a *vector processor* capable of 130 Mflops (*megaflops*, or millions of *flops*, floating-point operations per second). The eight-processor Cray Y-MP and the four-processor Cray-2 of the late 1980s are capable of Gflops (*gigaflops*, or thousands of Mflops) performance. These machines exhibit limited parallelism (as far as the user is concerned) with just a small number of powerful processing units operating in parallel. The race is on for the development of a Tflops (*teraflops*, or millions of Mflops) machine, something which is likely to be realised within the next few years. It is probable that such a machine will exploit more, perhaps considerably more, parallelism than the earlier supercomputers.

The second strand of parallel computer development has centred around the desire to produce machines (*minisupercomputers*, or *superminicomputers*) which are capable of performances approaching that of a supercomputer, but at considerably reduced cost. These machines achieve their performance either by using vector processing capabilities, or by including a number of parallel processing units (when we refer to a *multiprocessor system*), or both. Thus parallelism in computing is not only present in supercomputers, but also increasingly common in less powerful machines.

It is instructive to relate these developments in computer architecture to the computer solution of numerical problems. Throughout the decades 1950–1980 the architectural developments were, for the most part, invisible to the user. Programs which worked optimally on one machine were likely also to work well on some other system. The only hardware feature of which the programmer might need to be aware was the available memory. Virtual memory aided portability; programs simply had to minimise the amount of data transfer between main and secondary storage. The development of units possessing vector processing capabilities meant that the way a user chose to express his algorithm could have a significant effect on the performance of an implementation. By making the vector structure of the algorithm visible to the compiler significant improvements could be obtained over a corresponding code for which this structure was not apparent. The impact of the computer architecture on the user is considerably greater in the case of a multiprocessor system and is compounded if its memory is distributed over the individual processors. Such architectural considerations crucially affect the choice of algorithm and the way that an algorithm is expressed. It is this aspect of parallel processing that we take as our principal theme.

The demand for increased performance may result from a desire

- to decrease the execution time of certain programs so that results can be obtained in a reasonable time (possibly even in real time), or
- to run programs for the same amount of time but obtain higher accuracy or more accurate modelling of the underlying physical problem by increasing the problem size in some way, or
- to solve problems previously considered too large for existing architectures.

In the history of computing it has been largely possible to achieve these requirements by improved technology aimed at a uniprocessor. However, there are physical limits to just how far such a machine can be improved and the use of parallelism is seen to be one of the most attractive avenues to explore in the quest for increased performance. In Section 1.2.2 we introduce some terminology with which we can attempt to categorise parallel systems, and in Sections 1.2.3–1.2.5 we give a brief overview of some of the techniques employed by computer architects in the design of such systems. We describe three types of parallel system in what is loosely a reverse chronological order of the introduction of the first machine of the given type, considering first multiprocessors, then array processors and finally vector processors. This approach has been adopted to reflect the aims of this book. We make the assumption that the reader is familiar with the basic terminology of computer systems. The section is concluded with an overview of languages which the programmer can use to exploit the architecture (Section 1.2.6), and some ideas on how any achieved performance improvements can be measured (Section 1.2.7).

1.2.2   Flynn's taxonomy

One of the standard ways of classifying computer systems is that proposed by Flynn (1972). Machines are categorised according to the way that data and instructions are related. According to the Flynn classification there are four basic types:

- *Single instruction stream – single data stream* (SISD)
  The von Neumann model.
- *Single instruction stream – multiple data stream* (SIMD)
  Includes machines supporting array parallelism.
- *Multiple instruction stream – single data stream* (MISD)
  No systems have been built which fit this classification.
- *Multiple instruction stream – multiple data stream* (MIMD)
  Covers the multiprocessor systems supporting process parallelism on which we concentrate.

Whilst the categorisation itself is reasonably well defined, many modern systems are hybrids and there are arguments over the classification of certain systems. Nevertheless, it provides a useful starting point.

1.2.3   Multiprocessors

Multiprocessor systems consist of a number of interconnected processors each of which is capable of performing complex tasks independently of the others. According to Flynn's classification multiprocessors are MIMD machines. An individual processor, or node, may be a scalar or vector processor, or even a multiprocessor. Vector processors are specialised machines that are specifically

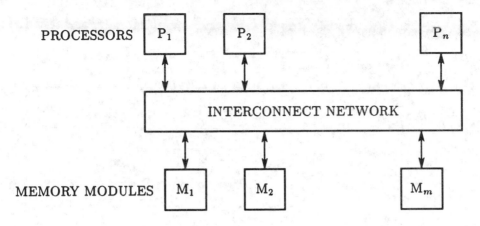

Figure 1.1 *Shared memory system*

targeted at certain types of problem, namely, those which involve a considerable amount of vector arithmetic. If a given problem does not exhibit such a property then the vector processing capability will be largely wasted. In this sense multiprocessors constructed of scalar units only are more general-purpose.

We distinguish between

- *shared memory systems*, in which the main memory is a global resource available to all processors, and
- *local memory systems*, in which the only main memory is that possessed by each processor.

These situations are illustrated in Figures 1.1 and 1.2. Of particular interest here are the Encore Multimax and the Intel Hypercube, shared and local memory systems respectively. In the case of the Multimax each processor has scalar characteristics only. For the Hypercube, processors may or may not possess vector capabilities depending on the particular model. A detailed study of multiprocessor systems, and the Multimax and Hypercube in particular, is deferred until Section 1.3.

1.2.4 Array processors

An array processor (or *data parallel machine*) is of SIMD type and consists of a number of identical and relatively elementary processors connected together to form a grid, which is often square. Each active processor (some may be idle) obeys the same instruction, in lockstep fashion, issued by a single control unit but on different, local, data. Because there is a single control unit there is a single program which operates on all the data at once, and thus there are no difficulties with synchronisation of the processors. Array processors are particularly well

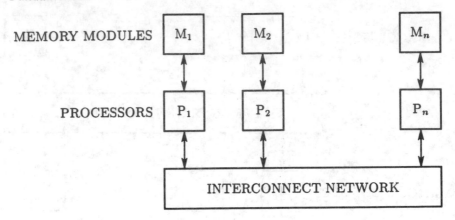

Figure 1.2 *Local memory system*

suited to computations involving matrix manipulations. Examples of array processors are the Active Memory Technology (AMT) Distributed Array Processor (DAP), Thinking Machines' Connection Machine, and the MasPar MP-1. Note that MIMD machines can behave in SIMD mode but this is not likely to lead to economical use of the resources available.

1.2.5   Pipelined vector processors

Many of the early attempts at exploiting parallelism in computer architecture were based on the use of pipelining. The analogy of the car assembly line given in Section 1.1 describes the situation well. The ATLAS computer, designed and built at the University of Manchester, UK, in the late 1950s, was one of the earliest computers to make use of pipelining. An instruction execution was divided into four phases (instruction fetch, operand address calculation, operand fetch, and arithmetic operation), and these phases could be pipelined, so that, in the ideal case, phase 4 of instruction 1, phase 3 of instruction 2, phase 2 of instruction 3 and phase 1 of instruction 4 could all be performed concurrently. To make good use of these facilities it was necessary to have some sort of *lookahead* in the instruction stream to identify which instructions could be pipelined without affecting the intended logic of the program.

Vector pipelines are simply the extension of the idea of pipelining to vector instructions with vector operands. The reason why vector operations are well suited to pipelining is that they form a predictable, regular, sequence of scalar operations on the components of the vectors. As a simple example, suppose that we wish to form the vector sum $c \leftarrow a+b$ and that the addition of two scalars can be divided into three stages, each requiring one clock cycle to execute. Then,

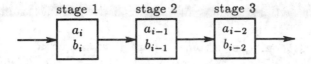

Figure 1.3 *Status of the pipeline on the ith clock cycle*

- on the first clock cycle
    - a_1 and b_1 are in stage 1 of the addition,
- on the second clock cycle
    - a_1 and b_1 are in stage 2 of the addition, and
    - a_2 and b_2 are in stage 1, and
- on the third clock cycle
    - a_1 and b_1 are in stage 3 of the addition,
    - a_2 and b_2 are in stage 2, and
    - a_3 and b_3 are in stage 1.

Thereafter all three stages of the pipeline are kept busy until the final components of **a** and **b** enter the pipe. At the end of the ith clock cycle, illustrated in Figure 1.3, the result $c_{i-2} \leftarrow a_{i-2} + b_{i-2}$ is produced.

A vector pipeline processor is an implementation, in hardware, of the pipelining just described. Vector pipelining forms the basis of many of today's vector supercomputers, such as the Cray family and the Japanese (Fujitsu, Hitachi and NEC) supercomputers. The Cray supercomputers have several processors and a number of pipeline units per processor, and hence they support both process and pipeline parallelism.

The classification of vector processors is a matter of some debate. For example, the Cray-1, which is a pipeline vector processor, is classified as an SIMD machine by Hockney and Jesshope (1988), and as an SISD machine by Hwang and Briggs (1984). Later vector processors, such as the Cray Y-MP and the Cray-2, which are multiprocessors, fit comfortably into the MIMD classification.

1.2.6 Programming languages

As computer systems developed, so did the means of exploiting them. The earliest machines had to be programmed in machine code, but it soon became clear that some higher level of abstraction was necessary. Since most of the early applications of computers were concerned with the solution of scientific problems, the principles underlying the first computer languages reflected the need to express computations defined by mathematical formulae. Hence the development of Fortran (*Formula translation*), Algol (*Algorithmic language*) and their derivatives. The facilities provided by these languages closely reflect the von Neumann model; a Fortran program consists of a sequence of instructions which the program writer expects to be obeyed in sequence unless he explicitly includes a jump instruction to some other part of the code.

More recent developments in computer architecture, particularly those associated with parallel systems, have, in turn, led to language developments designed to exploit the new facilities. This has been accompanied by the development of software tools designed to assist the transfer of a sequential code to a parallel environment.

There are very few programming languages in common use that explicitly support parallelism. Rather, extensions have been made to existing languages, such as Fortran and Pascal (and even Basic), to provide appropriate facilities. In the next chapter we consider in some detail two languages which explicitly support concurrency, namely, Ada and occam, and also extensions to Fortran which are appropriate for multiprocessor systems. If near-optimal performance is to be achieved then it is incumbent on the programmer to make full use of these facilities explicitly. Some compilers have the capability of automatically generating parallel object code from a sequential source (what is termed *autoparallelisation*). This is usually achieved by a pre-processor in one of two ways; parallel constructs are used to replace sequential constructs, or compiler directives are inserted to indicate the potential for concurrency. Whilst, in principle, we need not be aware of the fundamentals underlying the autoparallelisation process, we can certainly help by carefully writing the sequential source in an appropriate manner.

In the case of array processors, languages such as CM Fortran and DAP Fortran have been developed to permit exploitation of the underlying architecture. These languages contain array extensions to the Fortran 77 standard which have subsequently been incorporated into Fortran 90. For vector processors we can use a standard language, such as Fortran, and rely on a compiler to *vectorise* the code, that is, produce object code which makes use of the hardware pipelines. Nevertheless, as with autoparallelisation, we can again help the compiler by careful preparation of the original Fortran source.

1.2.7   Performance measurements

To compare the performance of various computer systems some means of measuring their speed is clearly required. We have already mentioned Mflops as a means of quantifying the speed of a floating-point unit. An alternative performance measure frequently employed is mips (millions of instructions per second). It should be noted that the conversion between Mflops and mips is machine-dependent; on a scalar processor a floating-point operation will typically involve between two and five instructions.

Manufacturers usually quote performance figures which refer to

- *peak speed*, calculated by determining the number of instructions (or flops) that theoretically could be performed in the cycle time of the processor, and

- *sustained speed*, a measure of the speed likely to be achieved for a 'typical' application.

The peak and sustained figures may be in terms of the system as a whole, or of individual components of the system. The former represents an upper bound on the performance of the system, whilst the latter is more likely to give a useful measure of actual system performance; it may be a fraction only of the theoretical peak speed.

In addition to the above, standard sets of software are frequently employed to provide *benchmarks* for computer system performance:

- *Whetstones*

  The Whetstone benchmark is a 'typical' program containing a rich variety of floating-point operations, procedure calls, array indexing and transcendental functions (Curnow and Wichman, 1976). Computer manufacturers will quote performance figures in terms of the number of Whetstones per second.

- *Dhrystones*

  The Dhrystone benchmark is also a 'typical' program, but with no floating-point arithmetic (Weicker, 1984).

- *Livermore loops*

  This is a set of 24 Fortran loops from actual production codes at the Lawrence Livermore National Laboratory. They are claimed to be typical of those present in computationally intensive scientific software (Feo, 1988). In particular, they make reference to components of two- and three-dimensional arrays. Manufacturers will quote Mflops figures for a particular loop in the set.

- *LINPACK*

  LINPACK is a collection of Fortran routines for solving systems of linear algebraic equations and least squares problems (Dongarra, Bunch, *et al.*, 1979). In particular, LINPACK includes the Level 1 BLAS (*B*asic *L*inear *A*lgebra *S*ubprograms, Section 3.2), a set of 'building blocks' from which many of the higher-level routines in the collection have been constructed. The figures usually quoted are the Mflops rates achieved in using the LINPACK routines SGEFA and SGESL in the solution of a 100×100 system of linear equations (Dongarra, 1988). The LINPACK benchmark figures which we quote in Section 1.3 for various computer systems are the *towards peak performance* figures from the report by Dongarra (1991a).

- *Level 1 BLAS*

  Manufacturers will often quote Mflops figures for a particular Level 1 BLAS routine, such as `saxpy` (single precision addition of a multiple of one vector to another).

In addition to quoting performance rates for the processing units of a system, computer manufacturers will also give figures for memory capacities in terms of kbytes (*kilobytes*, thousands of bytes), Mbytes (*megabytes*, millions of bytes), Gbytes (*gigabytes*, thousands of megabytes), or Mwords (*megawords*, millions of words, with each word typically consisting of four or eight bytes). They will additionally quote figures for the speed of transfer of data. This data communication may take place on the processor itself, or between the processor and some external device, such as a disk drive or another processor. In either case the figures quoted will be in terms of Mbytes/s (megabytes per second) or Gbytes/s (gigabytes per second), and this is referred to as the *bandwidth* of the communication device.

Besides the main memory, many modern processors possess a relatively small, high access speed, memory known as a *cache* which acts as a buffer between the processor and main memory. We can expect, therefore, manufacturers to quote figures relating to the capacities of both main and cache memory, and the transfer rate between the two.

Performance measurements give an indication of overall speed only. Whether one particular system is 'better' than another will almost certainly be dependent on a given application. Thus it is difficult, if not impossible, to make comparisons between systems which are universally valid.

1.3 Multiprocessor systems

By multiprocessor systems we mean systems consisting of a number of separate processing elements which are in some way connected. The distinction between shared and local memory machines has already been noted and we consider both types of system in some detail in this section. Whilst historically the division is well defined from a hardware point of view, many modern shared memory systems are constructed from processing elements which have sizeable caches. As such they exhibit some of the properties of local memory systems. Nevertheless the shared/local memory distinction from a software point of view remains valid. These factors influence algorithm development in subsequent chapters; they are considered further, in a more general context, in Section 8.3.

First we outline some of the mechanisms that are used to connect the processors to each other, or to connect the processors to memory.

1.3.1 Interconnection

The physical connection between processors and/or memory modules may take the form of a common bus, or the processors themselves may have communications capabilities which allow them to be directly connected together. Alternatively, some sort of switching mechanism may be employed.

A *crossbar switch* is one which is able to connect a number of inputs with a number of outputs according to specified permitted combinations. For example, a 2 × 2 crossbar switch has two inputs and two outputs (Figure 1.4). The switch

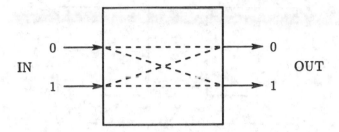

Figure 1.4 *A 2 × 2 crossbar switch*

will permit input i ($i = 0$ or 1) to be connected to output i, and also input i to be connected to output $(i + 1)$ mod 2. Further, a *broadcast* can be achieved by permitting either input to be connected to both outputs simultaneously.

By increasing the number, N, of inputs and outputs the number of possible interconnections grows rapidly. Building a single switch for large N can be very expensive, and a (current, in 1991) practical limit is around 2^7. An alternative approach is to let the number of inputs/outputs be large, but to limit the permitted available connections; see Hockney and Jesshope (1988) for details.

Another way of implementing an interconnection network is to construct a *multistage switch* in which each component is a crossbar switch. For example, four inputs can be connected to four outputs via a two-stage system in which the switches at each stage are 2×2 crossbars (Figure 1.5).

1.3.2   Shared memory systems

The building blocks for a shared memory system are processing elements, memory modules and some interconnect system linking processors to memory. Given this basic model there remains plenty of scope for variation. For example, each processing element may have scalar or vector processing capabilities; the interconnect network may be a crossbar or multistage switch, or a bus. We consider here a small number of examples of the more common systems. In particular we concentrate on the architecture of the Encore Multimax, since we exploit its facilities in some detail elsewhere in the book. Despite this specialisation, the Multimax provides a suitable model for a wide range of parallel machines.

Vector supercomputers
Foremost in the development of vector supercomputers is Cray Research Incorporated. The Cray-2 (1984/5) is a four-processor vector machine with 256 Mwords memory and a LINPACK benchmark performance of 1.4 Gflops. The Cray Y-MP (1987/8) is also a multiprocessor vector machine, with up to eight processors and a LINPACK benchmark performance of 2.1 Gflops.

In terms of ultimate performance the most powerful commercial machines at

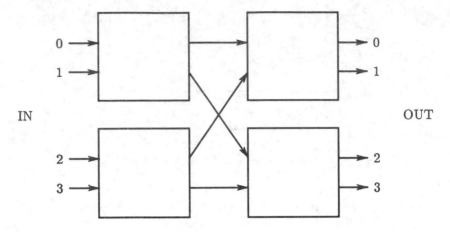

Figure 1.5 *A 4 × 4 multistage switch composed of 2 × 2 crossbars*

present are the Japanese vector supercomputers. The NEC SX-3/14 and the Fujitsu VP2600/10, with processor cycle times of 2.9 and 3.2 ns (*nanoseconds*) respectively, have LINPACK benchmarks of over 4 Gflops. (A nanosecond is one thousandth of a *microsecond*, which, in turn, is one millionth of a second.)

In 1991 Convex launched its C3800 supercomputer with gallium arsenide technology, up to eight processors, and 4 Gbytes of memory. With a one-processor LINPACK benchmark figure of 0.1 Gflop, the eight-processor LINPACK benchmark should be approaching a Gflop.

In the minisupercomputer league is the Alliant FX/80 (1987) possessing eight processing elements, each with vector processing capabilities. The processing elements are connected via a crossbar switch to a 512 kbyte cache with a bandwidth of 376 Mbytes/s. Communication between the caches and main memory (up to 256 Mbytes) is via two 188 Mbytes/s buses. Peak performance is 188 Mflops and the LINPACK benchmark figure is 69 Mflops.

Symmetric shared memory multiprocessors

The most common commercial shared memory MIMD systems are those based on attaching a number of standard scalar processors and memory modules to a bus. Of these, perhaps the best known are the Sequent Symmetry machines. A similar architecture is employed in the Encore Multimax which is described in detail here.

The central core of the Multimax is a nanobus supporting a bandwidth of 100 Mbytes/s. Any of the following four card types can be plugged into this bus:

- *System control card*
  Each system has a single card of this type which provides system control and nanobus coordination.
- *Dual processor card*

A system can have between one and ten of these cards. Initial systems were based on National Semiconductor 32032 processors possessing 8 kbytes of cache, each card capable of 1.5 mips.

- *Shared memory card*
  A system can have between one and eight cards, with each card having 4 or 16 Mbytes of memory.
- *Ethernet/mass storage card*
  A system can have between one and ten cards, each providing interfaces to Ethernet and to mass storage device controllers.

It should be noted that the Multimax is a truly symmetric multiprocessor; there is no master/slave arrangement. Nevertheless, the parallel programming paradigm is of a parent spawning child processes. Further, there are no terminal/printer cards attached to the nanobus; Ethernet-based terminal servers (Annex) are used as distributed front-end processors to support multiple users (up to 16 per Annex).

The Newcastle set-up

As an example of a shared memory computing facility we consider that installed at the University of Newcastle upon Tyne, UK, which houses the Centre for Multiprocessors, part-funded by the UK government's Department of Trade and Industry (DTI). The current equipment consists of two Encore Multimax machines, each running UMAX (Encore's version of UNIX):

- *Encore Multimax 320*
  The system has eight processing elements (NS 32332) and a total main memory of 48 Mbytes. Each processing element is rated at 2 mips and has 32 kbytes of *write back cache*, that is, as soon as an entry in the cache is modified the corresponding value in main memory is updated.
- *Encore Multimax 520*
  The system has fourteen processing elements (NS 32532) and a total main memory of 96 Mbytes. Each processing element is rated at 8.5 mips and has 256 kbytes of *write deferred cache*, that is, if an entry in the cache is modified the corresponding value in main memory is updated only when it is required elsewhere. The use of write deferred cache helps to reduce the amount of bus traffic.

In principle, therefore, the 320 is capable of 16 mips, whilst the 520 can perform at 119 mips.

The 320 is primarily used as a machine to support student teaching. Nearly all programs executed do not explicitly exploit parallelism; they are usually sequential Pascal programs. The user, who sees just a standard UNIX environment, is normally oblivious of the multiprocessor architecture.

The 520 is mainly employed as a research machine but is again available as a multi-user resource. Languages supported are C, Pascal and Fortran 77; Ada is also provided by Encore. Programs written in these languages can dynamically spawn concurrent tasks and, except for Fortran, the number of these tasks is not limited to the physical number of processors available. A multitasking library groups tasks into processes, and processes are allocated to processors by the system software. Process allocation is entirely outside the control of the user. He can indicate how many processors are required but exclusive use of those processors cannot be guaranteed; he must compete for resources with other users. For this reason, timing the performance of parallel programs can be difficult and needs to be done during periods of low activity. Even then, the chances of getting the same elapsed time for two separate runs of a program are remote due to system requirements, particularly if the user tries to employ all, or nearly all, of the processors available.

1.3.3   Local memory systems

The building blocks for a local memory system are multiple processing elements, each with its own substantial amount of main memory. Again, there is plenty of scope for variation, not least due to the different types of interconnect system which may be employed. Again, we give examples only.

Hypercubes

A simple way to construct a local memory multiprocessor system is to connect together a number of processors and to provide each with communication links with its nearest neighbours. One of the problems with such a naïve approach is that the number of links that information has to pass through to get from one extreme of the system to another can be large, and thus the time taken to communicate such a message can be quite high. (The delay incurred from the initiation of a request for data to the time when the transfer of that data item begins is referred to as *latency*.) One of the design philosophies behind hypercube systems is to minimise this worst-case communication cost.

We define the hypercube of dimension 0 to be a single processor. Hypercubes of higher dimensions are then constructed by taking two copies of the next lower dimension hypercube and joining corresponding nodes. We show in Figure 1.6 a diagrammatic representation of hypercubes of dimensions 0, 1, 2 and 3. It can be seen that the hypercube of dimension i has $p = 2^i$ processing elements, with each node connected to i neighbours, and that the maximum distance (number of edges) between any two nodes is i (that is, $\log_2 p$).

The prototype hypercube machines were the Caltech Cosmic Cube (Seitz, 1985) and the Caltech (JPL) Mark II (Tuazon, Peterson, *et al.*, 1985) which developed into the Intel iPSC/1 Hypercube and the AMETEK S14 Hypercube.

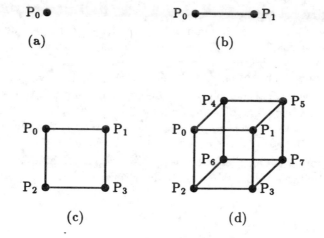

Figure 1.6 *Hypercubes of dimensions (a) 0, (b) 1, (c) 2, (d) 3*

Intel Hypercubes

The iPSC/1 supported up to 128 processing nodes, each consisting of an Intel 80286 processor with an 80287 numeric co-processor. Messages had to be routed from node to node to reach their destination, and thus the hypercube topology of the nodes was important.

The iPSC/1 evolved into the iPSC/2 Hypercube which can similarly have up to 128 nodes, each consisting of an Intel 80386 processor with an 80387 numeric co-processor and up to 16 Mbytes of memory. The co-processor can be replaced by an SX processor which is a high-speed scalar co-processor. A high-speed vector co-processor can also be added, although this limits the memory to 8 Mbytes. As far as the user is concerned the most significant improvement incorporated into the iPSC/2 is the provision, in hardware, of routing facilities, known as the *Direct-Connect Module*. These routing facilities enable a node to send a message to any other node; the message passes through the communications modules of intermediate nodes, but does not disturb those node processors. Essentially the user can treat the machine as an ensemble of fully connected nodes and does not need to refer to the hypercube topology of the nodes. Another significant change is that cube sharing is possible, so that a number of users can each have a part of the hypercube.

The iPSC/860 is similar to the iPSC/2 except that the node processors are Intel i860 processors which achieve high floating-point scalar and vector performance by using advanced pipelining techniques. The memory of each node is 8 Mbytes. Dongarra (1991a) reports a performance figure of 2.4 Gflops for a 128 processor iPSC/860 in the solution of a system of 13 000 linear equations.

Both the iPSC/2 and the iPSC/860 are front-ended by a *System Resource Manager* (SRM), which is an Intel 80386-based workstation. The SRM is used to compile node programs and to control the operation of the hypercube.

The Daresbury set-up

As an example of a local memory computing facility we consider the Intel Hypercube machines installed at the UK government's SERC (Science and Engineering Research Council) Laboratory, Daresbury, UK. Two machines are available, a 32 node iPSC/860 and a 32 node iPSC/2 with dual scalar/vector nodes. Each machine has its own SRM which is connected by a local area network to the other machines at Daresbury, one of which is a Convex C220. One of the functions of the Convex is to act as a remote host for the hypercubes to relieve some of the workload of the SRMs.

The hypercubes are employed exclusively as research machines and exploit Intel cube-sharing facilities to provide a multi-user service. Languages supported are Fortran 77 and C.

Transputer systems

All of the (shared and local memory) multiprocessor systems that we have considered so far are built from standard chips. In contrast, the Inmos *transputer* (*Trans*istor for multicom*puter*) is a chip which was designed specifically as a building block for parallel systems. There are, in fact, several types of transputer available, each possessing

- a central processing unit,
- a random access memory module,
- an external memory interface, and
- four link controllers,

all connected by a common bus (Figure 1.7). Each link controller supports an external link, and it is via these hardware links that transputers can be connected, either directly, or indirectly using, say, a switch network, to give a parallel machine. It should be noted that the link controllers can operate independently of, and concurrently with, the CPU, giving parallelism on the chip. The standard link speed for the first generation of transputers (represented by the T414, T212 and T800, see below) is 10 Mbits/s (megabits, or millions of bits, per second) although links on all currently available versions can also operate at 5 or 20 Mbits/s. Inmos quote a bi-directional data rate for each link of up to 2.35 Mbytes/s.

Currently available transputers include the following:

- *T414*
  Introduced in 1985, the T414 heralded the arrival of the transputer. A 32-bit machine, the T414 is available with 2 kbytes of on-chip memory. The T414-20 can operate at 20 MHz, having an internal clock speed of 50 ns. Inmos claim a peak performance figure of 10 mips for this machine.
- *T212*
  The T212 is similar to the T414, the essential difference being that it is a

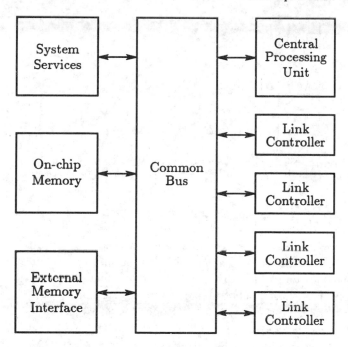

Figure 1.7 *Diagrammatic view of a transputer*

16-bit machine.

- *T800*

  A derivative of the T414 and developed within the European Commission ESPRIT-funded Supernode project (Harp, 1987), the T800 range of chips possess an additional on-chip 64-bit floating-point co-processor which is connected to the common bus and can operate independently of, and concurrently with, all the other units. 20 and 30 MHz versions of the T800 are available, with peak performance figures of 1.5 and 2.25 Mflops respectively.

For a more detailed description of these processors see Inmos (1988). See also McLean and Rowland (1985) for an interesting historical account of the establishment of Inmos.

A typical start-up transputer system consists of a front-end (say, a PC or UNIX workstation) acting as a host with a transputer board plugged into one of the expansion slots. A single, root, T800 transputer with its fast floating-point co-processor can be used to speed up existing sequential programs. Compilers are available in Fortran 77, C and Pascal. However, potentially greater performance is available if an application is written in occam (Section 2.5.1). Other, slave, transputers can be connected to the root transputer and the links configured so as to form a *linear chain* (Figure 1.8). If the last transputer in the chain has a link back to the root transputer then we refer to the topology as a *ring*.

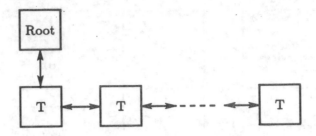

Figure 1.8 *Linear chain of transputers connected to a root*

Because the transputer has four links, other configurations are possible, such as the rectangular grid with wraparound (*torus*) shown in Figure 1.9. Note that, because the torus uses all the available links, an extra transputer is needed to connect the torus to the root. This architecture is similar to that employed in an array processor, although here the individual processing elements are considerably more powerful than those likely to be found in an array processor and can execute different operations concurrently (MIMD operation).

Partly as a result of development work within the ESPRIT-funded PUMA (*P*arallel *U*niversal *M*ultiprocessor *A*rchitecture) project (P2701), Inmos has recently launched the next generation of transputer chips, the T9000. The basic architecture is the same as the T800 series, although an increase in performance of the various units gives peak performance figures of 200 mips and 25 Mflops. The T9000 has an additional unit to support virtual channels for message multiplexing. Hardware throughrouting is provided on the T9000 and on the new C104 packet routing switch. For details of these developments see Inmos (1991).

1.3.4   Shared versus local memory systems

Since we later concentrate on the use of both shared and local memory multiprocessors it is, perhaps, useful to attempt some elementary comparisons between the two types of system:

- *Compilers*
  Compilers for many of the more common programming languages, such as Fortran 77, C and Pascal, with extensions to support parallelism, are readily available for both types of system. Ada compilers supporting true parallelism are also available for certain categories of multiprocessor. Further, compilers that automatically parallelise sequential codes are widely available for shared memory systems. For local memory systems it is necessary to partition not only the code but also the data; autoparallelisation for such systems is inherently more complicated and the science is correspondingly less well developed. Hence the transition from a uniprocessor to a multipro-

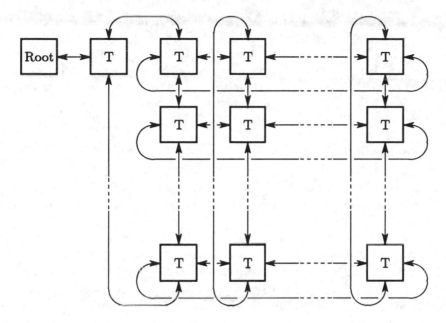

Figure 1.9 *Rectangular grid of transputers*

cessor is easier, from the point of view of the software writer, for a shared memory system, although how effective this transition is will depend on the autoparallelising compilers.

- *Start-up systems*

  The initial cost of a simple local memory system can be quite reasonable, and the cost of an upgrade involving the purchase of a number of additional processors can also be relatively cheap. In comparison the entry cost for shared memory systems is quite high, and upgrades are correspondingly costly.

- *Expansion*

  It will have been observed that the number of processing elements in each of the shared memory systems we have considered is fairly small, certainly considerably less than one would expect to find in a hypercube or transputer array. In compensation each processing element can be expected to be rather powerful. There is, in any case, a physical upper bound on the number of processing elements that can be used which arises from the limitations of the interconnection devices. For example, there are only enough slots on the nanobus of the Encore Multimax to accommodate 20 processors. In addition, if more than 20 processors were connected to the nanobus, system performance could degrade due to the sheer volume of bus traffic. For local memory systems there is, in principle, no theoretical limit to the number of processors that can be employed, although the distance between processors at the extremes of the system may become unacceptably high. We refer to systems built of very large numbers of processors as *massively parallel*.

For shared memory systems the software support and ease of programming are major attractions. Local memory systems win on low unit costs and potentially infinite expansion. Software support for such systems is rapidly improving, and advances are being made to develop local memory systems with interprocessor communications that are sufficiently fast for them to be regarded as shared memory systems. We look at some of these developments in the final chapter.

1.3.5   Processors and processes

In the above we have used 'multiprocessor' as a generic term for a machine which consists of a number of computing units, some form of interconnection and, possibly, a number of separate memory modules. It can be argued that we should apply the term to shared memory systems only and that, more properly, local memory systems should be termed 'multicomputers'. For the rest of this book we ignore such distinctions, but the reader should be aware of them.

More important is the need to differentiate between 'processors' and 'processes'. The former is a functional unit (hardware) whilst the latter is a combination of operations performed on data (software). On a uniprocessor we have only one process executing at any one time, but may have multiple processes time-shared. This can cloud the issue, particularly on a multiprocessor system. For the most part we make the simplifying assumption of a one-to-one correspondence between processors and processes; that is, on a multiprocessor with p processors there will be p processes with one process per processor and no time-sharing. On a local memory system, therefore, interprocess communication will involve messages being sent along interprocessor links.

1.4   Measuring program performance

In Section 1.3 we studied examples of multiprocessor systems and quoted figures indicating the performance that is achievable for certain benchmark programs. There is a danger that we may be lulled into believing that there is an inverse relationship between the execution time of a parallel program and the number of processors employed. In practice this is not usually the case and we now introduce some terminology which helps to explain why this is so.

1.4.1   Grain size

Besides categorising computer systems we can also categorise the inherent parallelism of an algorithm. The usual measure employed is that of *grain size*, which is an attribute of processes rather than processors. It refers to the number of instructions that can be performed concurrently before some form of synchronisation needs to take place. Grain size is a qualitative term but, typically, we might regard the parallelism as being

- *fine-grain,*
- *medium-grain,* or
- *coarse-grain,*

according to whether the number of instructions between synchronisation points is of the order of units, tens, or hundreds plus, respectively. SIMD machines are best suited to fine- to medium-grain parallelism, whereas MIMD machines tend to be more appropriate for medium- to coarse-grain parallelism.

Associated with grain size is the *compute/communication ratio*. With reference to software, this is a measure of the relative importance of arithmetic operations and access to the operands, with the latter involving access to shared memory or interprocess communication as appropriate. A large ratio implies large granularity.

1.4.2 Speed-up and efficiency

Let T_p be the execution time for a parallel program on p processors. We define the following terms:

- S_p, the *speed-up ratio* on p processors, is given by

$$S_p = T_0/T_p,$$

where T_0 is the time for the fastest serial algorithm on a single processor. It should be noted that this definition of speed-up ratio compares a parallel algorithm with the fastest serial algorithm for a given problem. It measures the benefit to be gained by moving the application from a serial machine with one processor to a parallel machine with p identical processors. Even if the same algorithm is employed we normally expect the time taken by the parallel implementation executing on a single processor (T_1) to exceed the time taken by the serial implementation on the same processor (T_0) because of the overheads associated with running parallel processes.

- $\bar{S}_p$, the *algorithmic speed-up ratio* on p processors, is given by

$$\bar{S}_p = T_1/T_p. \tag{1.1}$$

This quantity measures the speed-up to be gained by the parallelisation of a given algorithm. It thus directly measures the effects of synchronisation and communication delays on the performance of a parallel algorithm, and it is this definition of speed-up ratio that we use throughout the book without further qualification. Ideally we would like $\bar{S}_p$ to grow linearly with p, with

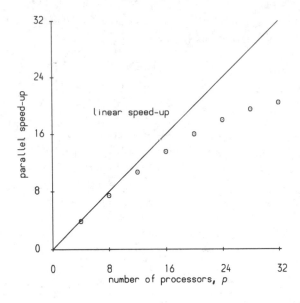

Figure 1.10 *Typical speed-up graph*

unit gradient. Unfortunately, even for a 'good' parallel algorithm we can at best expect the speed-up initially to grow at a close to linear rate, and then eventually to decline as the problem is saturated with processors, and synchronisation overheads, etc., start to dominate (Figure 1.10).

- E_p, the *efficiency* on p processors, is given by

$$E_p = 100 \times \bar{S}_p/p.$$

Note that efficiency is measured as a percentage and that in an ideal situation we would hope for 100% efficiency for all p. In practice we expect efficiency to decrease as we increase p (Figure 1.11).

It should be observed that, as defined here, both speed-up and efficiency are measured in terms of the number of processors, rather than processes, employed.

Speed-up and efficiency are clearly the driving forces behind parallel algorithm development. However, we must not lose sight of other characteristics, such as accuracy and reliability. Maintaining these attributes may well have a detrimental effect on the improvements that are potentially achievable.

1.4.3   Amdahl's law

As we have already remarked, the ideal situation is one in which the speed-up ratio increases linearly with p, with slope 1, giving an efficiency of 100%. Such a

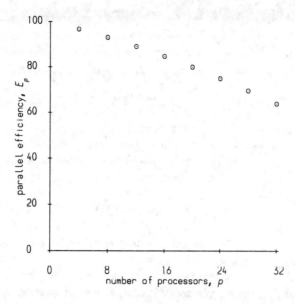

Figure 1.11 *Typical efficiency graph*

situation rarely occurs in practice, partly because of the need to synchronise parallel processes. There is a further consideration which arises from the inescapable fact that there are likely to be portions of a program which are inherently sequential. This result is encapsulated in *Amdahl's law* (Amdahl, 1967).

Amdahl's law: Suppose that r is that fraction of a program which is parallelisable and that $s = 1 - r$ is the remaining inherently sequential fraction. Then on p processors the algorithmic speed-up ratio $\bar{S}_p$ satisfies

$$\bar{S}_p \leq \frac{1}{s + r/p}.$$

The proof of this result is straightforward. We start with the definition of algorithmic speed-up ratio given by (1.1). Now, no matter how many processors we use to solve the problem the time for the sequential part remains constant (sT_1), whilst the best we can expect from the parallel part is for the time to decrease inversely with p and be given by rT_1/p. Hence

$$T_p \geq sT_1 + rT_1/p,$$

and

$$\bar{S}_p \leq \frac{T_1}{sT_1 + rT_1/p} = \frac{1}{s + r/p}.$$

Amdahl's law appears to have serious consequences as far as the parallelisation of algorithms is concerned, since it imposes an upper bound on the speed-up ratio. For example, if only 50% of a program is parallelisable, then $s = r = 1/2$, and

$$\bar{S}_p \le \frac{2}{1 + 1/p},$$

so that as $p \to \infty$, $\bar{S}_p \le 2$. Thus, for this particular program, the speed-up ratio is limited to 2, regardless of the number of processors used.

An alternative view of Amdahl's law, which assumes that the problem size increases with the number of processors, is arguably more appropriate. We assume that the part of the program which can only be executed sequentially, T_s, is fixed and that the part of the program that can be executed in parallel increases with the size of the problem, n, and is given by nT_p, where T_p is fixed. The assumption that the serial time is independent of the problem size is reasonable for some problems. For many other problems the dependence, on n, of the serial and parallel execution times is different, with, for example, the serial time given by nT_s and the parallel time given by $n^2 T_p$. In either case the sequential fraction, s, is given by

$$s = \frac{T_s}{T_s + nT_p},$$

with the parallel fraction r given by

$$r = \frac{nT_p}{T_s + nT_p}.$$

Then

$$\bar{S}_p \le \frac{T_s + nT_p}{T_s + nT_p/p} = \frac{1 + n\tau}{1 + n\tau/p},$$

where $\tau = T_p/T_s$ is fixed. Hence

$$\bar{S}_p \le p\,\frac{1 + n\tau}{p + n\tau},$$

and we conclude that $\bar{S}_p \le p$. Further, for a given choice of p, we can choose n sufficiently large that $n\tau \gg p$ and then $\bar{S}_p \to p$. On the other hand if, for large p, $n\tau = 1$ then $\bar{S}_p \le 2$. This view of Amdahl's law suggests that the reason for moving to a larger parallel machine should be to solve larger examples of a given problem rather than to solve same-sized problems more quickly.

1.4.4   Load balancing and throughput

If we wish to optimise the efficiency of a particular applications program then it is essential to ensure that all processors are doing useful work for as much time as possible. This means that, as far as possible, we must avoid processors being held at synchronisation points waiting for information from some other processor before they are able to proceed. Clearly, in this sense, a sequential program running on a single processor is 100% efficient. The aim is to implement an algorithm in such a way as to employ as many processors as possible whilst at the same time ensuring that all the processors are sufficiently usefully active; we refer to this as *load balancing*. An alternative way of expressing this is to say that for a given amount of time we wish to optimise the *throughput* of the system, that is, the amount of useful work that can be performed.

1.5 Preparing for parallelism

So, having decided that the future really does lie in parallel systems, we persuade some appropriate spending authority to purchase a machine which we believe best suits our needs. To convince the authority that a wise decision has been made the machine must be exploited to the full. For a given problem an appropriate algorithm must be chosen and, if possible, the potential speed-ups that might be achieved should be indicated beforehand. When programming the algorithm it is necessary to be aware of the additional pitfalls that can arise. We conclude this chapter with a brief look at each of these aspects in turn.

1.5.1 Matching the algorithm

The three main factors which affect the implementation of a numerical algorithm on a parallel machine are

- the granularity,
- the data distribution,
- the potential for vectorisation,

and it is arguable that all three are interrelated.

On a shared memory machine the granularity dictates whether or not the algorithm will efficiently exploit the machine. Where possible, we should aim for medium- to coarse-grain algorithms; high operation counts between synchronisation points will help to offset the processor idle time associated with the protected access to global data, a sequential section of code, etc. Another consideration on a number of the more powerful shared memory machines, such as those marketed by Alliant, Convex and Cray, is the potential to exploit the vector architecture of the processors. Thus we seek algorithms which parallelise well and within that parallel structure exhibit substructures which vectorise efficiently. The data distribution of the algorithm is important only when synchronised access to global data, and the way the cache memories of the processors are used, are taken into account.

On a local memory machine both the granularity and the data distribution of the algorithm are important since together they determine the communication costs of the algorithm. We again aim for a medium- to coarse-grain algorithm, with a data distribution which minimises the volume of communication traffic. The potential for vectorisation within the parallel strands of the algorithm becomes increasingly important as the power of the individual processors of the machine increases by, for example, the addition of a vector co-processor.

1.5.2   Timing models

In subsequent chapters of this book we propose a number of parallel algorithms, and it is natural to attempt to assess their performance in some way. One way to proceed is simply to monitor the behaviour of the implementation by gathering data in the form of elapsed timings. If the conclusion then is that the algorithm demonstrates only a limited speed-up (or even slow-down) we have wasted both time and effort.

A potentially better solution is to construct a mathematical *timing model* for the algorithm. That is, having determined the characteristics of the algorithm in terms of the number of arithmetic operations, the synchronisation points, and the data traffic, a mathematical formula, which purports to predict the execution time for the algorithm on some chosen machine, is derived. It is important to parametrise the model so that the variation in performance can be monitored as the problem size changes or the number of processors increases. It will almost certainly be necessary to build into the model a number of simplifying assumptions and 'worst-case' scenarios. It will therefore be necessary to validate the model by comparing predicted performance with that actually observed when the algorithm is implemented.

The construction of a timing model may be a less than trivial exercise, and its worth will depend on how accurately the model reflects the implementation. Nevertheless the time spent on the exercise can be of considerable benefit.

1.5.3   Run-time locks

We have now reached the point of having decided on the parallel machine to be used, the problem to be solved, an appropriate algorithm and a programming language, and are ready to develop our first parallel program. It is just possible that this will execute correctly. (Can you remember whether your first attempt at a sequential program succeeded?) Even if we are successful, at some future stage we will inevitably encounter compilation and run-time errors, and the latter type of error can cause problems an order of magnitude greater than in a serial environment. No longer do we have just one process to monitor, but many.

Further, we have an increased range of run-time errors to guard against. In particular, the program may *lock* for one reason or another.

We use the term *deadlock* to describe what happens when a number of processes are competing for the same resources but none can proceed. Effectively nothing happens; the system just halts. One of the classic examples of deadlock is provided by the dining philosophers' problem (Dijkstra, 1971). Here, a number of philosophers are seated around a table with a common resource of food available to all at the centre of the table. Between any two philosophers is a single fork, but each philosopher requires two forks to collect and eat food. The problem is to devise a scheduling algorithm which ensures that the philosophers can eat in a fair way, assuming that they all possess an equal appetite. It is clear that if each philosopher simultaneously picks up a fork on his right-hand side then none can have two forks, and so none can eat. The system is deadlocked as no further progress can be made.

Clearly deadlock is something which we need to guard against in our programs. Tracing it, when it inadvertently slips through to an implementation, can be a significant problem.

Exercises

1.1. An American football team provides illustrations of both process and array parallelism. Discuss.

1.2. Find out all you can about the architectures of the computer systems you regularly use. What are the memory hierarchies? How much memory is there? What is the data transfer rate? Are the systems networked and, if so, how? If you have access to a parallel system, what is the programming paradigm? What is the interconnection network?

1.3. What mechanism does your favourite language implementation support for program timing? Use it to work out the Mflops rate for a simple program (say, vector-vector addition) on a uniprocessor. Is this a measure of CPU or elapsed time? Is CPU time a useful measure on a multiprocessor?

1.4. Consider the respective merits of a grid and a torus local memory multi-processor topology. Draw a 4×4 array of processors in which the processors at each end of a row (or column) are directly connected to the next processor in the row (or column) and the next but one, and compare the resulting torus with a torus constructed as in Figure 1.9 (Seitz, 1990).

1.5. Consider a set of inputs labelled $0 \rightarrow 7$. A *perfect shuffle* takes the binary representation of each label and performs a unit circular left shift on it (so, for example, 110 becomes 101). Draw a diagram which shows how the input labels are transformed by a perfect shuffle. Draw a three-stage switch (an *omega network*, Lawrie, 1975) in which each stage consists of four 2×2 crossbars and the inputs are shuffled prior to the first stage, and the outputs are shuffled between stages but not after the last stage. Draw the switch system (an *indirect binary n-cube network*, Pease, 1977) which implements the reverse transformation, interchanging the position of the middle two switches in the middle stage, but none of the links.

1.6. Each node of a hypercube of dimension 2 possesses a single integer. It passes its current value to its neighbour within the first dimension, and then across the second dimension in turn (so that, for example, node 0 first passes a value to node 1, and then passes a value to node 2). After every data transfer each node computes the maximum of the two numbers it now possesses, and overwrites its own value with the result. What is the effect after two steps? Extend the idea to a hypercube of dimension 3.

1.7. A ring of transputers can be used to implement a coarse-grain process pipeline. Ignoring communication overheads, if n tasks arrive at a pipe of p processors, and the execution time of each task in each processor is t, determine the efficiency of the pipe.

1.8. Show that by a suitable node ordering a hypercube of dimension 4 can be regarded as (a) a linear chain, and (b) a 4×4 grid. To what depth can a binary tree be represented?

1.9. Draw the speed-up and efficiency graphs for a program in which (a) 5%, (b) 10%, (c) 20%, and (d) 50% of the code is inherently sequential and the rest of the code can be parallelised without overhead.

1.10. In terms of the dining philosopher's problem, what do you think is meant by *process starvation*? (Ben-Ari, 1990.)

Further reading

For comprehensive introductions to parallel systems from a hardware point of view, see Almasi and Gottlieb (1988), Hockney and Jesshope (1988), or Hwang and Briggs (1984) (in no particular order). Trew and Wilson (1991) have compiled an extensive survey of the development of parallel systems. Other books which concentrate on hardware issues of multiprocessors are DeCegama (1989), Stone (1990), and Tabak (1990). The collection of Suaya and Birtwistle (1990) contains a number of chapters which are of particular relevance to the foregoing discussion. The initial chapter by Seitz (1990) includes a historical account of the development of the Intel Hypercube, and the third chapter by Dally (1990) covers networks. For details of early developments in parallel systems, see Satyanarayanan (1980). Lazou (1986) concentrates on early supercomputer development.

Also worth consulting is the collection of Elliott and Hoare (1989), in which Kung (1989) looks at different models of parallelism. Several of the papers (Clementi, 1989; Fox and Furmanski, 1989; Wallace, 1989) deal with the solution of a variety of practical problems on various multiprocessor architectures. The papers by Hey (1989) and May (1989) cover transputer developments.

A list of books aimed more at software development is given at the end of Chapter 2. Several of these look at the principles underlying parallelism. In this category we mention Ben-Ari (1990), Krishnamurthy (1989) and Lakshmivarahan and Dhall (1990).

CHAPTER 2

System support

2.1 Software overview

As numerical problem solvers we have become accustomed to the provision on a uniprocessor of a number of sophisticated tools to aid program development. In a multiprocessor environment we expect to have these and more. At the very least we anticipate

- an operating system,
- high-level programming languages supporting parallelism,
- compilers which generate efficient code from programs written in these languages,
- libraries of numerical software, and
- tools to aid program development – particularly graphical aids for scheduling, debugging and profiling parallel codes.

From the operating system we might expect the capability to allocate processes to processors dynamically. From the programming languages we need to be able to express the parallelism of our problems in a clear and succinct manner. Ideally we would like to have available compilers which will help to identify parallelism inherent in existing sequential codes. If a numerical library is provided which itself exploits parallelism then we may be able to develop a parallel code by writing a sequential main program which simply makes calls to appropriate parallel routines. Finally, the problems of debugging and of understanding the performance of a parallel code should not be underestimated; the situation can be much more difficult to appraise than that obtaining in a serial environment. In particular, two consecutive runs of a given parallel program may not necessarily produce the same results. Simply adding output statements may provide insufficient information to allow the behaviour of the program to be determined. More importantly,

32

the output statements themselves may well affect the behaviour of the program. Hence, any tools which are likely to help in the area of debugging and profiling are extremely welcome.

In this chapter we look at some of these issues in general and indicate in detail the facilities that are likely to be available and how we expect to use them. Discussion of the provision of numerical libraries is left to the next chapter. We first indicate the approach to the development of parallel programs.

2.2   The programming paradigm

In Chapter 1 we were careful to distinguish between shared and local memory systems. This is a traditional way of classifying multiprocessors although, arguably, a better distinction would be to differentiate between systems with high- and low-latency communications. Whilst the difference between shared and local memory systems is fundamental, it is, arguably, not all that great and it is usually technically possible to program a local memory system as if it were a shared memory system, and vice versa. We return to the argument in Chapter 8.

We recall that the basic components of a typical shared memory multiprocessor are

- a number of processing elements, each with a relatively small amount of local memory,
- a number of main memory modules, and
- some appropriate interconnection network.

In principle, and in practice, there is no direct correspondence between a processor and a memory module. Indeed, the numbers of processors and memory modules may be considerably different. There is no need explicitly to pass information from one processor to another since all memory modules may be accessed from all processors (and it is assumed that the cost of a memory access between a processor and a memory module is constant). However, we have to be careful about multiple accesses to the same data, and the programming paradigm is therefore based on the synchronised use of data (*process synchronisation*) using spin locks, barriers, etc.

In contrast a local (or distributed) memory multiprocessor consists of

- a number of processing elements, each with a substantial amount of local memory, and
- some appropriate interconnection network.

Each processor acts independently but can receive data from, or send data to, other processors to which it is directly, or indirectly, connected. Here the programming paradigm is therefore that of *message-passing*.

Whichever programming paradigm is adopted the approach to algorithm design which is likely to lead to the best results is that of attempting to maximise granularity whilst endeavouring to ensure that all processors are kept usefully busy. In a shared memory environment there is the additional problem of attempting to minimise the amount of transfer of data between cache and shared memory; in this sense shared memory systems exhibit local memory system behaviour. The equivalent requirement in a local memory environment is that of minimising the amount of data transfer between processes running on different processors. In either case we will need to synchronise processes so that, on a shared memory machine, there is a correct sequence of writing to or reading from global shared memory, or, on a local memory machine, a data transfer takes place at an appropriate point. We need also to ensure that the wait time of any processes held at such synchronisation points is minimised.

2.3   Language requirements

In the previous section we indicated that there are essential differences which must be borne in mind when designing and implementing algorithms for shared and local memory multiprocessor systems. We recall that the two programming paradigms are

- process synchronisation (shared memory system), and
- message-passing (local memory system),

and that these two paradigms form the basis of many parallel languages. By a 'parallel language' we mean a programming language which is designed to exploit the facilities offered by a multiprocessor architecture. It may have been designed as a parallel language at the outset, or it may be simply a standard sequential language to which parallel facilities have been added. We give some examples of such languages in the next two sections. First we consider the two paradigms individually but in generality.

2.3.1   Process synchronisation

For shared memory systems we require facilities which enable us to

- create (spawn) processes (either statically or dynamically),
- destroy processes, and
- identify processes.

The reader might also expect the need to allocate processes to processors, but it is a feature of many shared memory systems that the placement of processes is outside the control of the programmer; it is left to the system software to perform this task dynamically. Similarly, the programmer is unaware of which memory modules will be used for his job, and how. However, we expect the ability to

- differentiate between
  - variables globally available to all processes, and
  - variables local to a given process.

In addition we expect the ability to

- establish synchronisation points

so that the operations of concurrent processes may be coordinated properly. For example, we might wish to accumulate a sum of values each of which is determined by a separate process. At any one time only one process should have both read and write access to the global variable in which the sum is accumulated. Failure to maintain this synchronised access is likely to lead to an incorrect result being formed. (Try it.)

The synchronisation facilities, which may be implemented in hardware or software, will employ some suitable locking arrangement based on stop and go flags. The utilities that the user can expect to see included are as follows:

- *Spin locks*
  A spin lock can be in one of two states: `locked` or `un_locked`. The status of a spin lock can be altered by any of three basic operations:
  - `spin_lock_init` initialises a spin lock to either the `locked` or the `un_locked` state.
  - `spin_lock` locks the spin lock before proceeding with the process if the lock is in the `un_locked` state, and performs a busy-wait (that is, repeatedly examines the lock until it is in the `un_locked` state) otherwise.
  - `spin_unlock` unlocks a spin lock.
  The use of a spin lock enables a section of code within a process to be protected in the sense that, at any one time, only one process is permitted to enter into a region protected by a given lock. In particular, it enables synchronised (guarded) use to be made of global data, and of a single output device.
- *Events*
  Events can be in the states `posted` or `cleared`. The notation employed is, arguably, a little confusing. The `posted` state is the event equivalent of the `un_locked` state for a lock; the `cleared` event state is equivalent to the `locked` lock state. There are three basic operations which can be performed on an event:
  - `event_post` sets an event to be in the `posted` state.
  - `event_clear` sets an event to be in the `cleared` state.

○ event_wait blocks a process (it performs a busy-wait) until the event is in the posted state, when it proceeds without altering the value of the event. Note that the setting of an event to the posted state does not guarantee that all processes blocked at an event_wait will proceed; the event may be reset to the cleared state before the value of the event is re-checked by a blocked process.

The essential difference between a spin lock and an event is, therefore, that the former permits one process, and only one process, to proceed when the flag is in the un_locked state, whereas the latter can be used to permit a number of processes to continue when the flag is in the posted state.

- *Barriers*
  A barrier can be used to ensure that all processes reach the barrier point before any is allowed to proceed.

A further type of synchronisation facility is the *semaphore*. This is a generalisation of the spin lock which permits groups of processes to enter a protected region at any one time. Typically, the semaphore flag will be an integer value. If a process is waiting on a flag whose value is greater than 1 the process proceeds and the flag is decremented by 1. If the value of the flag is 1 the process is blocked.

2.3.2   Message-passing

From any parallel language designed for use with a local memory system we again expect facilities for process creation, destruction and identification. In addition we require the ability explicitly to communicate between processes. Specifically, we may wish to

- send a message from one process to another, and
- receive a message in one process from another,

where a message involves some transfer of data. For example, we might again wish to accumulate a sum of values each of which is determined by a separate process (cf. Section 2.3.1). Assume that a single process is to form the sum; then all the other processes must communicate their values (pass messages) to that process. Besides these one-to-one message-passing primitives it is likely that we will also require facilities to

- send a message from one process to all other processes (a *broadcast* or *scatter*), and
- receive a message in one process from all others (a *gather*).

We distinguish carefully between two types of communication:

- *Synchronous communication*

  The sending process is held until the corresponding receiving process is ready. Similarly, the receiving process is held until the corresponding sending process is ready.

- *Asynchronous communication*

  The sending process is not held waiting for the receiving process. If the receiving process is not ready then the data to be transferred is put into a buffer until the message transfer can be completed.

Associated with the creation of processes is the placement of each process on a given processor. In a local memory environment this placement is usually undertaken by the programmer, rather than just left to the system software. Hence, additionally, we might expect facilities to

- associate a process with a specified processor, and, possibly,
- associate an interprocess communication link with a physical interprocessor link.

This process-to-processor allocation implicitly maps data requirements to local memory.

When developing parallel programs within a message-passing environment the physical structure of the computer system on which the code is to run may need to be borne in mind, although it is desirable to minimise the dependence of the code on the structure's topology. Communication between processes resident on physically adjacent processors is straightforward and takes place along the link between the processors. Communication between non-adjacent processors may require processes running on intermediate processors to pass the message on. Ideally we would like the need for such throughrouting to be hidden from the programmer, as is the case with the Intel iPSC/2 and the Intel iPSC/860 extensions to Fortran (Section 2.4.4); with throughrouting communications software the programmer can nominate a process to receive the message and then leave it to the system to ensure that the message arrives at the correct destination (and, where appropriate, to return an acknowledgement of the receipt of the message to the sending process).

2.4 Parallel extensions to Fortran

Having isolated the primitives that the user can expect to find in a parallel language the question arises as to whether to use a genuinely parallel language which incorporates these facilities (and which ensures some degree of portability across various systems, assuming that appropriate compilers are available), or to employ a sequential language to which parallel constructs have been added (and which is likely to be less portable). In the former category the obvious candidates are occam and Ada and we consider these languages in the next section, together with C which has facilities for parallel processing.

In this section we consider extensions to sequential languages which tend, by their very nature, to be non-standard, even though they are usually little more than variations on the mechanisms outlined in the previous section. Within the scientific community Fortran reigns supreme, despite the criticisms that it has frequently received. Consequently most multiprocessor systems come complete with their parallel extensions to the Fortran 77 standard. We provide details of some 'typical' extensions, two based on process synchronisation, the other on message-passing. We also mention an attempt to produce a standard for parallel Fortran. First we give a brief summary of some of the basic features of Fortran 77 itself. We assume that the reader has at least a passing knowledge of Fortran 77; for a more detailed account readers should consult one of the many texts available on the language, such as Ellis (1990), or the definitive reference, the American National Standards Institute standard (ANSI, 1978). We also look at the new standard, Fortran 90, with particular reference to the scope for parallelism inherent in its array handling facilities.

2.4.1   Fortran 77

A Fortran 77 program consists of one or more separately compilable units, one of which is a main program, with the others being subroutines, functions or block data. Each unit consists of a declaration part and an executable part. Standard types include `INTEGER`, `REAL`, `DOUBLE PRECISION` and `LOGICAL`, and multi-dimensional arrays of variables of these types may be declared. Whilst it is good programming practice explicitly to declare all variables used, this is not essential. Any undeclared identifier is assumed to have type `INTEGER` if its initial letter is in the range I–N, and type `REAL` otherwise. Values of constants are defined using the `PARAMETER` statement. Identifiers may be at most six alphanumeric characters long, with the first character being a letter. Many compilers accept identifiers longer than six characters with the underscore character employed as a separator; we make use of this extension in many of the codes which follow. Strictly speaking, lower-case alphabetic characters can be used in a Fortran program only as part of a comment line (see below), or as part of a character string used, say, in an output statement. Many compilers permit lower-case letters throughout the program in identifier names, etc., and again, we make use of this common extension. It should be noted that these extensions are case-insensitive. Fortran's history stems from the era of the punched card; statements appear one per line starting from column 7 and are terminated by the end of the line, which occurs after column 72. Continuation of a statement is possible by the inclusion of a non-space character in column 6 of each continuation line. Columns 1 through 5 are reserved for the insertion of numeric labels, with the exception that a C or * appearing in column 1 indicates a comment line.

Control structures in Fortran are somewhat limited; for repetition the language supports the DO loop, which takes the form

```
        DO <label> [,] <integer identifier> = <start>,<finish>[,<step>]
           <statements>
  <label> <the last statement in the loop>
```

where items contained in square brackets are optional. All statements down to, and including, that bearing the corresponding label (an integer in the range 1–99999) are executed for values of the control variable (with name specified by the integer identifier) equal to the start value and, in increments equal to the step, up to, but not beyond, the finish value. The default step is 1. Although not permitted by the standard, many Fortran 77 compilers accept a loop of the form

```
        DO <integer identifier> = <start>,<finish>[,<step>]
           <statements>
        END DO
```

and throughout this book we make use of this extension. (This loop construct is a feature of the new Fortran 90 standard; see Section 2.4.2.)

The IF...THEN...ELSE...END IF statement supports conditional branching and takes the form

```
        IF (<Boolean expression>) THEN
           <statements for THEN branch>
        ELSE
           <statements for ELSE branch>
        END IF
```

with the ELSE branch being optional.

A subroutine is a unit of the form

```
        SUBROUTINE <identifier>(<formal parameters>)
           <declaration part>
           <execution part>
        END
```

in which variables and constants specified in the declaration part are in scope until the end of the unit. Unlike block-structured languages such as Pascal, identifiers declared in the main program are not in scope in the subroutine; they must be passed explicitly either via the parameter list, or via a COMMON block. Hence the danger of inadvertently altering the values of variables in a non-local block can be minimised. A Fortran function has a form similar to that for the subroutine and is given by

```
        FUNCTION <identifier>(<formal parameters>)
           <declaration part>
           <execution part>
        END
```

with the type of the result being defined either in the declaration part, or prior to the word FUNCTION. The execution part must contain at least one statement which assigns a value to the function name.

Unlike many block-structured languages Fortran is not strongly typed. In particular the types of the formal parameters and the actual parameters in a subroutine or function call may not be checked by the compiler. This is a frequent source of program error, but it is also a feature which can be exploited by a programmer who is aware of the consequences. It should be borne in mind that, effectively, all parameter passing is implemented *by reference*. That is, the actual parameter is merely an address in store which will be treated within the routine as the start address of an element of the type as specified in the declaration part of that routine. Where this is particularly important is in the passing of array parameters. If, say, a routine has specification

```
SUBROUTINE BOB_DYLAN(X)
REAL X(*)
```

then, if B is a one-dimensional array in the unit calling this routine, the actual parameter can be B, or B(I) for some value of the integer I. In the former case X and B are, effectively, one and the same during the routine call. In the latter case the value of I need not be the lower bound of B; all that happens is that the start address of X is taken to be the address of B(I).

For two- and higher-dimensional arrays the situation is equivalent but slightly more complicated. First it must be remembered that, unlike most other high-level languages, Fortran stores two-dimensional arrays by columns. Thus, if C is declared as

```
REAL C(3,4)
```

then, beginning at the start address of C, the stored values are in the order C(1,1), C(2,1), C(3,1), C(1,2), C(2,2), etc. Now, we can use C(I,J), for some integer values I and J, as an actual parameter to the routine BOB_DYLAN, in which case the formal parameter X is just a vector of values starting at the address of this matrix element. In particular, if the actual parameter is C(1,J), then X can be regarded as the *j*th column of C.

To permit BOB_DYLAN to regard a row of a matrix as a vector we must include an additional integer parameter which is the *stride* length to be employed. We write the specification as

```
SUBROUTINE BOB_DYLAN(X,INCX)
INTEGER INCX
REAL X(*)
```

For positive INCX we refer inside the routine to the *i*th element of X as X(1+(I-1)*INCX). Now, for X to be the *j*th column of C the actual parameters should be C(1,J) and 1, whilst for X to be the *i*th row of C the actual parameters should be C(I,1) and LDC, where LDC is the declared leading (that is, first) dimension of C (3 in our particular case). (But what if INCX is negative? All is revealed in Section 3.2.1.)

2.4.2 Fortran 90

The Fortran 77 standard has recently undergone a revision, culminating in the specification of the Fortran 90 standard. As with all new versions of Fortran, the 90 standard is designed to be upwards compatible; no features of Fortran 77 have been removed and so any existing Fortran 77 program should perform in exactly the same way whether it is compiled by a Fortran 77 or a Fortran 90 compiler. However, some features (such as the arithmetic IF statement) have been marked as *obsolescent*, that is, recommended for deletion at the next revision of the language.

It should be recognised that Fortran 90 represents a major enhancement to the language. Program layout is more flexible, WHILE loops and user-defined types are supported, there is a more flexible sub-program parameter passing mechanism, and many other features peculiar to the 77 standard outlined in the previous subsection have been superseded; see Metcalf and Reid (1991) for a detailed description of the language. Unfortunately to date there are few compilers available even for uniprocessors. Although we anticipate that the situation will change rapidly in the coming years, recognition of the current state of the availability of Fortran compilers dictates that elsewhere in this book we restrict our attention to the 77 standard. Nevertheless, a study of the new standard in the context of parallel architectures is of interest. There are no explicit constructs in Fortran 90 to support parallelism, but the ability to treat arrays as objects permits compiler writers to parallelise the various array operations and intrinsics. We concentrate on these facilities here.

The treatment of multi-dimensional arrays in Fortran 90 is similar to that of DAP Fortran (AMT, 1990). In particular, the standard operators +, -, * and / can be applied to such structures, but in a point-wise way. Hence, for example, if A and B are two-dimensional arrays of the same size, then the (i, j) component of the result of the product C = A*B is A(I,J)*B(I,J), rather than the (i, j) component of the mathematical product of the two matrices. Clearly, the computation of each component of the result is independent of every other, and hence there is scope for parallelism. In a shared memory environment the work can be apportioned using the simple loop spreading ideas that we discuss in the next subsection. If, in a local memory environment, the arrays are distributed amongst the processors in a corresponding fashion (in the sense that the (i, j) component of each of the matrices is assigned to the same processor), then each processor can compute its portion of the result array.

As well as being able to make use of operators, the programmer in Fortran 90 has available a library of intrinsic functions which can be applied to arrays. In particular, the standard mathematical functions, such as `SIN` and `EXP`, accept array arguments. Again, the functions are applied point-wise so that, for example, if `A` and `B` are conforming two-dimensional arrays the (i, j) component of the result of `B = SIN(A)` is the sine of `A(I,J)`. Note that it is not necessary to give the magnitude of either of the dimensions of `A`.

In addition to extending the definition of standard intrinsics, Fortran 90 has available a number of routines implementing frequently used operations on arrays. All are functions returning results of the appropriate type and can be classified in the following manner:

- *Vector and matrix multiply functions*
  There are two routines in this category:
  - `DOT_PRODUCT(VECTOR_A,VECTOR_B)` forms the dot product of two vectors. We consider possible parallel implementations of this operation later in this chapter and in the next.
  - `MATMUL(MATRIX_A,MATRIX_B)` forms the result of matrix multiplication (that is, 'true' matrix multiplication, as opposed to the point-wise operation defined by the `*` operator). We also consider possible parallel implementations of this operation in the next chapter.
- *Array reduction functions*
  There are basically two different types of these functions:
  - *Functions which operate on a logical array*
    `ALL(MASK,DIM)` returns a `.TRUE.` value if all components of the logical array `MASK` in dimension `DIM` (an integer) are `.TRUE.`, and `.FALSE.` otherwise. For example, suppose that `MASK` is a two-dimensional array, then `ALL(MASK,2)` will return a vector of logical values, each corresponding to the `.AND.` operation applied to every element of the corresponding row of `MASK`. The parameter `DIM` is optional; if omitted the whole of the array is considered and a single logical result returned. Similar functions are `ANY` (`.TRUE.` if any value is `.TRUE.`) and `COUNT` (number of `.TRUE.` elements).
  - *Functions which extract values from an array*
    `MAXVAL(ARRAY,DIM,MASK)` returns the maximum values of the array `ARRAY` in the dimension `DIM`, with only those values of `ARRAY` which correspond to a `.TRUE.` value in the logical array `MASK` being considered. Both `DIM` and `MASK` are optional. When present, `MASK` must have the same dimensions as `ARRAY`. Other functions in this category are `MINVAL` (minimum value), `SUM` (sum of all elements) and `PRODUCT` (product of all elements).
- *Array inquiry functions*
  These functions can be used to determine the attributes of an array. Examples are `SIZE(ARRAY,DIM)` (the total number of elements in an array in dimension `DIM`), `LBOUND(ARRAY,DIM)` (lower bound in dimension `DIM`) and `UBOUND(ARRAY,DIM)` (upper bound in dimension `DIM`).

- *Array construction and manipulation functions*

  There are four routines in this category:

  o SPREAD(SOURCE,DIM,NCOPIES) replicates an array by adding a dimension, with NCOPIES being the number of copies of the array in that dimension.

  o MERGE(TSOURCE,FSOURCE,MASK) merges two arrays under the control of a mask; that is, a new array with the same shape as TSOURCE and FSOURCE is formed whose entries are the entries in TSOURCE or FSOURCE, according to whether the corresponding entries in MASK are .TRUE. or .FALSE..

  o RESHAPE(SOURCE,SHAPE,PAD,ORDER) returns a reshaped array with the same values as contained in SOURCE, but with elements in permuted dimension order and padded with elements specified by PAD. Here, SHAPE gives the size of the new array and ORDER (a permutation of $1, 2, \ldots, n$, with n the length of SHAPE) defines the order in which the array is filled. If PAD is absent then the size of the result must not exceed the size of SOURCE. If ORDER is missing then the result, in array element order as specified by SHAPE, is SOURCE, in its array element order, followed by PAD, in its array element order. For example, suppose that SOURCE is the one-dimensional array with entries $(1, 2, 3, 4, 5, 6)$. If SHAPE $= (2, 3)$, and PAD and ORDER are absent, then the result is the 2×3 array $\begin{pmatrix} 1 & 3 & 5 \\ 2 & 4 & 6 \end{pmatrix}$; that is, the result is filled in normal, column, order. If, however, SHAPE $= (2, 4)$ and PAD $= (0, 7)$, and ORDER is absent, then the result is the 2×4 array $\begin{pmatrix} 1 & 3 & 5 & 0 \\ 2 & 4 & 6 & 7 \end{pmatrix}$. If, in addition, ORDER $= (2, 1)$ then the result is the 2×4 array $\begin{pmatrix} 1 & 2 & 3 & 4 \\ 5 & 6 & 0 & 7 \end{pmatrix}$; that is, the array is filled in permuted, row, order.

  o TRANSPOSE(MATRIX) is applicable to two-dimensional arrays only and forms the transpose.

- *Array shifting functions*

  This set of functions can be regarded as a subset of the array construction and manipulation functions. Here the array retains its shape but the elements are reordered in appropriate way. For example, CSHIFT(ARRAY,DIM,SHIFT) performs a circular shift in the direction DIM, with SHIFT indicating the extent of that (not necessarily uniform) shift. The other routine which fits this classification is EOSHIFT(ARRAY,DIM,SHIFT,BOUNDARY) (end-off shift, introducing elements specified by BOUNDARY).

- *Array location functions*

  MAXLOC(ARRAY,MASK) determines the location of the maximum element of an array (cf. the Level 1 BLAS operation isamax, Section 3.2.1). Similarly, MINLOC locates the position of the minimum element.

With the exception of the array inquiry functions, all of these intrinsics offer scope for parallelisation.

2.4.3   Parallel Fortran for shared memory systems

In this subsection we describe two particular extensions to Fortran 77 which support parallelism on shared memory multiprocessors. Both follow the general pattern set out in Section 2.3.1 and hence the differences are, for the most part, minor and associated with syntax only. Elsewhere in this book we outline code development on shared memory systems with respect to one of these language extensions only, namely Encore Parallel Fortran, but most of what we have to say is applicable to programs written in other languages, whether dialects of Fortran or not, which conform to the common model.

Encore Parallel Fortran

Encore Parallel Fortran (EPF) (Encore, 1988b) is a superset of Fortran 77 possessing many of the common sequential extensions to the standard, such as the DO...END DO and WHILE loops. In addition it possesses a number of facilities for both medium- and coarse-grain parallelism. Before executing a compiled EPF program the user must indicate the number of processors required by issuing appropriate UMAX commands. For example,

```
EPR_PROCS=5
export EPR_PROCS
```

asks for five processors, and this situation is applicable to any job subsequently issued unless a new value is given to EPR_PROCS. It should be noted that this does not necessarily reserve five processors exclusively for any job which follows. The number of processors employed, and how the resources available are shared with any competitors (other user jobs, operating system interrupts, etc.) is left to the system software. This makes it difficult to obtain consistent and meaningful timings for a parallel program.

The extensions to standard Fortran 77 provided within EPF allow for the creation and synchronisation of concurrent processes. Most are programming constructs or statements (compiler directives) rather than calls to Fortran subroutines, although a small number of Fortran functions are provided. We begin with process creation. The constructs and statements available are as follows:

- PARALLEL
    <declarations>
    <statements>
  END PARALLEL
  This construct identifies a parallel block (otherwise execution is sequential). The first line should be viewed as indicating the activation of a number

of parallel tasks which can then have work apportioned to them, or the reawakening of tasks which already exist. There is a one-to-one correspondence between tasks (or, equivalently, processes) and processors. Hence the number employed is as specified by EPR_PROCS. All statements within the block are obeyed by all concurrent tasks, including the 'parent' task which was executing the sequential code immediately prior to the start of the parallel block. Sections of code can be protected and the control flow can be made dependent on the value of the task identifier (an integer value in the range $[0, p-1]$, with p the number of active tasks, see TASKID). In particular, some of the tasks can be given no work at all if we so choose. Any items declared within a parallel block are local (that is, PRIVATE, see below) to each concurrent task. END PARALLEL acts as a BARRIER (see below). On leaving a parallel block all tasks except the parent are idled until a new parallel block is encountered, while the parent continues by executing the code which immediately follows the end of the parallel block. Thus, the overheads associated with task creation are incurred once only, that is, when the first parallel block in a program is encountered. Nevertheless, there will be overheads associated with allocating work to those tasks. It should be noted that EPF does not support nested parallel blocks.

- DO ALL (<integer identifier> = <start>:<finish>[:<step>])
    <statements>
  END DO ALL
  When used within a parallel block this construct apportions the iterations of a loop to currently executing tasks using loop spreading. Unlike a conventional loop there is no ordering imposed on the iterations. Hence if this statement takes the form DO ALL (I = 1:10) and $p = 3$, then process 0 might obey the statements within the loop corresponding to values of I equal to 1, 4, 7 and 10, whilst processes 1 and 2 might execute the statements for I equal to 2, 5 and 8, and 3, 6 and 9, respectively. The precise way in which the loop is spread is outside the control of the programmer; there is no guarantee that a given process will be given a consecutive block of the loop, or that, as suggested here, the iterations of the loop will be allocated to the processes cyclically, or even that two separate runs will result in the same allocation. The integer identifier (which must be PRIVATE and therefore declared within the parallel block) takes values in the range start to finish in increments of step (which is optional). END DO ALL does not act as a BARRIER so that, for example, a new DO ALL can be activated before a previous one is complete. Nested DO ALLs are not supported by EPF.

- CRITICAL SECTION
    <statements>
  END CRITICAL SECTION
  Only one of the active tasks within a parallel block is permitted to execute the statements within a critical section at any one time. Thus, for example, this construct can be employed as a simple locking mechanism to restrict, to

one process at a time, access to global data, or to an output device.

- BARRIER
  Within a parallel block each active task is held at the barrier until all tasks have arrived at that point.
- BARRIER BEGIN
    `<statements>`
  END BARRIER
  This is similar to the BARRIER statement, but only one task is permitted to execute the statements protected by the construct. All other tasks are suspended until that task is complete, and then continue execution with the code which appears immediately after END BARRIER.

All variables and constants declared in the unit containing a parallel block are globally available to all concurrent tasks, whilst those declared within the parallel block are local to each task. Within a parallel block variables may be declared to be of type PRIVATE or SHARED, with obvious consequences. In addition, variables may be declared as type VOLATILE which means that they may be modified without notification to the other tasks.

Besides the barrier mechanism outlined above, synchronisation between tasks can be achieved using the SEND SYNC, WAIT SYNC, SEND LOCK and WAIT LOCK statements. The first two statements operate on items of type EVENT, the last two on items of type LOCK. EVENT items can have the values .WAIT. (which corresponds to the cleared state of Section 2.3.1) or .GO. (the posted state of Section 2.3.1), whilst LOCK items can be assigned the values .LOCKED. or .UNLOCKED.. As well as appearing as parameters to the appropriate SEND and WAIT statements, variables of these two types can be given suitable values in simple assignment statements, or may be employed in .EQ. or .NEQ. comparisons.

The SEND SYNC and WAIT SYNC statements have the following specifications:

- SEND SYNC(`<event list>`)
  The event list is a list of variables of type EVENT separated by commas. This statement assigns to each variable in the event list the value .GO..
- WAIT SYNC(`<event list>`)
  A task is held in a waiting state until all events in the event list have the value .GO..

Thus the execution of a task may be suspended at a WAIT SYNC point until the appropriate event(s) are put in the .GO. state by some other task.

The lock mechanism is similar, except that only one task is permitted to proceed beyond a WAIT statement. The specifications of SEND LOCK and WAIT LOCK are as follows:

- SEND LOCK(`<lock variable>`)
  The lock variable is set to the .UNLOCKED. state.

- WAIT LOCK(<lock list>)

  The lock list is a list of variables of type LOCK separated by commas. A task is suspended until all locks in the lock list have the value .UNLOCKED.. One, and only one, task is then permitted to proceed and all variables in the lock list are reset to the .LOCKED. state.

Additional to the facilities so far outlined are two parameterless integer functions:

- NTASKS() returns the number of tasks in the active set (the value given to EPR_PROCS).
- TASKID() returns the task identifier (a value in the range $[0, p-1]$, where p is the number of active tasks).

Contrary to the Fortran 77 standard, neither must be typed or declared in the unit from which they are called.

Cray Fortran

Outlined here are some of the extensions to Fortran provided for exploiting coarse-grain parallelism on the Cray range of supercomputers. Unlike EPF, parallelism is exploited via subroutine calls, and the mechanism for initiating tasks is also somewhat different. However, Cray Fortran does use locks and events to synchronise tasks.

The extensions to Fortran 77 provided in Cray Fortran are divided into three categories:

- *Tasks software*

  Two subroutines are provided:
  - TSKSTART(<taskid>,<a subroutine name>,<arguments>)

    This routine can be used to spawn a task which can run concurrently with the parent task. The task identifier is used to identify the task whose executable statements are defined by the given subroutine name. The formal parameters of this subroutine are given the actual values specified in <arguments>. It should be noted that the number of tasks in existence at any one time is not limited to the number of processors available. The placement of tasks on processors is left to the multitasking software which also controls the scheduling of those tasks.
  - TSKWAIT(<taskid>)

    After a parent task has spawned a child task using TSKSTART it may proceed with its own executable statements. The parent can be made to wait at any subsequent point until the child has terminated by a call to TSKWAIT, nominating the task identifier of the child.
- *Locks software*

  Four subroutines are provided, each with a single parameter, the name of a

lock:
  o LOCKASGN(<lock>) creates a lock and sets it to the .UNLOCKED. state.
  o LOCKON(<lock>) acts in a similar fashion to EPF's WAIT LOCK except that only one lock is involved. That is, if the lock is in the .UNLOCKED. state then the task sets it to the .LOCKED. state and proceeds. If the lock is in the .LOCKED. state then the task performs a busy-wait.
  o LOCKOFF(<lock>) sets a lock to the .UNLOCKED. state.
  o LOCKREL(<lock>) releases the identifier associated with the lock.

- *Events software*
  Five subroutines are provided, each with a single parameter, the name of an event:
  o EVASGN(<event>) creates an unassigned event.
  o EVWAIT(<event>) waits for an event to be in the .POSTED. state before proceeding (cf. EPF's WAIT SYNC).
  o EVPOST(<event>) sets an event to the .POSTED. state.
  o EVCLEAR(<event>) sets an event to the .CLEARED. state.
  o EVREL(<event>) releases an event.

The ability to spawn more tasks than the number of processors available can be a useful feature. It is possible to implement a similar scheme in EPF by forming a pool of processes, with each task in a PARALLEL block selecting a process from this pool when it has completed its current process; scheduling is therefore a matter for the program writer.

2.4.4   iPSC/2 Fortran

The extensions to the 77 standard in iPSC/2 Fortran are based on the message-passing paradigm and are similar to the Argonne/GMD message-passing macros which are designed to be portable across a wide range of machines (Bomans, Roose, *et al.*, 1990). In iPSC/2 Fortran there are a number of predefined subroutines and functions which enable process handling and message-passing to be performed. We consider a representative sample of the routines provided in iPSC/2 Fortran and refer to Intel (1991) for a complete description of the language. We divide the routines into three separate groups depending on their functionality; unless otherwise noted all parameters are input parameters.

Process handling (cube control) and identification

These routines are concerned with handling the hypercube itself, loading processes (in the form of compiled Fortran code) on to the hypercube nodes, specifying the identifier of a process and obtaining the identifier of a node processor:

- SUBROUTINE GETCUBE(CUBENAME,CUBETYPE,SRMNAME,KEEP)
  GETCUBE is called from the host process and is employed to make available

a number of hypercube node processors for its use. The parameters of this routine are

  CUBENAME – a character string which specifies the name of the cube,

  CUBETYPE – a character string which specifies the size of the cube (number of nodes) and, optionally, the type (co-processor and memory size) of the nodes to be allocated by the GETCUBE call,

  SRMNAME – a character string which specifies the name of the host processor,

  KEEP – an integer to indicate the lifetime of the cube allocated by the GETCUBE call. KEEP = 0 indicates that the cube is released when the calling process exits or is killed; if KEEP = 1 the cube is only released by an explicit call to RELCUBE (see below).

The number of nodes allocated by a call to GETCUBE is always a power of 2; if the size specified is not a power of 2 then it is rounded up appropriately.

- INTEGER FUNCTION NUMNODES()

The parameterless function NUMNODES returns the number of nodes in the allocated cube.

- SUBROUTINE SETPID(PID)

SETPID is called from the host process and is used to set the process identifier of that process. Its single parameter is

  PID – an integer which specifies the process identifier of the host process.

- SUBROUTINE LOAD(FILENAME,NODE,PID)

LOAD is used to load a node process on to a node processor; once loaded, the process begins execution immediately. The parameters of LOAD are

  FILENAME – a character string which specifies the pathname of the file which contains the compiled code of the node process which is to be loaded,

  NODE – an integer which identifies the node on to which the code is to be loaded; −1 is used to indicate that the code is to be loaded on to all of the nodes of the allocated cube. Nodes are numbered in the range $[0, p - 1]$ where p is the number of nodes in the allocated cube,

  PID – an integer which specifies the process identifier of the node process which is currently being loaded.

The NODE and PID parameters of the load operation are used in the communications routines described below. Note that several processes can be loaded on to a single processor and these different processes can be uniquely identified by the PID.

- SUBROUTINE RELCUBE(CUBENAME)

RELCUBE is called from the host process and releases the cube allocated by an earlier call to GETCUBE. Its single parameter is

CUBENAME – a character string which specifies the name of the cube (as given by a previous call to GETCUBE).

- INTEGER FUNCTION MYHOST()
  The parameterless function MYHOST returns the node identifier of the caller's host process.
- INTEGER FUNCTION MYNODE()
  The parameterless function MYNODE returns the node identifier of the calling process.

Note that MYHOST and MYNODE return node identifiers, not process identifiers. Process identifiers need be unique only with respect to a given node; it follows that if a one-to-one mapping of processes to nodes is employed, then all processes can be given the same process identifier.

Message-passing

We describe here those routines which enable synchronous message-passing to be performed:

- SUBROUTINE CSEND(TYPE,BUF,LEN,NODE,PID)
  CSEND is used to send a message to a nominated process. Its parameters are

  TYPE – an integer which identifies the message to be sent,

  BUF – a buffer containing the message to be sent,

  LEN – an integer which specifies the length of the message in bytes,

  NODE – an integer to specify the particular node to which the message is to be sent, a value of -1 indicating that the message is to be sent to all nodes,

  PID – an integer to specify the identifier of the process to which the message is to be sent.

  CSEND sends the message of length LEN identified by TYPE and contained in BUF to the process PID running on the processor NODE and waits until this message has been sent. In other words the calling process is blocked until the CSEND operation has been completed. The completion of a CSEND operation does not imply that the message has been received at the destination, only that the send buffer is available for reuse.
- SUBROUTINE CRECV(TYPESEL,BUF,LEN)
  CRECV is used to receive a message. Its parameters are

  TYPESEL – an integer which identifies the message which is awaited. If TYPESEL is a non-negative integer then a message of the corresponding type will be received, whilst if TYPESEL is -1 then the first message (of any type) to arrive for the given process will be received. Otherwise, any message for which the type is one of a set of message types will be received,

> BUF – an output parameter, a buffer in which the message is received,
>
> LEN – an integer which specifies the size of BUF in bytes.

CRECV receives the message sent by CSEND, or by the corresponding asynchronous send function ISEND, whose TYPE is identified by TYPESEL. It stores the message in BUF and waits until the desired message has been received. That is, the calling process is blocked until the CRECV operation has been completed.

- SUBROUTINE CPROBE(TYPESEL)

  CPROBE is used to block a process until a message of the specified type is sent to it; that is, it performs a busy-wait. Its single parameter is

  > TYPESEL – an integer which identifies the message which is awaited. TYPESEL can be set as in CRECV.

  CPROBE blocks the calling process until a message whose TYPE is identified by TYPESEL is available for receipt. When CPROBE returns the desired message can be received by using the receive subroutine CRECV described above, or the asynchronous receive subroutine IRECV. Alternatively, more information about the pending message could be obtained using the INFO functions (INFOCOUNT, INFONODE, INFOPID, INFOTYPE) described below.

- INTEGER FUNCTION INFOCOUNT(), INFONODE(), INFOPID(), INFOTYPE()

  The parameterless INFO functions return information about pending messages detected when CPROBE returns, or about messages which have been received on return from CRECV. The calls to the INFO functions must immediately follow the corresponding message-passing routine:

  o INFOCOUNT() returns the length (in bytes) of the message.

  o INFONODE() returns the node identifier (NODE) of the node which sent the message.

  o INFOPID() returns the process identifier (PID) of the process which sent the message.

  o INFOTYPE() returns the TYPE of the message.

Routines of a similar form to the above but supporting asynchronous communication are also available.

Global communications utilities

The communications routines described above provide facilities for sending a message from an individual process to some other individual process. iPSC/2 Fortran also supports higher-level communications routines which provide, either implicitly or explicitly, communications between a number of processes. These routines do not include the process identifier (PID) as a parameter and hence cannot be used when multiple processes are executing on a node; in fact they assume that all participating processes have the same PID. We include here two examples of these more general routines:

- SUBROUTINE GSSUM(X,N,WORK)

  GSSUM is used to sum the components of a distributed vector. Its parameters are

  > X – an input/output parameter, a real one-dimensional array containing the input for the operation. When the call to GSSUM is complete, X contains the final result which is the componentwise sum of X across the participating nodes,
  >
  > N – an integer which is the length of X,
  >
  > WORK – a real one-dimensional array which is used to receive the contributions from the other participating nodes.

  All participating nodes call GSSUM with different data values. GSSUM calculates the sum of each component of X across all the nodes and the final result is returned, in X, to the nodes.

- SUBROUTINE GSENDX(TYPE,X,LEN,NODENUMS,NLEN)

  GSENDX is used to broadcast a message to a number of specified processors. Its parameters are

  > TYPE – an integer which identifies the message being sent,
  >
  > X – a buffer containing the message to be sent,
  >
  > LEN – an integer which specifies the length of the message in bytes,
  >
  > NODENUMS – a list of node identifiers to which X is to be sent,
  >
  > NLEN – the number of node identifiers in the list NODENUMS.

  GSENDX sends the message contained in X to the set of nodes specified in NODENUMS. All the nodes which are to receive the message must call an appropriate receive routine.

Note that GSSUM is a particular instance of the Fortran 90 intrinsic SUM (Section 2.4.2).

2.4.5   An example – the dot product

To illustrate the implementation of a parallel program using both EPF and iPSC/2 Fortran (and, later, occam and Ada) we consider the calculation of the dot product of two vectors. Formally, if $\mathbf{x}$ and $\mathbf{y}$ are both vectors of length n then their dot product, a scalar value, is $\mathbf{x}^T\mathbf{y} = \sum_{i=1}^{n} x_i y_i$. Thus, a straightforward sequential algorithm in pseudo Fortran for computing this quantity is

$$
\begin{aligned}
&sum \;\leftarrow\; 0 \\
&do\ i = 1\ to\ n \\
&\quad sum \;\leftarrow\; sum \;+\; x_i y_i \\
&end\ do
\end{aligned}
$$

where $\leftarrow$ denotes the assignment operator. Code 2.1 gives a single-precision implementation of this algorithm in Fortran 77, with the value of the dot product being returned through the function name `sdot`. We observe here that the dot product described in this example is a simplified version of the Level 1 BLAS operation `sdot` which we consider in some detail in the next chapter and exploit elsewhere in this book.

```
      real function sdot(n,x,y)
c Computes a single precision dot product of
c length n of the vectors x and y
      real zero
      parameter(zero = 0.0e0)
      integer n
      real x(*),y(*)
      integer i
      real sum
      sum = zero
      do i = 1,n
        sum = sum+x(i)*y(i)
      end do
      sdot = sum
      end
```

Code 2.1 *Sequential* sdot

We should make it clear at the outset that it is unlikely that in practice we would attempt to implement the calculation of the dot product as a parallel computation. Unless n is large the total amount of work (and, hence, the amount of work that we are likely to be able to apportion to each parallel process) is small, and therefore the overheads that we are likely to incur may well outweigh any potential gains. In terms of the notation introduced in Chapter 1, the grain size is too small. Nevertheless, the study of this calculation serves as a useful illustration of the issues involved in developing a parallel implementation of a numerical algorithm. We defer a more detailed discussion of the merits of our algorithms until the next chapter.

Suppose that we wish to construct a simple parallel implementation of the dot product using p processes with, for simplicity, n divisible by p. Then we can subdivide the vectors **x** and **y** so that each process gets n/p elements of each vector, whose products can be summed independently to form partial dot products. The aggregate of these partial sums (the computation of which involves some form of synchronisation) is the required dot product. It should be noted that the solution computed may not be invariant under such a transformation of the algorithm since the effect of round-off errors will depend on the order in which the components of the dot product are accumulated.

The case in which n is not divisible by p can also be handled, but it is slightly more complicated. Let $r = \lfloor n/p \rfloor$, where $\lfloor x \rfloor$ denotes the greatest integer less than or equal to x (that is, the result of integer division), and let $q = n \bmod p$ be the remainder of the integer division of n by p (so that $q = n - p\lfloor n/p \rfloor$). Then we allocate $r + 1$ components of each vector to q of the processes and r

components to the remaining $p - q$ processes. In the following we assume that n is divisible by p.

The dot product in EPF

To implement the dot product in parallel using EPF we need to make use of features to

1. create new processes,
2. apportion work to those processes,
3. distinguish between local (the local sums and loop counters) and global (the two vectors and the final result) variables,
4. synchronise the formation of the aggregate sum, and
5. destroy processes.

For (1) and (5) we use EPF's `PARALLEL...END PARALLEL` construct. There are two obvious ways of implementing (2) in EPF:

- Use the `DO ALL...END DO ALL` construct for which the subdivision of the two vectors and consequent load balancing is left to the system software. This solution has the advantage that the non-divisibility of n by p causes no additional problem.
- Use the task identifier to determine which portion of the arrays each task should work on. With this solution we have to recognise as a special case that in which n is not divisible by p.

The first of these options appears to be, and is, the easier. For (3) above we know that, by default, all variables declared at the beginning of the unit containing the parallel constructs are global, whilst those declared within a parallel block are local, and that we can override this convention by making use of `PRIVATE` and `SHARED` declarations. To facilitate the implementation of (4) we could use locks or events, but the simplest approach is to employ the critical section mechanism. Our proposed solution is given in Code 2.2. Here we have implicitly declared a number of copies of the same code (the section between the keywords `parallel` and `end parallel`) to be executed by a number of parallel tasks, one of which is the task which spawns the others, with each task being given a portion only of the `do all` loop by the system software. All tasks enter the critical section and read from and write to the global variable `sum`, but they do so one at a time.

Using the task identifier to apportion work is slightly trickier but illustrates a more general point. The vector subdivision is now left to the programmer's discretion, with the most obvious choice being consecutive blocks of elements. This is the approach employed in Code 2.3. First we need to determine the number of processors available (as previously set by an assignment to `EPR_PROCS` and determined by `ntasks`), and `r`, the length of the subvectors assigned to each task. The start position of each subvector is computed as a function of the task

```
      real function par_sdot(n,x,y)
c Computes a single precision dot product of length n of the vectors
c x and y using epf's parallel constructs and loop spreading
      real zero
      parameter(zero = 0.0e0)
      integer n
      real x(*),y(*)
      real sum
      sum = zero
      parallel
         integer i
         real locsum
c Compute the local sum
         locsum = zero
         do all (i = 1:n)
            locsum = locsum+x(i)*y(i)
         end do all
c Update the global sum
         critical section
            sum = sum+locsum
         end critical section
      end parallel
      par_sdot = sum
      end
```

Code 2.2 *Parallel dot product in EPF using* do all

identifier (as determined by **taskid**), with the length of each subvector providing a suitable offset. We have made use of the sequential **sdot** function of Code 2.1 to compute partial dot products, which are again accumulated in a critical section.

```
      real function par_sdot(n,x,y)
c Computes a single precision dot product of length n, a multiple
c of the number of tasks, of the vectors x and y using epf's parallel
c constructs and task allocation
      real zero
      parameter(zero = 0.0e0)
      integer n
      real x(*),y(*)
      integer r
      real sum,sdot
      external sdot
      sum = zero
      r = n/ntasks()
      parallel
         integer start
         real locsum
         start = 1+r*taskid()
c Compute the local sum using a sequential dot product routine
         locsum = sdot(r,x(start),y(start))
c Update the global sum
         critical section
            sum = sum+locsum
         end critical section
      end parallel
      par_sdot = sum
      end
```

Code 2.3 *Parallel dot product in EPF using the task identifier*

A third variation, given in Code 2.4, illustrates the use of locks to achieve global memory synchronisation rather than a critical section. Initially the lock is set to the unlocked state. The first process to reach the protected section sets the lock to the locked state before updating the global sum. Any other process to reach this section whilst the lock is in the locked state is held. When the global sum has been updated the lock is released so that some other process can add its partial sum to the global sum.

```
      real function par_sdot(n,x,y)
c Computes a single precision dot product of length n, a multiple
c of the number of tasks, of the vectors x and y using epf's parallel
c constructs, task allocation and an explicit lock
      real zero
      parameter(zero = 0.0e0)
      integer n
      real x(*),y(*)
      integer r
      real sum,sdot
      lock sum_lock
      external sdot
      sum = zero
      r = n/ntasks()
      sum_lock = .unlocked.
      parallel
         integer start
         real locsum
         start = 1+r*taskid()
c Compute the local sum using a sequential dot product routine
         locsum = sdot(r,x(start),y(start))
c Update the global sum
         wait lock(sum_lock)
            sum = sum+locsum
         send lock(sum_lock)
      end parallel
      par_sdot = sum
      end
```

Code 2.4 *Parallel dot product in EPF using a lock*

Of the three possible solutions offered here, the first (Code 2.2) is the most straightforward. In particular, we avoid the need to take special account of those cases for which n is not divisible by p. The second solution (Code 2.3) requires more care. It does, however, possess an important advantage. For suitably large n, Code 2.2 may involve a greater amount of data movement from main memory into cache as the loop iterations are apportioned to processes according to some, unknown, algorithm. By contrast, Code 2.3 works with blocks of consecutive elements and so the data movement is likely to be less. For this particular example this is unlikely to be a serious consideration but it is a factor which needs to be borne in mind when more substantial problems are tackled. Code 2.4 has been included purely for illustrative purposes and, in terms of efficiency, has nothing additional to offer compared to Code 2.3.

The dot product in iPSC/2 Fortran
This simple example illustrates well the extra complication involved in implementing the dot product algorithm (and, indeed, most other algorithms) on the

local memory Intel iPSC/2 Hypercube. Again the dot product is formed by dividing the calculation into a number of partial dot products and accumulating these in separate processes. The features used are similar to those employed in implementing the algorithm in EPF, with iPSC/2 message-passing replacing the synchronised use of data. The major difference is that it is now necessary explicitly to write two codes,

- a *host*, or *root*, *process* which acts as the master process whose principal role is to pass data to and receive results from the slave processes, and
- a *slave process* whose role is to compute a partial dot product.

The model we adopt here is, therefore, that of $p + 1$ processes: a host process which takes care of the data distribution and result accumulation, and runs on the SRM, and p slave processes which perform the shared work and run on nodes of the hypercube.

The host process takes overall control of the computation. In detail, it

1. requests (allocates) the node processors of the hypercube,
2. loads the slave processes (compiled Fortran code in `sdotslave.out`) on to the nodes,
3. sends data to and receives results from the slave processes,
4. accumulates the complete dot product from the partial dot products, and
5. releases the node processors allocated in (1) above.

Our solution is given in Code 2.5. This shows only the form of the routine `par_sdot`, which may be called from a main program, or any other unit, of the host process. Note that in order to send out the blocks of the x and y vectors, pairs of blocks are placed in a temporary vector `temp`. Thus, the sending of such a pair involves a single call of `csend` (and `crecv` in the slave process) as opposed to the two which would otherwise be necessary. This is a small, but important, point as it reduces the total communication start-up time (Section 3.3).

The (replicated) slave process accumulates a partial dot product. It must

1. receive from the host process any initialisation information, together with its portions of the two arrays,
2. form the dot product of the elements of these two arrays, and
3. return a partial sum to the host process.

These activities are implemented in Code 2.6.

Adapting the code to deal with the case that n is not divisible by p requires changes to the host process, but the slave process is unchanged since it merely deals with the partial dot product of vectors of whatever length is specified by the host process.

```
      real function par_sdot(n,x,y,cube)
c Host code for the dot product using iPSC/2 Fortran
c
c n     - an integer which is the length of the vectors - unchanged
c           on exit
c x,y   - the vectors - unchanged on exit
c cube  - a character string which is the number and type of nodes of
c           the hypercube to be allocated - unchanged on exit
      real zero
      integer maxtemp
      parameter(zero = 0.0e0,maxtemp = 500)
c Assume that an integer variable requires 4 bytes of storage and
c that a real variable also requires 4 bytes of storage
      integer intlen,reallen
      parameter(intlen = 4,reallen = 4)
      integer n
      real x(*),y(*)
      character*4 cube
      real temp(maxtemp),sum,partsum
      integer numslave,numnodes,number,itype,ibytes,pid,ipointer,i,j
      external getcube,setpid,load,numnodes,csend,crecv
c Allocate a cube
      call getcube('sdotcube',cube,' ',0)
c Set the process id of the host process to be 1
      call setpid(1)
c Load each slave with the same code and give each the process id 1
      call load('sdotslave.out',-1,1)
c Set numslave equal to the number of nodes in the allocated cube
      numslave = numnodes()
c Calculate the number of elements to be handled by each slave.  We
c assume that n is exactly divisible by numslave
      number = n/numslave
c Send the number of elements to be handled to the slave processes
      itype = 1
      ibytes = intlen
      pid = 1
      call csend(itype,number,ibytes,-1,pid)
c Send the next number elements of x and y to the next slave process using
c the temporary vector temp
      itype = 2
      ipointer = 0
      ibytes = 2*number*reallen
      do i = 0,numslave-1
        do j = 1,number
          temp(j) = x(ipointer+j)
          temp(number+j) = y(ipointer+j)
        end do
        call csend(itype,temp,ibytes,i,pid)
        ipointer = ipointer+number
      end do
c Receive partial sums from all the slave processes
      itype = 3
      ibytes = reallen
      sum = zero
      do i = 0,numslave-1
c Receive a message from the next calling process
        call crecv(itype,partsum,ibytes)
        sum = sum+partsum
      end do
      par_sdot = sum
      call relcube('sdotcube')
      end
```

Code 2.5 *Parallel dot product in iPSC/2 Fortran – host process*

```
      program slave
c Slave code for the dot product using iPSC/2 Fortran
      real zero
      integer maxtemp
      parameter(zero = 0.0e0,maxtemp = 500)
c Assume that an integer variable requires 4 bytes of storage and
c that a real variable also requires 4 bytes of storage
      integer intlen,reallen
      parameter(intlen = 4,reallen = 4)
      integer itype,ibytes,num,i,node,myhost,pid
      real sum,temp(maxtemp)
      external crecv,myhost,csend
c The first message received is the number of elements, num, to be
c handled by this slave
      itype = 1
      ibytes = intlen
      call crecv(itype,num,ibytes)
c The second message received contains num elements of x followed
c by num elements of y
      itype = 2
      ibytes = 2*num*reallen
      call crecv(itype,temp,ibytes)
c Accumulate the partial dot product in sum, noting that x is stored
c in the first num elements of temp and y is stored in the next num
c elements of temp
      sum = zero
      do i = 1,num
        sum = sum+temp(i)*temp(num+i)
      end do
c Return the partial dot product to the host process
      itype = 3
      ibytes = reallen
c Determine the node id of the host process
      node = myhost()
      pid = 1
      call csend(itype,sum,ibytes,node,pid)
      end
```

Code 2.6 *Parallel dot product in iPSC/2 Fortran – slave process*

Two important points about this local memory implementation example should be borne in mind:

- The example is unusually simple, although it does illustrate the essence of the parallel implementation of a numerical algorithm. The simplicity results from the fact that the slave processes are completely independent of, and do not need to communicate with, each other – the only communication is between the host and the slave processes.

- With such a simple computation the costs of loading the slave processes on to the node processors and of communicating the data to and from the slave processes far outweigh any benefits that result from parallelising its solution. In a more typical application the dot product will be calculated as part of a larger computation and it should then be possible (and necessary for efficiency) to arrange for the data (the vectors **x** and **y**) to be predistributed on the appropriate node processors.

Both points illustrate the importance of granularity. The lack of interprocess communication helps to maximise this, but the amount of computation per process is small unless n is very large compared to p.

2.4.6   Parallel Computing Forum

The Parallel Computing Forum (PCF) has attempted to establish a standard for parallel extensions to Fortran targeted at shared memory systems. It is hoped that the standard will be adopted by manufacturers (particularly those who contribute to PCF) in order to facilitate portability. We give here a flavour of some of the language facilities.

Many of the PCF extensions follow the spirit of EPF and Cray Fortran. For example, a parallel loop construct is provided of the form

```
PARALLEL DO [<qualifiers>,] <identifier> = <start>,<finish> [,<step>]
    <declarations>
    <executable statements>
END PARALLEL DO
```

(This is not quite the same as EPF's `DO ALL` construct, see later.) Here, the list of qualifiers may include any of the following:

- `ORDERED`
  If absent, the executable statements must be such that they impose no ordering on the loop iterations. If present, an executable statement can employ synchronisation which explicitly requires the completion of some previous iteration of the loop.
- `MAX PARALLEL = <number of tasks>`
  As with Cray Fortran, PCF Fortran can support more tasks than the number of processors available.
- `HARD EXITS`
  This specifies the action to be taken to terminate the loop in the event of an explicit jump out. In the event of a `HARD EXIT`, all threads of execution terminate as soon as possible, even if they are waiting at synchronisation points. For a `SOFT EXIT`, which is the default setting, all currently executing threads complete before termination; in particular, all synchronisations are completed.

PCF Fortran carefully distinguishes between this type of parallel loop and that which appears within a parallel region, so as to minimise the start-up effects associated with spawning tasks (see below).

PCF Fortran supports a `PARALLEL SECTIONS` construct of the form

```
PARALLEL SECTIONS [<qualifiers>]
    <section specifications>
END PARALLEL SECTIONS
```

in which the specification of each section takes the form

```
SECTION [<section name>] [WAIT(<section names>)]
  <declarations>
  <executable statements>
```

Here, a number of sections of code are defined as being concurrently executable, with the WAIT qualifier being used to indicate that a section will be initiated only when certain other, named, sections have been completed. The construct is not dissimilar to EPF's PARALLEL...END PARALLEL construct, with the action taken wholly dependent on the task identifier. It is assumed that the number of tasks is not limited to the number of available processors.

PCF Fortran distinguishes carefully between parallel constructs and work sharing constructs. In the case of PARALLEL DO and PARALLEL SECTIONS these two functions have been combined. However, the construct

```
PARALLEL [<qualifiers>]
  <declarations>
  <executable statements>
END PARALLEL
```

is a parallel construct only, defining a parallel region (cf. EPF). Permitted qualifiers are HARD EXITS and MAX PARALLEL. Within this region work sharing constructs may be employed, such as the loop

```
PDO [<qualifiers>] <identifier> = <start>,<finish>[,<step>]
  <executable statements>
END PDO
```

with the qualifiers including those for PARALLEL DO. Hence, it is the PDO construct which is equivalent to EPF's DO ALL.

The work of the PCF has largely been subsumed by the ANSI standards committee X3H5, which is examining language support for parallelism in a context wider than just Fortran.

2.5 Parallel languages

Many other variants of Fortran than those outlined in the previous section exist for various multiprocessor architectures (for example, 3L Fortran for transputer systems). In addition, modifications to many other languages have been made, such as Pascal Plus (Welsh and Bustard, 1979). In contrast to this proliferation of variants of standard languages the number of programming languages which were designed from the outset with multiprocessors in mind is relatively small. Perhaps the two best known are occam and Ada and we now pay some attention to these languages, together with C which has certain parallel processing capabilities.

2.5.1   occam

occam is a block-structured language developed from CSP (Communicating Sequential Processes) (Hoare, 1978; 1986). It is a compact language with few constructs and standard data types. Our aim here is to describe only the essential features, making particular reference to those which support parallelism, leading to the description of an occam implementation of the dot product. Despite its block-structured origins the language contains a number of idiosyncrasies which make it rather different to other high-level languages and so we spend some time looking at the essential building blocks. For a complete description of the occam language, see Burns (1988) or Pountain and May (1987).

Program layout and types

Programs written in occam consist of blocks, each of which has a declaration part and an executable part, with a declaration preceding any use of an identifier. The usual rules of scope apply, so that a variable declared within a block is in scope from its point of declaration until the end of that block, and variables declared in an outer block are in scope in an inner block unless redefined. Where occam is a little idiosyncratic (there are those who would put it stronger than this) is that the blocking structure is defined not in terms of reserved words (such as BEGIN and END), but by the use of indentation. A block of statements (or, more properly, processes) which are to be obeyed sequentially is indicated by the use of the keyword SEQ. The statements then appear, one after the other, each indented from the start of the word SEQ by at least two spaces. Any declarations of variables used locally in the block appear before SEQ. Thus, a section of code to add two integers i and j is

```
INT i,j,isum:
SEQ
  i := 3
  j := 2
  isum := i+j
```

Declarations are terminated by a colon. The end of an executable statement is indicated by the end of a line. A statement may be continued on the next line down provided that the continuation part is indented by at least a further two spaces. Additionally, the break must appear at a 'sensible' place. By this we mean not only that a break cannot appear in the middle of an identifier name, but also that the break must be such that it indicates that a continuation must follow. In an arithmetic expression, therefore, an appropriate place for a break would be immediately after an arithmetic operator. It should be noted that occam assumes no operator priority and so this must be imposed, explicitly, by the use of brackets. In addition, occam is strongly typed, to the extent that mixed-mode arithmetic is not permitted without the explicit use of type-conversion functions. A block is terminated by a process appearing at the same indentation level as that which obtained before the start of the block. A comment may appear on the same line as, but after, a process, or on a separate line (and is then subject to the usual indentation rules); it is indicated by the use of --.

Identifier names are chosen according to the usually accepted rules and can be of any length. In addition, full stops can be used as separators. occam is case-sensitive so that for example, sum and Sum refer to two entirely different objects. Amongst the standard data types are INT, REAL32 (single-precision, 32-bit, real), REAL64 (double-precision real) and BOOL, and both variables and constants of these types can be declared. For example,

```
VAL REAL32 pi IS 3.14159(REAL32):
```

declares a single-precision real constant pi holding an approximation to π. One-dimensional arrays of these types may be declared and ranges of array indices always begin at 0. Thus,

```
VAL INT n IS 10:
[n] REAL32 vector:
```

declares vector to be a one-dimensional array of length 10 with indices $0, 1, \ldots, 9$. The length of an array may be determined using the standard operator SIZE (so that SIZE vector has the value 10). Slices of arrays are permitted; for example, [vector FROM 3 FOR 2] can be used to represent the vector of two elements [vector[3],vector[4]]. Slices may appear on both the left- and right-hand sides of an assignment provided that the lengths of the slices are the same. Multi-dimensional arrays are declared as arrays of arrays.

Loops and conditionals
As with all high-level languages, occam, viewed as a sequential language, has loop and conditional statements. The equivalent of a DO loop is the replicated SEQ, for example,

```
SEQ i=0 FOR n
```

Here, i is declared (to have type INT) simply by its appearance in the replicated SEQ. The value of n indicates the number of cycles of the loop, not the upper limit of i. The 'body' of the loop (a process) must appear on the next line, indented by two spaces. If the body consists of several processes to be obeyed sequentially, then they must be grouped into a single process using a further SEQ. occam also possesses a WHILE statement, for example,

```
WHILE i<n
```

Again, the body of the loop must appear on the following line, indented by two spaces.

The IF construct in occam is more like a standard CASE statement (although there is a separate occam CASE construct). After the reserved word IF there may be any number of branches, each guarded by a conditional expression, and each indented by the usual two spaces. The first `TRUE` conditional in the list determines which process is to be obeyed. For example,

```
IF
  i>j
    maxval := i
  TRUE
    maxval := j
```

sets `maxval` equal to the maximum of the two integers i and j. The use of the Boolean constant `TRUE` effectively makes the above process equivalent to Fortran's `IF...THEN...ELSE...ENDIF` construct. If none of the conditions specified in an `IF` statement yields a `TRUE` result then the statement is equivalent to the process `STOP`. The `STOP` process starts, never proceeds, and never finishes (that is, it just hangs). One way to overcome this problem, and thus simulate a Fortran `IF...THEN...ENDIF` construct, is to include a `TRUE` branch, whose process is the `SKIP` process which starts, does nothing, and then finishes, as the final branch of an `IF` construct.

The keyword used in occam to denote a subprogram is `PROC`. Parameters are passed by reference unless the keyword `VAL` precedes the type specification. An occam `FUNCTION` may return more than one result. In line with good programming practice, any attempt to alter the value of a parameter or global variable inside a function is expressly forbidden.

Communication

Interprocess communication is synchronous and unbuffered. To send a message a process of the form `chan0 ! <list.of.variables>` is used, where `chan0` is a channel name. Channels may be declared in the same way as variables, and may be universal or specific. For example, `CHAN OF ANY` is a channel along which variables of any number and type may be sent or received; `CHAN OF REAL32;REAL32` is a channel which can be used to transmit a pair of single-precision reals only. The format of a channel can be specified using a `PROTOCOL` statement. First the protocol is declared and then channels of that protocol are declared, as in

```
PROTOCOL tworeals IS REAL32;REAL32:
CHAN OF tworeals channel:
```

Thus, a process to send the values of the two single-precision real variables x and y is

```
channel ! x;y
```

Corresponding to this send process there must be a receive process

```
channel ? a;b
```

waiting to receive the values into two single-precision real variables, here a and b. These two processes may be on the same processor, in which case they must be concurrent processes (see below), or on separate processors. If the processors are transputers then the communication takes place down one of the transputer links, and, in order to achieve this, the channels have to be placed on to the links.

The idea of a protocol can be generalised to a tagged, or variant, protocol which enables a channel to possess a finite number of protocols, each possessing a tag value. Whenever communication takes place the tag (a normal identifier) indicates precisely which protocol is being employed. For example, a protocol for the transmission of a single-precision real or, separately, an array of reals is

```
PROTOCOL tagged.protocol
  CASE
    number;REAL32
    vector;INT::[]REAL32
  :
CHAN OF tagged.protocol tagged.channel:
```

Here, in the second tagged protocol, the integer gives the length of the array to be transmitted. Using these declarations a process sending a real value takes the form

```
tagged.channel ! number;x
```

On the other hand, if a vector is to be sent we would use

```
tagged.channel ! vector;n::[a.vector FROM 0 FOR n]
```

A receive process might be of the form

```
tagged.channel ? CASE
  number;a
  vector;len::receive.vector
```

with **receive.vector** of length **len**. If it is known that only certain tag values are possible at a particular point in a computation the remaining tags may be omitted from the receive process.

occam permits a variant on the receive process which enables the programmer to receive the first of several alternative messages. For example,

```
ALT
  channel.1 ? variable.list.1
    process.1
  channel.2 ? variable.list.2
    process.2
```

will receive a message on **channel.1** or **channel.2**, whichever is ready first. If both are ready simultaneously then, according to the language definition, either message is equally likely to be received. This indeterminacy can be avoided by using the **PRI ALT** construct which gives priority to the first branch in the sequence. Having received the message, the appropriate process is scheduled. Boolean guards can be attached to ensure that a channel is considered a candidate for availability only if some appropriate Boolean condition is satisfied.

Process creation

We have already seen that the beginning of a block whose processes are to be obeyed sequentially is indicated by SEQ, and that the processes themselves appear on the following lines, each indented from the start of the SEQ by at least two spaces. Similarly, the keyword PAR can be used to indicate a block whose processes can be executed concurrently (and there can be parallel blocks within parallel blocks). Thus the assignments

```
a := x
b := y
```

where a, b, x and y are all of type REAL32 could be preceded by PAR (with the two assignments then each indented by a further two spaces). There is no advantage in doing this (in fact there is a disadvantage, since the two concurrent assignments will be time-shared), but it illustrates a point. As a further illustration, we can replace the assignments by

```
PAR
  channel ! x,y
  channel ? a,b
```

where channel is a channel for the transmission of two reals, although we again note that there is no advantage in doing this. (What would happen if PAR were replaced by SEQ? Remember, communication is synchronous.) Any attempt to assign values to the same variable in two parallel processes is an error. In particular, this applies to an array, even if different array components are involved. Often this is over-restrictive and so occam compilers permit such *usage checking* to be disabled.

A replicated PAR creates a number of replicas of the same process, for example,

```
PAR i=0 FOR 5
  a.process.replicated.5.times
```

The parallel processes so created can be sequential processes (in which case they are grouped together by a SEQ), or parallel (grouped by PAR), or, indeed, any combination of these. Note that current implementations of occam permit a static number of replications of a PAR process only.

If all of the processes to be executed in parallel reside on the same processor then there would appear to be no real scope for parallelism. However, as we saw in Chapter 1, the transputer has, on a single chip, a number of independent functional units, namely the central processing unit, the floating-point unit (on a T800), and the link engines. Thus, in particular, it is possible to perform some floating-point computation whilst, at the same time, obtaining data from a process executing on another processor. Further, we can give priority to the communication, on which some other process may be waiting, by using a `PRI PAR` which prioritises in a similar way to `PRI ALT`.

Process and channel placement

It is normal to implement a distributed occam program in terms of a single host and a number of slave processes. Since channels can exist between any two processes, whether they reside on the same or physically separate processors, it is a relatively straightforward matter to develop the code, first in 'pseudo parallel' on a single processor, and then transfer it to a distributed environment. To achieve this transition we require the ability to place the processes on to physical processors (cf. the iPSC/2 Fortran `LOAD` routine) and assign channels to inter-processor links. For this placement we employ a `PLACED PAR`.

The `PLACED PAR` construct takes the form

```
PLACED PAR
  PROCESSOR <processor.number> <processor.type>
    <placement statements>
```

Here, the processor type can be T2, T4 or T8. A placement statement may consist of the process which is to be placed on that processor. It is normal to make this a `PROC` with parameters corresponding to the channels being employed for communication. Other possible placement statements are those which associate a channel with a physical link; these take the form

```
    PLACE <channel.identifier> AT <link.number>
```

where the link number is an integer in the range [0, 7]. (Remember, each transputer has four bi-directional links.)

The dot product in occam

As a simple example of an occam program we again consider the dot product of two vectors. The approach is similar to that employed for the iPSC/2 Fortran code given in Section 2.4.5. An important difference, however, is that in occam the sender and receiver of a message are identified by the channel along which the message is sent, rather than via a node identifier and process identifier. Further, the processors are not restricted to any particular connection topology, and so the code for one particular arrangement of processors (say, a linear chain) is unlikely to be the same as that for some other arrangement (say, a square grid). Here we consider an implementation on a linear chain in which the host process (running on the host transputer) passes to the first slave in the chain all of the relevant data and receives back the final result. Each slave has to

1. receive from its predecessor in the chain the data for its successors and pass this on to its immediate successor,
2. receive from its predecessor its own data,

3. compute a partial dot product,
4. receive from its successor the sum of the partial dot products of all of its successors and add this to its own partial dot product, and
5. send this sum to its predecessor.

First we need to establish a variant protocol capable of dealing with

- initialisation information (two integers, the identifier of a slave process and the total number of slave processes),
- two arrays (blocks of the vectors whose dot product is required), and
- a real (a partial dot product).

A suitable protocol is given in Code 2.7.

```
PROTOCOL dot.protocol
  -- Protocol for initialisation, sending out the vectors and
  -- returning partial sums
  CASE
    init;INT;INT
    arrays;INT::[]REAL32;INT::[]REAL32
    dot.product;REAL32
:
```

Code 2.7 *Protocol for occam dot product*

```
PROC par.sdot(VAL[]REAL32 x,y, VAL INT n,n.slaves, REAL32 result,
  CHAN OF dot.protocol from.slaves,to.slaves)
  -- Host code for the dot product in occam
  INT block.size:
  SEQ
    -- Work out the portion of work per slave
    block.size := n/n.slaves
    -- Send out the initialisation information
    to.slaves! init;1;n.slaves
    -- Send out the data
    SEQ i=0 FOR n.slaves
      to.slaves ! arrays;block.size::[x FROM i*block.size FOR block.size];
        block.size::[y FROM i*block.size FOR block.size]
    -- Get the result back
    from.slaves ? CASE dot.product;result
:
```

Code 2.8 *Parallel dot product in occam – host code*

The host code passes data (in blocks) to the slaves and receives back the result. We give, in Code 2.8, the form of a procedure which handles this data transfer. It makes use of two channels, `to.slaves` and `from.slaves`, of our chosen protocol `dot.protocol`. The use of the channels changes their state; hence we need to use a procedure for `par.sdot`, rather than a function. Note that it is necessary to have a separate channel for each direction of data transmission and that the channels will need to be associated with the physical link joining the host and the first slave transputer. Unlike the iPSC/2 Fortran code, the protocol employed permits blocks of the vectors x and y to be transmitted together without the need to employ a temporary vector. `par.sdot` can be called anywhere in the host process from the block in which is declared.

```
PROC slave.code(CHAN OF dot.protocol from.left,to.left,
                                      from.right,to.right)
  -- Slave code for the dot product in occam
  INT my.number,n.slaves,block.size:
  REAL32 my.dot,right.dot:
  VAL INT maxn IS 30:
  [maxn]REAL32 x,y:
  REAL32 FUNCTION sdot(VAL[]REAL32 a,b)
    -- A sequential function for determining the dot product of two vectors
    VAL zero IS 0.0(REAL32):
    REAL32 dot.prod:
    VALOF
      SEQ
        dot.prod := zero
        SEQ i = 0 FOR SIZE a
          dot.prod := dot.prod+(a[i]*b[i])
        RESULT dot.prod
  :
  SEQ
    -- Get the initialisation information
    from.left ? CASE init;my.number;n.slaves
    -- Pass this on to any other processes in the chain
    IF
      my.number<n.slaves
        to.right ! init;my.number+1;n.slaves
      TRUE
        SKIP
    -- Get and pass on the data for remaining processes in the chain
    SEQ i = 0 FOR n.slaves-my.number
      SEQ
        from.left ? CASE arrays;block.size::[x FROM 0 FOR block.size];
          block.size::[y FROM 0 FOR block.size]
        to.right ! arrays;block.size::[x FROM 0 FOR block.size];
          block.size::[y FROM 0 FOR block.size]
    -- Get my own data
    from.left ? CASE arrays;block.size::[x FROM 0 FOR block.size];
      block.size::[y FROM 0 FOR block.size]
    -- Work out my partial dot product
    my.dot := sdot([x FROM 0 FOR block.size],[y FROM 0 FOR block.size])
    -- Add on partial dot products from remining processes in the chain
    IF
      my.number < n.slaves
        SEQ
          from.right ? CASE dot.product;right.dot
          my.dot := my.dot+right.dot
      TRUE
        SKIP
    -- Return the accumulated result
    to.left ! dot.product;my.dot
  :
```

Code 2.9 *Parallel dot product in occam – slave code*

The slave code (Code 2.9) does the computation part. It makes use of four channels, two linking it to the preceding processor in the chain and two linking it to its successor. Note that of the four links on each transputer, only two are being employed for this neighbour-to-neighbour communication. Note also that the performance of the code could be improved by overlapping the input of data from the left with output to the right.

As it stands the code is a little inelegant, since it explicitly addresses the need to forward data which is destined for some other process. Software packages are available (for example, CS Tools, Meiko, 1989) which take care of the throughrouting. Further, the latest developments in transputer technology will mean that systems built from the T9000 transputer and the C104 communications chips will deal with throughrouting in hardware.

2.5.2   Ada

Ada is a powerful high-level language with a tasking structure to facilitate the implementation of concurrency. Because Ada is such a comprehensive language we can only briefly describe some of its features here. In particular, we concentrate on the features provided for parallelism. Rather than give examples of each of the facilities we describe, an implementation of the dot product is used to illustrate the relevant features. For a more general introduction to the language, see Barnes (1989).

Program layout

Ada is also a block-structured language with, arguably, fewer idiosyncrasies than occam. In particular, Ada is a free-format language with blocks defined by the use of the reserved words `begin` and `end`, `if` and `end if`, etc. Statements are terminated by semi-colons. Identifiers may be of any length, with the underscore character employed as a separator. Ada is not case-sensitive although it is a convention to use lower-case letters for all reserved words and upper-case for everything else. Standard types include `INTEGER`, `FLOAT`, `LONG_FLOAT` and `BOOLEAN`, and the language supports subranges, and user-defined enumerated types and records. Note that these are predefined entities which may be redefined if required. Hence, `FLOAT`, etc., are not reserved words. Ada supports multi-dimensional arrays, with the attributes `'FIRST` and `'LAST` giving the index values of the first and last elements of the first (or only) dimension.

Ada has `for`, `while` and `loop...end loop` constructs with an `exit` statement which can be used to terminate a loop at any point of its execution. The `if` statement is similar to its Fortran equivalent. Ada has, additionally, a `case` statement.

Subprograms in Ada are denoted using the reserved word `procedure`. Parameters may be denoted as `in`, `out` or `in out`, indicating in which direction the information denoted by the identifier is to be passed. The matching of actual parameters to formal parameters may obey the usual positional convention, as in Fortran 77. Alternatively, the match may be made explicit at the point of call by using the notation

```
<actual parameter> => <formal parameter>
```

(a scheme which has been incorporated into Fortran 90). Now the ordering of the actual-to-formal parameter assignments is unimportant; further, if certain parameters are not required they may be omitted completely. The declaration of a subprogram is much as would be expected, except that the identifier may be a standard operator symbol. This *overloading* of the operator means that it can be employed with operands of any type, as long as the effect of that operation has been defined. The technique can be extended to procedures as well, and the terminology of *generics* is employed. That is, a subprogram can be written in terms of some abstract generic type and an *instantiation* of that subprogram can be created which deals with items of some actual type.

Process creation

Ada was not designed with any particular parallel architecture in mind. Thus, its parallel constructs are general and encompass both the message-passing and synchronised use of shared data programming paradigms. The basic construct is the Ada **task**. A given module (procedure, package, task, etc.) acts as a parent which can spawn a number of tasks. Communication between tasks (and the parent) is achieved using **entry** points. The specification of an entry point is similar to that for a procedure; in particular, it has a name, a number of parameters, and it can have associated with it a number of executable statements. Within the task in which an entry point is defined a reference to that entry point, an **accept** statement, suspends the task until a releasing call is made elsewhere. The calling task makes what is, essentially, a procedure call using the task name and entry name, and is itself suspended until the corresponding **accept** statement is ready (that is, communication is synchronous). This mechanism is referred to as a *rendezvous*, and it permits data to be passed between the calling and called tasks via the entry parameter list. The rendezvous is complete only when the statements of the entry have been completed. When the rendezvous is complete both tasks can continue.

The dot product in Ada

Our approach to the development of an Ada dot product routine owes much to the EPF Code 2.3 which employs the task identifier. It also owes something to the iPSC/2 Codes 2.5 and 2.6 since we have both a master/slave model of parallelism and explicit communication. The role of the master process is to

- start up a number of parallel tasks,
- assign a task identifier to each task, and
- obtain a partial dot product from each task and update the global sum,

whilst the role of each slave process is to

- receive the task identifier from the master process,
- work out a partial dot product, and
- return the partial dot product to the master process.

Our implementation is given in Code 2.10.

```
function PAR_SDOT(X,Y : in VECTOR; NTASK : in INTEGER) return FLOAT is
  -- Computes the dot product of the two vectors X and Y, of the
  -- pre-defined type VECTOR, using NTASK slave tasks
  SUM,RESULT : FLOAT;
  N          : INTEGER := X'LENGTH;
  -- Specification for slaves
  task type SUB_TASK is
    entry SYNC(NO : in INTEGER);              -- My id
    entry ANSWER(PARTTOT : out FLOAT);        -- Partial sum
  end SUB_TASK;
  DOT_TASK : array (1..NTASK) of SUB_TASK;
  -- Body for slaves
  task body SUB_TASK is
    LOCSUM : FLOAT := 0.0;
    MYID,OFFSET : INTEGER ;
  begin
    -- Get my id
    accept SYNC(NO : in INTEGER) do
      MYID := NO;
    end;
    -- Work out partial sum
    OFFSET := (N/NTASK) * (MYID-1) + X'FIRST;
    for COUNT in OFFSET..(OFFSET + N/NTASK - 1) loop
      LOCSUM := LOCSUM + X(COUNT)*Y(COUNT);
    end loop;
    -- Return partial sum
    accept ANSWER(PARTTOT : out FLOAT) do
      PARTTOT := LOCSUM;
    end;
  end SUB_TASK;
-- Body of master
begin
  -- Give each slave its id
  for I in 1..NTASK loop
    DOT_TASK(I).SYNC(I);
  end loop;
  -- Get the partial sums back in order
  SUM := 0.0;
  for I in 1..NTASK loop
    DOT_TASK(I).ANSWER(RESULT);
    SUM := SUM+RESULT;
  end loop;
  return SUM;
end PAR_SDOT;
```

Code 2.10 *The dot product in Ada*

In Code 2.10 the master process is identified by the function name PAR_SDOT. The declaration part of this subprogram includes the introduction of an array of items of type SUB_TASK. The definition of this **task type** is in two parts:

- The *specification* part describes the interface between this and other tasks.

Here there are two entry points, SYNC and ANSWER, with which rendezvous can be established.

- The *body* part describes the behaviour of the task. Here an accept statement is used to establish the rendezvous with the master process by which the task obtains its identifier. Using this, the task is able to work out which portion of the arrays to use when forming a partial dot product. A second accept statement establishes the rendezvous with the master process by which the partial dot product is returned.

The activation of the tasks is effected at the beginning of the executable portion of PAR_SDOT. Within this executable part a loop is used to perform the rendezvous with each sub-task in order. A second loop performs the rendezvous for the return of the partial dot products, again in order, with these values being accumulated in SUM.

2.5.3   C

We have already commented on the continuing large scale support for Fortran within the scientific community. However we also recognise a growing interest in C and here briefly describe those features of the language which support parallelism. For a detailed study of the language see Kernighan and Ritchie (1988), or Banahan, Brady, *et al.* (1991).

C is a traditional free-format block-structured language and is the basis for the UNIX operating system. Standard types include int, short int, long int, float, and double. The only sub-program unit provided in C is the function. Subroutines, or procedures, are implemented by returning a void result from a function. Of particular relevance here is the UNIX system routine fork. This is a parameterless function which returns an integer result and can be used to spawn a concurrent process. On the issue of a fork an exact copy is made of the parent process. That is, after the fork call both the parent and the child obey the same instructions (although the route taken through those instructions can be made dependent on the task identifier). In particular, each child inherits its own copy of any variables available to the parent. The value returned by fork to the child is zero, whilst the value returned to the parent is the task identifier of the child. Once a process has been spawned it may be terminated using the exit function, which returns a void result. The parent process can be made to wait for the termination of a child using the wait function, which returns an integer value, the process identifier of the terminated process.

Armed with the primitives fork, exit and wait we can implement a number of the constructs that exist in other parallel languages, or extensions to sequential languages. Exactly how this may be achieved is outside the scope of this book. We refer the interested reader to Brawer (1989) who shows how to construct parallel blocks using fork, exit and wait, how memory may be shared between the parent and child processes, and how UNIX V semaphores may be employed in the implementation of spin locks. Encore supports a C *threads* (lightweight processes) multitasking system library (Encore, 1988a) which is callable from Encore's Fortran 77 (f77) and Pascal (pc) compilers.

2.6   Program development tools

The tools provided to help in the development of parallel programs vary according to the degree to which they try to automate the process. We introduce three categories:

- *Autoparallelisers (parallelising compilers)*

  The objective of an autoparalleliser is to detect and exploit parallelism in a code without requiring the user explicitly to write the code using a (parallel) language like those described in Sections 2.4 and 2.5. Such compilers are capable of detecting a limited degree of parallelism in fairly standard 'dusty deck' Fortran codes and are thus attractive to users who do not wish to spend their time rewriting a large code for parallel execution.

- *Parallel programming environments*

  We have already mentioned that the major high-level programming language supported by any parallel computer is likely to be an extension of Fortran, although there are important exceptions. The language is often a component of a wider parallel programming environment which includes other tools to aid parallel program development.

- *Parallel debugging aids and parallel performance profilers*

  Because the task of debugging a parallel program, with the additional complications of process synchronisation or message-passing, can be considerably more difficult than debugging a sequential code, tools to aid debugging are an important component in the armoury of the parallel programmer. It is one thing to write a bug-free piece of software that employs parallelism; it is quite another to write codes which efficiently exploit a given architecture. In a sequential environment profiling tools play an important part in the development of highly efficient software. In a parallel environment we naturally expect such tools to be available also. However, we additionally require tools to aid in the determination of

  - synchronisation points, and
  - processor utilisation,

  with graphical representation being almost essential.

Note that it may not be necessary for any of these tools themselves to operate in a parallel environment.

2.6.1 Autoparallelisers

In the dot product example of Section 2.4.5 we saw that simple loop spreading can be used to produce parallel codes. Since it is relatively easy to automate this process autoparallelisers can be used to remove this chore from the programmer, thus giving parallelism with no effort. A direct consequence of this autoparallelising capability is the potential ability to speed up 'dusty deck' Fortran programs. Unless the programmer has some detailed knowledge of the underlying algorithms on which the program is based, the starting point is the acquisition of profiling information to determine precisely those areas of the program where the most significant gains are likely to be achievable. These sections of the program can then be compiled by the autoparalleliser, the output inspected, and its performance monitored. Speed-up is not guaranteed, let alone optimal performance, so some further hand-modification of the program may be required. Note that here we are referring to autoparallelisation at the coarse-grain level (process parallelism), in which loop spreading is applied at the outermost level possible. Contrast this with vectorisation which tends to be applied to innermost loops (pipeline parallelism).

In Section 2.4.3 we described the extensions to the Fortran 77 standard supported by the EPF compiler. In addition to providing these facilities EPF has autoparallelising capabilities. A pre-processor scans the Fortran source code and inserts EPF parallel constructs where appropriate. The result, which may be saved for subsequent examination, is then compiled in the normal manner.

As a simple example of the performance of the autoparallelising capabilities of EPF we examine what happens when the sequential dot product code of Code 2.1 is compiled with the 'concurrentise' option switched on. Using a further option to save the results, compilation produces two files. The first, with a `.lst` extension, is shown in Code 2.11. The C at the beginning of a line is not a Fortran comment but an indication that the compiler recognises that the loop can be spread. The second file produced by EPF, with a `.E` extension, is the parallel version of the original code and is shown in Code 2.12. The original code remains in lower case, whilst the upper-case statements indicate the modifications introduced by EPF. We note that, apart from a few details, the code is as given in Code 2.2.

It should be noted that Code 2.12 will not necessarily produce identical results on separate runs using different numbers of processors, or even the same number of processors. Further, we cannot even guarantee that it will ever produce results identical to Code 2.1. The reason for this is that, as we remarked earlier, the effects of round-off errors will cause different results depending on how the separate partial dot products are determined and accumulated. To guard against this variation in behaviour the EPF compiler contains an option which disables all transformations of this form. Using this option to compile Code 2.1 produces a `.lst` file as shown in Code 2.11, but with the C (concurrentized) on lines 11 and 12 replaced by DD (data dependence). A comment at the end of the `.lst` file indicates that this is associated with the variable **sum**. Of course, we know

```
                   1        real function sdot(n,x,y)
                   2      c Computes a single precision dot product of
                   3      c length n of the vectors x and y
                   4        real zero
                   5        parameter(zero = 0.0e0)
                   6        integer n
                   7        real x(*),y(*)
                   8        integer i
                   9        real sum
                  10        sum = zero
c          +---------  11        do i = 1,n
c          *           12          sum = sum+x(i)*y(i)
           *_____  13        end do
                  14        sdot = sum
                  15        end
```

```
1 loops total

1 loops concurrentized
```

Code 2.11 *Concurrentisation detected by EPF*

```
      real function sdot(n,x,y)
c Computes a single precision dot product of
c length n of the vectors x and y
      real zero
      parameter(zero = 0.0e0)
      integer n
      real x(*),y(*)
      integer i
      real sum
      sum = zero
      PARALLEL
      INTEGER I
      REAL SUM1
      PRIVATE SUM1, I
      SUM1 = 0.
      DOALL (I=1:N)
        SUM1 = SUM1 + X(I) * Y(I)
      END DOALL
      CRITICAL SECTION
      SUM = SUM + SUM1
      END CRITICAL SECTION
      END PARALLEL
      sdot = sum
      end
```

Code 2.12 *Dot product parallelised by EPF*

that EPF is capable of removing this data dependence, but we have explicitly prevented it from so doing. The .E file is now identical to the source code.

The autoparalleliser is likely to be able to perform limited optimisations only, so programmer assistance (by, for example, reordering loops) should be given wherever possible. In particular, the autoparalleliser may be limited in its ability to determine data dependencies; they may be known at run-time only. As a consequence, loops which incorporate calls to subroutines may be ignored. For example, if we implement a matrix-vector product $\mathbf{y} = A\mathbf{x}$ as a sequence of dot products involving the rows of A we obtain, in sequential Fortran, the code

```
do i = 1,m
  y(i) = sdot(n,a(i,1),lda,x,1)
end do
```

The dot product function used here is the Level 1 BLAS sdot to be described in Section 3.2.1, rather than the simplified dot product considered in Section 2.4.5. Each vector has associated with it a stride (Section 2.4.1). A stride of LDA (the leading dimension of A) for A means that the *i*th row of A is regarded as a vector. The computation of each component of Y is completely independent of that of every other component. However, an autoparalleliser may have difficulty in recognising this since it requires knowledge of the way that the matrix A and the vector X are modified by the function call sdot. In fact, of course, they are not modified at all, but the autoparalleliser may not be clever enough to realise this. If the function call is replaced by in-line code, to give

```
do i = 1,m
  sum = zero
  do j = 1,n
    sum = sum+a(i,j)*x(j)
  end do
  y(i) = sum
end do
```

then the autoparalleliser has full knowledge of the access to data and hence may have greater success at parallelisation. This can be applied to either the outer loop or the inner loop (or even both). The general strategy is to apply parallelism to the outermost loop whenever possible. In this case this is highly desirable since the parallel version of the inner loop contains a synchronisation construct (accumulating the dot product). Hence, EPF replaces the outer loop by a DO ALL construct. We return to the matter in Section 3.3.2.

As a further example of the possible limitations of autoparallelisation we consider the Level 1 BLAS operation scopy, which copies one vector into another (Section 3.2.1). A simplified implementation of this routine is

```
subroutine scopy(n,x,y)
integer i,n
real x(*),y(*)
do i = 1,n
  y(i) = x(i)
end do
end
```

Superficially it appears that the assignment of a component of x to the corresponding component of y is independent of every other such operation. Thus EPF replaces the loop by a DO ALL construct. However, there are dangers in this. Suppose that, in the unit which calls scopy, a is a single-precision real array of dimension 1:10. Then a call to scopy of the form

```
call scopy(5,a(6),a(1))
```

can be used to copy the second half of a into the first half. Suppose, now, that the second actual parameter a(6) is replaced by a(2). The sequential code will give a shift by one element of components 2 through 6. However, the performance of the parallelised code is indeterminate (can you explain why?). Hence, we cannot ensure complete compatibility of the parallelised code with the original Fortran source without imposing very stringent, and unrealistic, constraints on the autoparalleliser.

If we modify the routine slightly and permit non-unit increments for x and y then, for positive incx and incy, the loop becomes

```
ix = 1
iy = 1
do i = 1,n
  y(iy) = x(ix)
  ix = ix+incx
  iy = iy+incy
end do
```

In this case the autoparalleliser may have difficulty with the updating of the indices. For example, EPF spreads the loop but retains global variables for the indices and synchronises the updating of their values using events. The effect is, unfortunately, to degrade the performance of the code. If, however, the code is rewritten so that the assignment to y is given by

```
y(1+(i-1)*incy) = x(1+(i-1)*incx)
```

an autoparalleliser might be expected to produce somewhat better code (although, sadly, EPF leaves it entirely unchanged).

A further example illustrates how loop reordering can help parallelisation. We consider the Fortran 90 intrinsic EOSHIFT but restricted to a two-dimensional array, shifting columns one place to the right and introducing zeros in the first column. Ignoring the fill-in of zeros in the first column, the natural way to write this in Fortran 77 is

```
do j = n,2,-1
  do i = 1,n
    a(i,j) = a(i,j-1)
  end do
end do
```

where the j loop is outermost to reflect Fortran's column storage. As it stands parallelisation can be applied to the innermost loop but not, directly, at the outer level. Switching the order of the loops leaves the meaning of the code unchanged, but parallelisation can now be applied at the outermost loop. The lesson, therefore, is that we should attempt to help the parallelising compiler by reordering loops wherever appropriate. We may even be fortunate enough to have a compiler which will do the reordering for us (and, for the code above, EPF does, indeed, switch the loops).

We have illustrated here some of the limitations of autoparallelisation. However, we wish to stress that autoparallelisers can be invaluable tools in the development of parallel codes from dusty deck Fortran programs. It is important to remember that autoparallelisation alone may not necessarily produce significantly faster execution times on a multiprocessor, and that some assistance, either before or after the code has been transformed, may well be necessary if significant speed-up is to be achieved.

2.6.2 Parallel programming environments

A number of parallel programming languages have been developed in recent years and in Sections 2.4 and 2.5 we gave some examples. Some, such as Ada, are essentially an attempt to provide a uniform framework in which to develop parallel programs. Others form part of a sophisticated parallel programming environment which also includes debugging and profiling tools. We briefly describe two such programming environments: SCHEDULE, developed by Dongarra and Sorensen (1987a) (see also Dongarra, Sorensen, *et al.*, 1988) and implemented on a number of machines (both shared and local memory); and the Transputer Development System, a system for the development of occam programs on transputer arrays.

SCHEDULE

SCHEDULE is based on the assumption that a parallel program can be divided into a number of units of computation or processes with *execution dependencies* amongst the processes. These dependencies describe the order in which computations should occur and are based not only on data dependencies but also on other considerations, such as load balancing. SCHEDULE is Fortran-based and it is natural to associate a process with a Fortran subroutine and its data; in this way SCHEDULE allows the exploitation of existing sequential Fortran subroutines in parallel code.

The starting point for writing a SCHEDULE program is to construct a *dependency graph* which represents the processes and their execution dependencies – this enables the coarse-grain parallelism of the problem to be exploited. From this dependency graph a Fortran 'control flow' subroutine describing the processes and their execution dependencies can be written. For ease of use the dependency graph can be defined using a mouse-driven graphics device, and from the graph the control flow subroutine can be generated automatically.

The facilities described so far enable SCHEDULE to express the static partitioning of a problem, that is, a partitioning of the problem into parallel processes which can be performed *before* any computation is performed. However, for some problems it is not known until the program is executing that a particular process could, and should, be further subdivided into a number of parallel processes. The requirement here is the facility dynamically to spawn processes during the execution of a process which has been defined statically. SCHEDULE provides such dynamic spawning capabilities.

As noted earlier, SCHEDULE is much more than a language for expressing parallelism. It also includes graphical output to trace the progress of a parallel program by following the flow of execution through the dependency graph. This enables a user to identify serial bottlenecks and load balancing problems in his current parallel implementation, and to redefine his dependency graph to attempt to eliminate these problems.

Transputer Development System

The Transputer Development System (TDS) (Inmos, 1990) does not have the full functionality of SCHEDULE, nevertheless it represents an interesting attempt at providing an integrated program development package. The first point to note is that it is not menu- or command-driven; rather, individual keystrokes (or combined keystrokes) are employed to drive the package. If the host machine is an IBM PC, then TDS is invoked by issuing a DOS command (TDS2). TDS itself then runs on the root transputer in the host machine.

Having entered TDS the programmer can edit, compile and debug programs, and interact with the filestore of the host machine. One of the peculiar, but much admired, facilities of the system is the screen editor, which is known as a *folding editor*. The analogy is with a piece of paper which is folded so that some middle portion is unseen, but labelled. On opening the fold the contents of that hidden portion come into view. Thus, in program development the programmer is encouraged to take a top-down approach. It is natural to make each block or subprogram into a fold, and there may be folds within folds for well-defined sub-problems. Cursor keys can be used to move around the screen, and to move up and down within a fold. Folds are created, deleted, entered, exited, etc., using simple keystrokes. The usefulness of the folding editor is not restricted to occam, or even as a support tool for parallel programming, and several stand-alone versions (for example, *origami*) have proved popular.

TDS has three sets of *utilities*:

- *Compiler*
  The compiler is invoked on a fold containing, amongst other things, the occam source. In the event of a compilation error the folding editor is auto-matically entered at the point of error.
- *Debugger*
  From a core dump of the program the debugger can be invoked and attributes of the program at the point of a run-time error can be determined. In particular, the values of variables can be inspected, as can the status of channels. It should be noted that the debugger can be applied to each processor.
- *File handling*
  TDS files are stored, on the host machine, in a special format. This par-ticularly applies to occam source files constructed using the folding editor.

The file handling utilities permit, amongst other things, the conversion of the occam source file to a standard host (for example, DOS) file.

Also available from Inmos is the Transputer Development Toolkit (TDT), a more conventional command-driven development environment.

2.6.3   Parallel debugging aids and parallel performance profilers

The problems of debugging a parallel program are the same as those of debugging a sequential program with the added (severe) complication of synchronisation and message-passing bugs. We briefly describe the DECON Concurrent Debugger provided on the Intel iPSC/2. As would be expected, DECON provides all the facilities usually provided in a symbolic sequential debugger. In addition it has facilities to find 'lost' messages – messages which have been sent but not received at their destination, and requests to receive messages which have not yet been satisfied. DECON has the ability to 'zoom' in on a particular node of the hypercube and apply a debugger to a particular process; alternatively, its debugging capabilities can be applied to the complete parallel environment of the hypercube.

We have already referred to the profiling capabilities of the SCHEDULE package. A further set of profiling tools worthy of mention and intended as aids to the programmer of a local memory multiprocessor such as an Intel Hypercube or an NCUBE is that which has been developed at Tufts University (see Krumme, Couch, *et al.*, 1989, and the references therein). This supports a real-time performance monitor program, SEEPLEX, which displays graphically the state of each node processor and the activity levels of all the hardware links between processors. The tool is useful in the initial debugging of a parallel program as well as in the subsequent stage of making the program more efficient by eliminating synchronisation delays and communication bottlenecks. Also provided is an off-line post-mortem trace analyser, SEECUBE. Again, this can be used to produce a graphical display of processor and communications activity. Being an off-line tool it uses profiling data which has been collected during the execution of the parallel program and it can provide a more detailed analysis than SEEPLEX. SEECUBE supports multiple views of a single execution and backward and forward replay of executions at variable processing speeds. Both SEEPLEX and SEECUBE make extensive use of high-level colour graphics, using different colours to denote different levels of activity of a processor or of a communications link.

With all debugging and monitoring tools it should be remembered that the collection of measurements (of activity levels, for example) can itself interfere with the behaviour being monitored. The greater the amount of data being collected, the greater the potential for this interference.

Exercises

2.1.   If you have access to a parallel system find out what software support is provided. What (parallel) languages are available? Implement the dot product and time its performance (a) for a fixed number of processors, p, with the vector length, n, increasing, and (b) for a fixed n and p increasing. What support is there for debugging? Force a run-time error (say, division by zero) and examine the information provided.

2.2.   How can the coarse-grain parallel facilities in Cray Fortran be used to implement the dot product? Indicate how to construct an EPF code which operates in a similar way, with tasks being allocated from a list.

2.3.   Modify Codes 2.3 and 2.5 so that they deal with the case that n is not divisible by p.

2.4.   Show how EPF's BARRIER construct may be implemented using events. What modification would you employ to provide the utility of the BARRIER BEGIN...END BARRIER construct?

2.5.   Use the dot product codes of Section 2.4.5 to produce codes in EPF and iPSC/2 Fortran which, given a real n-vector $\mathbf{x}$, find the smallest integer, k, such that $|x_k| \geq |x_i|$, $1 \leq i \leq n$.

2.6.   Why is it not possible to employ a single event in place of a lock in Code 2.4? How could you use events to guard the update of the global sum?

2.7.   The elements of the matrix A are defined by $a_{ij} = a_{i-1,j} + a_{i,j-1}$, for $i, j > 1$, and $a_{ij} = 1$ for $i = 1$ or $j = 1$. Write a section of Fortran code which initialises the first row and column of A and then computes the remaining elements. Parallelise this code using EPF's DO ALL and BARRIER constructs.

2.8.   Use EPF's parallel extensions to parallelise the loop

```
do i=1,n
  x(i) = x(i+1)
end do
```

(Hint: Assume that n is divisible by p and divide the vector into blocks, taking care with edge effects.)

2.9.   Modify the occam code of Code 2.9 so that input from the left is overlapped with output to the right.

2.10.   Write a sequential loop which searches for a particular value (for example,

zero) in an integer vector (it need not find the lowest index in the event of a tie).
How would you parallelise this loop using EPF? How would you parallelise a
general WHILE loop in EPF?

Further reading

For comprehensive introductions to parallel systems from a software point of view see Ben-Ari
(1990), Bustard, Elder, *et al.* (1988), Carriero and Gelernter (1990), Chandy and Misra (1988),
Lakshmivarahan and Dhall (1990) or Perrott (1987). Ben-Ari (1990) uses Ada, occam and Linda;
Bustard, Elder, *et al.* (1988) use Pascal Plus; and Carriero and Gelernter (1990) use C and
Linda as vehicles for expressing parallel processes. Linda is a parallel communication paradigm
developed at Yale University (Ahuja, Carriero, *et al.*, 1986). Fox, Johnson, *et al.* (1988) consider
program development on a hypercube, with algorithms presented in pseudo code. Cok (1991)
considers practical aspects of programming current generation transputer systems. The book
deals with various configurations and programming methodologies, and develops occam software
supporting inter-process communication.

In Section 2.6 we highlighted some of the limitations of the EPF autoparallelisation pre-processor,
particularly with respect to data dependencies. However, rather more sophisticated autoparallelisers for shared memory systems are currently under development. See Polychronopoulos, Girkar,
et al. (1989) and the references therein, for details of the techniques that underpin these tools.

Dongarra, Brewer, *et al.* (1990) describe SHMAP, a tool for indicating array accesses involved in
multiprocessor implementations of linear algebra algorithms.

CHAPTER 3

Numerical libraries

3.1 The need for libraries

Applications programmers in the standard serial environment have become accustomed to the use of software packages to determine solutions to their subproblems. Rather than, say, write their own code to solve a system of linear equations, they reach for an off-the-shelf subroutine which is known to be efficient and reliable. Of the commercially available libraries of numerical software the best known, and most widely used, are those marketed by NAG (Numerical Algorithms Group Limited, Oxford, UK) and IMSL (International Mathematical and Statistical Libraries, Houston, Texas, USA). In addition to these there are a number of packages which are 'public domain'; that is, they are available for the cost of a handling charge only. Examples of these packages are the 'PACK' software which includes

- EISPACK (eigenvalue problems, Smith, Boyle, *et al.*, 1976),
- LINPACK (matrix factorisations and the solution of systems of linear equations, Dongarra, Bunch, *et al.*, 1979),
- MINPACK (the solution of nonlinear equations and nonlinear least squares problems, Moré, Sorensen, *et al.*, 1984),
- QUADPACK (numerical quadrature, Piessens, de Doncker-Kapenga, *et al.*, 1983), and
- LAPACK (linear algebra, Anderson, Bai, *et al.*, 1992).

Many of the routines in these packages have been incorporated into either the NAG or IMSL libraries, or both. As noted in Section 1.2.7, certain of the LIN-PACK routines are frequently used as a standard benchmark by which to measure the comparative performance of different computer systems. Also worthy of mention here are three further sources:

- SLATEC (Sandia, Los Alamos and Air Force Weapons Technical Exchange Committee, Vandevender and Haskell, 1982) which combines the resources of several US Department of Energy laboratories.
- NETLIB (Dongarra and Grosse, 1987), an electronic mail repository which is a source of routines covering many numerical areas (for example, the PACK software described above is available via NETLIB). This software is freely available but comes with the warning that the reliability of any individual routine is not guaranteed.
- The Numerical Recipes software (Press, Flannery, *et al.*, 1986). Again, the reliability of this, and other, software made available in conjunction with published texts on numerical methods is not guaranteed and the quality can be variable, but the codes can be a useful aid to software development. The work of Kahaner, Nash, *et al.* (1989) covers much of the PACK and SLATEC software.

Most of the numerical software available commercially, or in the public domain, is written in Fortran although other languages are also used. For example, NAG now markets its main numerical library in Fortran 77, with subsets in C, Pascal and Ada.

For multiprocessor architectures the choice of true parallel numerical libraries is, at present, rather limited. Part of the reason for this is portability. To maximise the potential benefits to be gained from a multiprocessor system it is necessary to exploit fully the architectural features of that machine. Failure to do this can significantly degrade performance. Thus an algorithm designed, say, for a local memory machine is unlikely to perform optimally on a shared memory machine, and vice versa. Some manufacturers have relied on parallelising compilers to produce parallel versions of standard serial library codes. In the local memory environment such tools are in their infancy and this has led to the development of a limited number of tailor-made parallel numerical libraries.

Central to the solution of many numerical problems is the application of linear algebra. It is not surprising, therefore, that much of the work undertaken in providing both serial and parallel numerical libraries has centred around the provision of efficient codes to perform such tasks. Further, there has been, in recent years, an attempt to categorise the most common components in linear algebra computations. This has resulted in the specification of the BLAS (Basic Linear Algebra Subprograms). The intention is that linear algebra computations should be performed, as far as possible, by making calls to these BLAS 'building blocks'. This development has important repercussions for the applications programmer working in either a serial or a parallel environment. In the next section we define the BLAS (summary specifications are included as an Appendix) and in Section 3.3 we consider possible implementations, with the emphasis on parallel versions. In Section 3.4 we consider the motivation for the development of block algorithms for linear algebra computations which can be expressed in such a way that they make extensive use of the BLAS. We conclude with an overview of operation models for parallel numerical libraries.

3.2   The BLAS

The BLAS are relatively low-level linear algebra operations which are intended as basic building blocks from which higher-level linear algebra routines can be constructed. Because numerical techniques used in other problem areas, such as optimisation, approximation theory and the solution of partial differential equations, make extensive use of these higher-level linear algebra routines, the BLAS can be introduced implicitly in many areas of numerical computation.

The three levels of BLAS are categorised according to their floating-point operation counts:

- Level 1 BLAS (Lawson, Hanson, *et al.*, 1979a; 1979b) are vector-vector operations involving $O(n)$ operations on $O(n)$ data.
- Level 2 BLAS (Dongarra, Du Croz, *et al.*, 1988a; 1988b) are matrix-vector operations involving $O(n^2)$ operations on $O(n^2)$ data.
- Level 3 BLAS (Dongarra, Du Croz, *et al.*, 1990a; 1990b) are matrix-matrix operations involving $O(n^3)$ operations on $O(n^2)$ data.

For each pair of references cited above, the first gives a specification of the routines as to their functionality and parameter lists; the second defines a model Fortran implementation and test software. Routines are defined for single- and double-precision real and complex data types (and, in a few cases, for mixed types). In subsequent chapters we exploit the BLAS as a convenient mechanism with which to express high-level linear algebra algorithms and a detailed description of their functionality is therefore in order.

3.2.1   Level 1 BLAS

The single-precision real Level 1 BLAS, involving vector-vector operations, have the following specifications:

- SUBROUTINE SSWAP(N,X,INCX,Y,INCY)
  sswap interchanges two vectors $(\mathbf{x} \leftrightarrow \mathbf{y})$ of length n. Here
    o the first character of the routine name denotes the data type of the vectors (S, as in the example, and D for single- and double-precision real, C and Z for single- and double-precision complex), and
    o the parameters INCX and INCY are the strides for the vectors X and Y. If, for example, INCX is positive the vector X is regarded as having components X(1), X(1+INCX), X(1+2*INCX), and so on. If INCX is negative then the components are X(1-(N-1)*INCX), X(1-(N-2)*INCX),

etc., that is, the vector is assumed to be stored in reverse order. The use of a non-unit stride, with INCX equal to the declared leading (first) dimension of a matrix, permits a row of that matrix to be regarded as a vector. A column of a matrix can be regarded as a vector by setting INCX equal to 1.

- SUBROUTINE SSCAL(N,ALPHA,X,INCX)
  sscal scales a vector by α ($\mathbf{x} \leftarrow \alpha\mathbf{x}$).
- SUBROUTINE SCOPY(N,X,INCX,Y,INCY)
  scopy copies one vector into another ($\mathbf{y} \leftarrow \mathbf{x}$).
- SUBROUTINE SAXPY(N,ALPHA,X,INCX,Y,INCY)
  saxpy adds a multiple of one vector to another ($\mathbf{y} \leftarrow \alpha\mathbf{x} + \mathbf{y}$).
- FUNCTION SDOT(N,X,INCX,Y,INCY)
  sdot forms the dot product of two vectors (SDOT $\leftarrow \mathbf{x}^T\mathbf{y}$).
- FUNCTION SNRM2(N,X,INCX)
  snrm2 forms the two-norm of a vector (SNRM2 $\leftarrow \|\mathbf{x}\|_2 \equiv \sqrt{\sum_{i=1}^{n} x_i^2}$).
- FUNCTION SASUM(N,X,INCX)
  sasum forms the one-norm of a vector (SASUM $\leftarrow \|\mathbf{x}\|_1 \equiv \sum_{i=1}^{n} |x_i|$).
- FUNCTION ISAMAX(N,X,INCX)
  isamax finds the lowest index of the element of maximum modulus in a vector (the first k such that $|x_k| \geq |x_i|$, $\forall\ i$). Note that it is the second letter here which denotes the type and precision of the argument X.

There are, in addition, a few routines which deal with plane rotations, a discussion of which is deferred until Section 5.2.3. It should be noted that, according to their specification, there is no error reporting associated with the Level 1 BLAS. Thus, for example, if sscal is called with a negative value for N, the vector length, the routine simply returns with X unchanged.

The computers available when the Level 1 BLAS were first conceived in the early 1970s were traditional serial machines and it is for such machines that the Level 1 BLAS are especially appropriate. Because the Level 1 BLAS require only $O\,(n)$ floating-point operations they have too small a granularity fully to exploit the capabilities of vector and parallel computers. With this in mind the larger-grain ($O\,(n^2)$ floating-point operations) Level 2 BLAS were proposed.

3.2.2   Level 2 BLAS

The single-precision real Level 2 BLAS, involving matrix-vector operations, are more numerous but have a small number of basic forms:

- SUBROUTINE SGEMV(TRANS,M,N,ALPHA,A,LDA,X,INCX,BETA,Y,INCY)
  sgemv forms the matrix-vector product $\mathbf{y} \leftarrow \alpha\mathrm{op}(A)\mathbf{x} + \beta\mathbf{y}$, where $\mathrm{op}(A) = A$ or A^T. Here

o the declared first dimension of the matrix A is LDA, and the portion of A to be used in the matrix-vector product is assumed to have dimensions M × N, and

o TRANS is a character string. If the first letter of the string is N or n (that is, no transpose) then op$(A) = A$; if the first letter is T (transpose) or C (conjugate transpose), or their lower-case equivalents, then op$(A) = A^T$ or op$(A) = A^H$, where H denotes conjugate (Hermitian) transpose. The sizes of **x** and **y** must correspond accordingly. Note that it is only the first character of the string TRANS which is significant, so that, for example, the strings 'N', 'No transpose' and 'Nicole' are all equivalent and indicate that op$(A) = A$.

- SUBROUTINE STRSV(UPLO,TRANS,DIAG,N,A,LDA,X,INCX)
  strsv forms the matrix-vector product $\mathbf{x} \leftarrow \text{op}\left(A^{-1}\right)\mathbf{x}$ with A triangular (that is, solves the system of linear equations op$(A)\mathbf{y} = \mathbf{x}$, with the solution **y** overwriting **x**). Here

  o UPLO is a character string which indicates whether A is lower or upper triangular. If the first letter of the string is L or l then A is assumed to be lower triangular; if U or u then the matrix is assumed to be upper triangular, and

  o DIAG is a character string. If the first letter of the string is U or u then A is assumed to have a unit diagonal, and the entries on the diagonal are not accessed; if the first letter is N or n then A is assumed to be non-unit diagonal.

- SUBROUTINE SGER(M,N,ALPHA,X,INCX,Y,INCY,A,LDA)
  sger forms the general rank-one update $A \leftarrow \alpha\mathbf{x}\mathbf{y}^T + A$.

- SUBROUTINE SSYR(UPLO,N,ALPHA,X,INCX,A,LDA)
  ssyr forms the symmetric rank-one update $A \leftarrow \alpha\mathbf{x}\mathbf{x}^T + A$.

- SUBROUTINE SSYR2(UPLO,N,ALPHA,X,INCX,Y,INCY,A,LDA)
  ssyr2 forms the symmetric rank-two update $A \leftarrow \alpha\mathbf{x}\mathbf{y}^T + \alpha\mathbf{y}\mathbf{x}^T + A$.

Again, the initial letter of the routine indicates the data type of the matrices and vectors employed in the operation.

Variants of **sgemv** exist for the case that A is

- banded (**sgbmv**),
- symmetric (**ssymv**),
- symmetric and banded (**ssbmv**),
- symmetric and stored in packed form (**sspmv**),
- triangular (**strmv**),
- triangular and banded (**stbmv**), and
- triangular and stored in packed form (**stpmv**),

with the operation for triangular matrices simplifying to $\mathbf{x} \leftarrow \text{op}(A)\mathbf{x}$. If a matrix is symmetric but stored in normal form, only the upper or lower triangle is accessed. If the lower (upper) triangle is stored in packed form the matrix is represented by a one-dimensional array, with the first n (1) elements containing the first column, the next $n - 1$ (2) elements containing the second column, and so on. For a banded matrix only the co-diagonals containing non-zeros are stored, with rows of the stored array corresponding to diagonals of the original matrix and columns of the array corresponding to columns of the original matrix. Thus, for example, the principal diagonal of a non-symmetric tridiagonal matrix is stored in the second row of the array, starting at position 1, the co-diagonal below is stored in the third row of the array, starting at position 1, and the co-diagonal above is stored in the first row of the array, starting at position 2. For a symmetric, banded matrix either the lower or upper triangle must be provided, but with rows again representing diagonals.

Variants of `strsv` exist for the cases that A is banded (`stbsv`), or is stored in packed form (`stpsv`), and there are variants of `ssyr` and `ssyr2` for the case that A is stored in packed form (`sspr` and `sspr2`).

Unlike the Level 1 BLAS, the specifications for the Level 2 routines require that certain errors should be flagged. These are associated with the correctness of the calling sequence only. In particular, all character strings that begin with an invalid letter (for example, anything other than N, T, C, or their lower-case equivalents, for `TRANS`) should be reported as an error and the execution of the routine terminated. What the specification does not define is precisely how this error reporting should take place. The model implementations in Dongarra, Du Croz, *et al.* (1988b) make calls to the error handling routine `XERBLA` which has the specification

```
SUBROUTINE XERBLA(SRNAME,INFO)
```

Here,

- `SRNAME` (a character string) is the name of the routine which called `XERBLA`, and
- `INFO` (an integer) is the position of the invalid parameter in the parameter list of the calling routine.

It should be noted that only simple checks on the parameters are made; no checks are made in `strsv` and its variants that A is invertible.

The greater granularity of the Level 2 BLAS makes them better suited for implementation on computers which have a vector pipeline unit. However the Level 2 BLAS involve $O\left(n^2\right)$ floating-point operations on $O\left(n^2\right)$ data items and they thus have a compute/communication ratio of $O\left(1\right)$. This makes an efficient implementation of the Level 2 BLAS on a parallel computer difficult (in the sense of achieving a close to linear parallel speed-up), and this is particularly so on a local memory system. Because of this the Level 3 BLAS, which have even greater granularity and, more importantly, have a compute/communication ratio of $O\left(n\right)$, have been proposed.

3.2.3 Level 3 BLAS

The single-precision real Level 3 BLAS involving matrix-matrix operations are small in number:

- SUBROUTINE SGEMM(TRANSA,TRANSB,M,N,K,ALPHA,A,LDA,B,LDB,BETA,C, LDC)

  sgemm forms the matrix matrix product $C \leftarrow \alpha\text{op}(A)\text{op}(B) + \beta C$. Here
  - TRANSA and TRANSB indicate whether A and B are to be transposed,
  - M and N are the numbers of rows and columns of C,
  - K is the inner dimension of the multiplication, so that if, for example, A and B are not to be transposed then A is an M × K matrix, B is a K × N matrix, and
  - LDA, LDB and LDC are the declared leading dimensions of A, B and C respectively.

- SUBROUTINE SSYMM(SIDE,UPLO,M,N,ALPHA,A,LDA,B,LDB,BETA,C,LDC)

  ssymm forms the matrix-matrix product $C \leftarrow \alpha AB + \beta C$ or $C \leftarrow \alpha BA + \beta C$. Here
  - SIDE is a character string which indicates the order of the matrix-matrix multiplication. If the first letter of that string is L or l then A multiplies B on the left; if the letter is R or r then multiplication by A takes place on the right.

- SUBROUTINE SSYRK(UPLO,TRANS,N,K,ALPHA,A,LDA,BETA,C,LDC)

  ssyrk forms the symmetric rank-k update $C \leftarrow \alpha AA^T + \beta C$ or $C \leftarrow \alpha A^T A + \beta C$, where C is an N × N matrix.

- SUBROUTINE SSYR2K(UPLO,TRANS,N,K,ALPHA,A,LDA,B,LDB,BETA,C,LDC)

  ssyr2k forms the symmetric rank-two update $C \leftarrow \alpha AB^T + \alpha BA^T + \beta C$ or $C \leftarrow \alpha A^T B + \alpha B^T A + \beta C$.

- SUBROUTINE STRMM(SIDE,UPLO,TRANSA,DIAG,M,N,ALPHA,A,LDA,B,LDB)

  strmm forms the matrix-matrix product $B \leftarrow \alpha\text{op}(A)B$ or $B \leftarrow \alpha B\text{op}(A)$, where A is a triangular matrix.

- SUBROUTINE STRSM(SIDE,UPLO,TRANSA,DIAG,M,N,ALPHA,A,LDA,B,LDB)

  strsm forms the matrix-matrix product $B \leftarrow \alpha\text{op}(A^{-1})B$ (solves a system of equations with the same coefficient matrix but several right-hand sides) or $B \leftarrow \alpha B\text{op}(A^{-1})$, where A is a triangular matrix.

It can be seen that all Level 3 BLAS routines, in one form or another, represent a matrix-matrix product.

3.3 Implementation of the BLAS

The BLAS constitute specifications of the functionality of routines, but not the way in which they arrive at a particular result. The model implementations mentioned in the previous section perform well in most circumstances. All code is in-line; that is, the Level 3 routines do not make calls to Level 2 or Level 1 routines, and Level 2 routines do not make calls to Level 1 routines, thus avoiding the associated subroutine call overheads.

Since it is intended that the BLAS should form the basis of, and be called from, a large proportion of numerical codes, it is important that as near optimal versions as possible are provided. By using the BLAS to write higher-level numerical codes, portability of the calling software is maintained, whilst a particular implementation of the BLAS may be strictly non-standard if it is written to exploit fully the architectural features of a given machine. One way of achieving this optimal performance is to provide assembler-coded versions. Many computer manufacturers have recognised the importance of the BLAS approach to the development of numerical codes and now provide very efficient versions of these routines.

In the following subsections we consider possible implementations of the three levels of BLAS in turn. In each case we consider the implementation of a particular example routine on a serial machine and then describe how that routine may be adapted to exploit the facilities of both shared and local memory multi-processors.

3.3.1   Level 1 BLAS

Serial implementation
Implementation of the Level 1 BLAS is straightforward. For example, the operation sscal, $\mathbf{x} \leftarrow \alpha\mathbf{x}$, may be expressed as shown in Code 3.1 when the vector has unit stride (INCX $= 1$).

```
do i = 1,n
  x(i) = alpha*x(i)
end do
```

Code 3.1 sscal *for unit stride*

The model implementations referred to at the beginning of Section 3.2 take separate account of the case that the vector has unit stride, in which case they use *loop unrolling* (Exercise 3.1) for increased efficiency, and simply return if $\alpha = 1$ or n is non-positive.

Parallel implementations
Parallelisation of the sscal operation is also, in principle, straightforward and can be achieved by the simple process of loop spreading, whereby each process is given a portion of the vector to scale. If this is done equitably, then, as with the simplified dot product operation outlined in Section 2.4.5, the number of elements of the vector to be dealt with by each process is either $\lfloor n/p \rfloor + 1$ or $\lfloor n/p \rfloor$, where

p is the number of processes (and we assume a one-to-one mapping of processes to processors).

It is instructive to develop a simple timing model of this parallel implementation of sscal. The number of floating-point operations performed by each process concurrently is at most $\lceil n/p \rceil$ multiplications, where $\lceil n/p \rceil$ is n/p if n is divisible by p, and $\lfloor n/p \rfloor + 1$ otherwise. (We ignore the integer arithmetic associated with array access and loop overheads.) Thus, if T_f is the time taken to perform a single floating-point operation, the time to compute the sscal operation using p concurrent processes is $\lceil n/p \rceil T_f$. As p increases, therefore, the computation time decreases approximately linearly. If the components of the vector $\mathbf{x}$ (and the multiplier α) are already available locally (that is, on the processor on which the computation is to be performed), and the scaled vector $\alpha \mathbf{x}$ is to remain only locally available, then we can achieve linear speed-up.

Unfortunately, the data may not be immediately accessible. In a shared memory environment it may be necessary to move the required elements of the vector $\mathbf{x}$ from the shared global memory into the local caches of the processors; in a local memory environment it may be necessary to distribute the required vector elements from a host processor to the slave processors. Let T_c be the time taken to move a single floating-point value. (We conveniently make the simplifying assumption that all data movements take the same amount of time.) Then, in the worst case, the time required to distribute and gather $\mathbf{x}$ is $2nT_c$ (and to this we should add the time associated with the distribution of α, although since this is less significant by an order of magnitude we omit it from consideration). Hence the total time required to complete the parallel sscal operation is

$$total\ time = \lceil n/p \rceil T_f + 2nT_c. \tag{3.1}$$

This simple timing model indicates that, as p increases, the computation time decreases but the communication time remains unchanged. We are therefore unlikely to obtain any significant parallel speed-up.

For a local memory system it is more realistic to assume that the time taken to move r floating-point numbers between any two processors is $T_s + rT_c$, where T_s represents a *start-up time*. The total time required to complete the parallel sscal operation is now

$$total\ time = \lceil n/p \rceil T_f + 2pT_s + 2nT_c. \tag{3.2}$$

This model indicates that as p increases the computation time decreases, but the communication start-up time increases. The net effect is that the total time may well also increase with p. For the iPSC/2, typical values for T_f, T_s and T_c are approximately 5.95, 660.0 and 1.44 μs respectively (Bomans and Roose, 1989). Assuming that we wish to scale a vector with n components the execution times predicted by (3.2) for $n = 1000$ and $n = 10\,000$ and for various values of p are given in Figure 3.1. We see that the parallel speed-up is not substantial, even for a vector with $10\,000$ components. For local memory systems, therefore, parallel

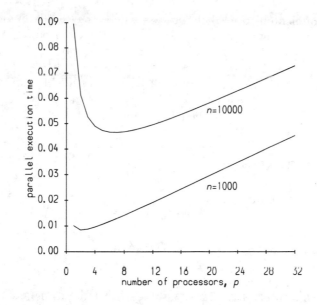

Figure 3.1 *Predicted times for* sscal *on the iPSC/2*

implementation of Level 1 BLAS operations involving distribution of data is not recommended. On some systems (for example, transputers) it may be possible to exploit *excess parallelism*, whereby the distribution of data takes place at the same time as some other, unrelated, computation. If this computation is significant then it may be possible to 'hide' the communication in this way.

In a shared memory environment the situation is more hopeful, but (3.1) indicates that the speed-up achievable is still likely to be somewhat limited. Additional to the costs indicated are the overheads associated with spawning tasks and allocating work to those tasks. We recall that, in EPF, the costs associated with task creation are incurred once only, so that if, as is likely, the evaluation of sscal is but one of a number of computations to be performed in parallel, these overheads may not be significant. The work allocation costs, however, may overwhelm any potential speed-up unless the vector length is very large.

The difficulty in obtaining a parallel speed-up for the Level 1 BLAS can be summarised by noting that the compute/communication ratio of the routines is $O\left(1\right)$. If T_c and T_s are significant compared to T_f then a marked parallel speed-up is unlikely to be achievable.

As a further example of a Level 1 BLAS we consider the sdot operation SDOT$\leftarrow \mathbf{x}^T \mathbf{y}$. A parallel version of a simplified form of this operation was considered in Section 2.4.5. Compared to sscal there is an added complication in a parallel implementation of sdot in that a global sum must be accumulated. In a shared memory environment this means that it is necessary to synchronise the updating of a global variable; in a local memory environment partial sums must be communicated to a single process where they are accumulated. Thus the communication costs of sdot are even greater than those for sscal. We are therefore even less likely to observe good parallel performance. Thus, despite the fact that we have used the sdot operation as an example of a parallel code to run on the iPSC/2, we stress here that this code is not ideally suited to a parallel implementation (and, in fact, the overall execution time often increases as more processors are used to calculate the dot product).

3.3.2   Level 2 BLAS

Serial implementation

Since the Level 2 operations are fairly elementary their implementation is, again, straightforward. However, there is the added complication that various alternative implementations are possible. For example, one way of expressing the sgemv operation $\mathbf{y} \leftarrow \alpha \mathrm{op}(A)\mathbf{x} + \beta \mathbf{y}$ with $\mathrm{op}(A) = A$ is to write it as $y_i \leftarrow \alpha \bar{\mathbf{a}}_i^T \mathbf{x} + \beta y_i$, where $\bar{\mathbf{a}}_i^T$ denotes the ith row of A. (To avoid confusion with our notation for columns of matrices, we use barred vectors to denote rows of matrices.) If we separate out the scaling of $\mathbf{y}$ and assume that TRANS = 'N' and that INCX = INCY = 1, then the code which implements this version of the sgemv operation is given by Code 3.2. In practice we would consider accumulating the sum in the inner loop in a real variable to avoid the overheads associated with repeated array accesses, and multiply this by α once only outside the loop. Even if we did not do this, we might hope that an optimising compiler would do it for us. Further, we would put conditional guards on the first loop and the double loops to avoid unnecessary computation when $\alpha = 0$ and/or $\beta = 1$.

```
do i = 1,m
  y(i) = beta*y(i)
end do
do i = 1,m
  do j = 1,n
    y(i) = y(i)+alpha*a(i,j)*x(j)
  end do
end do
```

Code 3.2 sgemv *implemented in terms of dot products*

Taking Code 3.2 as it stands we see that the first loop is an sscal operation. In addition, of the two nested loops, the inner one is the dot product of $\bar{\mathbf{a}}_i^T$ with $\mathbf{x}$ – an sdot operation. Hence we may express the sgemv operation as shown in Code 3.3. The use of the increment lda, the declared first dimension of a, as the third actual parameter of sdot ensures that the ith row of a is regarded as a vector. Writing the code in this way, with calls being made to Level 1 BLAS routines, means that subroutine call overheads are incurred. However, if particularly efficient versions of sdot and sscal are available then this may more than compensate for such overheads. Here, and elsewhere, in expressing our algorithms we make full use of the BLAS notation in the interests of clarity, whilst recognising that in certain circumstances the use of in-line code may be more efficient.

```
call sscal(m,beta,y,1)
do i = 1,m
  y(i) = y(i)+alpha*sdot(n,a(i,1),lda,x,1)
end do
```

Code 3.3 sdot *version of* sgemv

An alternative approach to the **sgemv** operation is to recognise that, again for a non-transposed matrix, it may be written $\mathbf{y} \leftarrow \alpha \sum_{j=1}^{n} x_j \mathbf{a}_j + \beta \mathbf{y}$, with $\mathbf{a}_j$ being the jth column of A. To implement the operation in this form we simply rearrange the order of the loops in Code 3.2. The inner loop is now a **saxpy** operation, and hence this implementation can be written as shown in Code 3.4.

```
call sscal(m,beta,y,1)
do j = 1,n
  call saxpy(m,alpha*x(j),a(1,j),1,y,1)
end do
```

Code 3.4 saxpy *version of* sgemv

A sensible modification of this code would be to avoid the **saxpy** call if x_j is zero. For this, and other reasons, a Fortran implementation of the **saxpy** variant of the **sgemv** code is to be preferred to the earlier **sdot** variant on a uniprocessor. By making references to columns we may hope to minimise the amount of data transfer to/from a local cache and/or secondary storage. In addition, if the processor on which this code is to be executed possesses a pipeline unit, then the **saxpy** variant is again likely to perform better.

We have considered here the case $\text{op}(A) = A$ only. If $\text{op}(A) = A^T$ then the columns of A are the rows of A^T, and vice versa. In this case the **sdot** variant may turn out to be the more efficient.

Parallel implementations

We now turn to the parallel implementation of the Level 2 BLAS **sgemv** operation. The parallelisation of the **sdot** variant is straightforward since the dot products are independent operations and loop spreading can be applied to the outermost loop. In a shared memory system the algorithm will achieve its best performance if the whole of A can be stored in the cache of each processor. This follows from Fortran's storage of two-dimensional arrays by columns; since the independent dot product operations access the array by rows, paging overheads will be incurred if the array is too large for the cache. This will require a large cache and/or a small value for n. In a local memory system the algorithm will perform well if A is pre-distributed by rows, say, process $i, 0 \leq i \leq p-1$, has rows $ib+1, ib+2, \ldots, (i+1)b$, where $b = n/p$ is some chosen block size, and we assume, for simplicity, that n is divisible by p, the number of processes. To avoid any communication each process will require a local copy of the whole of **x** and local access to the components of **y** which correspond to its rows of A.

The situation with regard to the `saxpy` version is less clear since the operations of the outermost loop are not independent; each iteration is dependent on the partial column sum calculated by the previous iteration. One way to proceed would be for each process to be allocated a number of columns of A. (In a shared memory environment, if these are consecutive we are likely to reduce the amount of data traffic to and from the local cache.) Each process performs a number of `saxpy` operations corresponding to its columns in local vectors and these partial results are then aggregated. What we have, in essence, is a vector accumulation of partial sums and the situation is analogous to the parallel `sdot` considered in Section 3.3.1. We have the problem of synchronising access to a global sum (for a shared memory system) or sending partial sums to an accumulating process (for a local memory system with A distributed by columns). An alternative approach would be to perform a sequence of independent `saxpy` operations, dealing with shorter vectors of length $b = n/p$ in each process. This suggests a row-based array distribution, either implicitly in the case of a shared memory system, or explicitly in the case of a local memory system. In the former case paging overheads may again be a problem.

Whichever approach is adopted, we observe that the number of operations involved is $3n^2$ for the matrix-vector multiplication and n for the multiplication of **y** by a scalar. If, in a local memory environment, the data needs to be distributed, this involves the transmission of $n^2 + 2n$ floating-point numbers (and the two scalars, α and β), with the return transmission of a further n floating-point numbers defining the result. Thus the compute/communication ratio is, again, $O(1)$; for a local memory system we face the same problem as was observed for the Level 1 BLAS and are therefore unlikely to achieve a significant parallel speed-up.

3.3.3   Level 3 BLAS

Serial implementation

As with the lower-level routines, the Level 3 BLAS are easily implemented, although the number of different possibilities (corresponding to different orderings of the loops) is even greater than was the case for the Level 2 BLAS. As an example, we consider the `sgemm` operation $C \leftarrow \alpha AB + \beta C$, where C is an $m \times n$ matrix, A is $m \times k$ and B is $k \times n$. This operation can be carried out using Code 3.5, in which we have ordered the `i` and `j` loops so that the elements of the assigned matrix C are calculated in column order. Essentially, each element of C

```
do j = 1,n
  do i = 1,m
    c(i,j) = beta*c(i,j)
    do l = 1,k
      c(i,j) = c(i,j)+alpha*a(i,l)*b(l,j)
    end do
  end do
end do
```

Code 3.5 sgemm *in terms of dot products*

is computed as a dot product.

With the ordering of the loops shown in Code 3.5 it is straightforward to replace the two innermost, i and l, loops by a single call to the Level 2 BLAS sgemv, and this yields Code 3.6. For $j = 1, 2, \ldots, n$, this code updates c_j, the jth column of C, by adding to its scaled value α times the product of the matrix A and b_j, the jth column of B; that is,

$$c_j \leftarrow \alpha A b_j + \beta c_j.$$

We refer to this implementation as the column-based sgemv version of sgemm.

```
do j = 1,n
  call sgemv('n',m,k,alpha,a,lda,b(1,j),1,beta,c(1,j),1)
end do
```

Code 3.6 *Column-based* sgemv *version of* sgemm

An alternative implementation stems from expressing the sgemm operation as $C^T \leftarrow \alpha B^T A^T + \beta C^T$. To express the operation in this form we simply reorder the j and i loops, so that the elements of C are calculated in row order. If $\bar{c}_i^T$ and $\bar{a}_i^T$ denote the ith rows of C and A respectively, then the innermost, j and l, loops of the code so transformed represent

$$\bar{c}_i^T \leftarrow \alpha \bar{a}_i^T B + \beta \bar{c}_i^T,$$

or, equivalently,

$$\bar{c}_i \leftarrow \alpha B^T \bar{a}_i + \beta \bar{c}_i.$$

Thus, remembering that $\bar{c}_i^T$ and $\bar{a}_i^T$ represent rows of the matrices C and A respectively, we see that the sgemm operation may be expressed as shown in Code 3.7. This code updates the ith row of C, for $i = 1, 2, \ldots, m$, by forming α times the product of B^T with the ith row of A. We therefore refer to it as the row-based sgemv version of sgemm.

```
do i = 1,m
  call sgemv('t',k,n,alpha,b,ldb,a(i,1),lda,beta,c(i,1),ldc)
end do
```

Code 3.7 *Row-based* sgemv *version of* sgemm

A third way of expressing **sgemm** in terms of Level 2 BLAS results from writing the **sgemm** operation as $C \leftarrow \alpha \sum_{l=1}^{k} \mathbf{a}_l \bar{\mathbf{b}}_l^T + \beta C$, where $\mathbf{a}_l$ and $\bar{\mathbf{b}}_l^T$ are the lth column and the lth row of A and B respectively. To implement the operation expressed in this form we need to reorder the loops so that the 1 loop is outermost, which gives Code 3.8.

```
do j = 1,n
  do i = 1,m
    c(i,j) = beta*c(i,j)
  end do
end do
do l = 1,k
  do j = 1,n
    do i = 1,m
      c(i,j) = c(i,j)+alpha*a(i,l)*b(l,j)
    end do
  end do
end do
```

Code 3.8 sgemm *with* 1 *loop outermost*

The innermost, i and j, loops of Code 3.8 implement the rank-one update

$$C \leftarrow \alpha \mathbf{a}_l \bar{\mathbf{b}}_l^T + C.$$

Thus the **sgemm** operation can be implemented by first scaling the matrix C, using n calls to the Level 1 BLAS operation **sscal**, and then making k calls to the Level 2 BLAS **sger** (Code 3.9).

```
do j = 1,n
  call sscal(m,beta,c(1,j),1)
end do
do l = 1,k
  call sger(m,n,alpha,a(1,l),1,b(l,1),ldb,c,ldc)
end do
```

Code 3.9 sgemm *in terms of calls to* sger

In a Fortran environment, where matrices are stored by columns, there would seem to be nothing to be gained by implementing the row-based **sgemv** version of **sgemm** rather than the column-based version. The comparison between the column-based **sgemv** version and the **sger** version is less clear-cut and may, to some extent, depend on the relative dimensions of the matrices. Again, we have considered the non-transposed case only. If $op(A) = A^T$ and/or $op(B) = B^T$ then our conclusions may need to be modified.

Parallel implementations
The column-based **sgemv** version of **sgemm** decomposes the multiplication into n independent processes. Thus loop spreading enables parallelism to be applied

at the outermost level. For a shared memory system the allocation of work to block columns will help to minimise the amount of data traffic between caches and global memory. For a local memory system the algorithm will be appropriate if B and C are distributed by columns, whether blocked or scattered. In either case it is desirable for the whole of A to be available locally. This involves much replication of data and may preclude the multiplication of very large matrices on a local memory system without simulated paging.

The **sger** version of **sgemm** decomposes the multiplication into k processes. However, these are not independent since each is updating the result matrix C. Each process could update a local matrix C; these partial updates of C would then need to be accumulated. In essence we have a matrix accumulation of partial sums, and so we have the problem of synchronising access to the global sum (for a shared memory system) or sending partial sums to an accumulating process (for a local memory system with A distributed by columns and B distributed by rows). However, we note that this local memory **sger** version does not require replication of either of the matrices A or B. In contrast, because of Fortran's column storage, accessing B by rows in a shared memory environment may result in significant data traffic to and from the caches of the processors.

Whichever variant is adopted we observe that the operation count of **sgemm** is $3mnk$, whilst the amount of data on which these operations are performed is $mn + nk + mk$. Assuming $m = n = k$, we have a compute/communication ratio of $O(n)$, and we can therefore expect good parallel speed-up.

3.3.4 Cannon's algorithm for matrix-matrix multiplication

The aim of this book is to investigate the derivation and implementation of parallel numerical algorithms. As with the Levels 1, 2 and 3 BLAS, our starting point will often be the best serial algorithm whose potential for parallelism we then attempt to exploit. There are good reasons for taking this approach; the serial algorithms have a good pedigree and have shown themselves to be reliable and robust. However, the blinkered approach (a uniprocessor as the target machine) taken in the development of the serial algorithm may make parallelisation at best difficult, and occasionally impossible, and Amdahl's law (Section 1.4.3) may mean that we obtain very little parallel speed-up. Moreover, it is frequently the case that we can achieve improved results by taking a step backwards and developing, without preconceived ideas, new algorithms which are specifically targeted at multiprocessors.

The first such new algorithm we consider in this book, due to Cannon (1969), performs matrix-matrix multiplication (**sgemm**) on a local memory multiprocessor without replication of data. To describe the algorithm we assume that there are p^2 processors available, connected in the form of a torus, with the processors labelled as shown in Figure 3.2. We further assume, for simplicity, that A, B and C are $n \times n$ matrices, with n divisible by p, and that there are no transposes

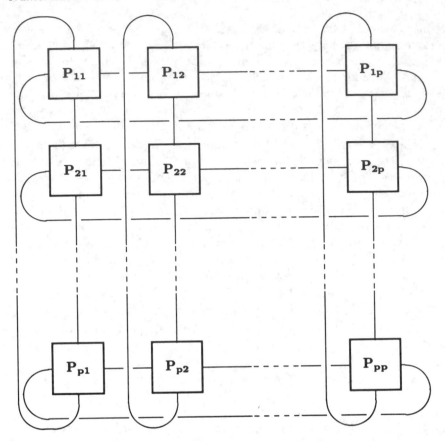

Figure 3.2 *Processor configuration for Cannon's algorithm*

to be performed. Then we can write A, B and C as $p \times p$ block matrices with each square block A_{ij}, B_{ij}, C_{ij} being of order b, where $b = n/p$. Matrix-matrix multiplication is the same, whether the matrix elements are real values or blocks (submatrices) of real values. Here, we wish to form

$$C_{ij} = \alpha \sum_{k=1}^{p} A_{ik} B_{kj} + \beta C_{ij},$$

for $i, j = 1, 2, \ldots, p$.

The start position for the algorithm is that the (i, j) block of each matrix is initially stored on processor $\mathbf{P}_{ij}$. The process, P_{ij}, resident on processor $\mathbf{P}_{ij}$ then executes the following steps:

- Step 0:
  Scale my block of C by β.

- Step 1:
  - (a) Shift my block of A i places to the West.
  - (b) Shift my block of B j places to the North.
  - (c) Update my block of C by adding to it α times the product of my new blocks of A and B.
- Step k, for $k = 2, 3, \ldots, p$:
  - (a) Shift my block of A one place to the East.
  - (b) Shift my block of B one place to the South.
  - (c) Update my block of C by adding to it α times the product of my new blocks of A and B.

Here, North means the next processor up in the grid, etc. Using the wrap-around connectivity of the processors the shifting operation is circular. Hence, for example, if process P_{pp} shifts a block one place East then the receiving process is P_{p1}, whilst it shifts a block one place South to process P_{1p}. Each step of the algorithm is executed concurrently, with synchronisation between steps; that is, all processes execute step 0 in parallel, synchronise, execute step 1 in parallel, synchronise, and so on. Note that in principle each processor has in store only one block (which we refer to as 'my block') of each of the matrices A, B and C, at any one time, although there may be a certain amount of overlap as blocks are shifted.

To see how the algorithm implements the **sgemm** operation we consider in detail a 3×3 torus of processors; for simplicity the processor connectivity is omitted from the diagrams which follow. Each diagram shows the data distribution after the shifts have taken place and, where appropriate, the **sgemm** operation required to update the blocks of C.

- Step 0:
  Scale my block of C.

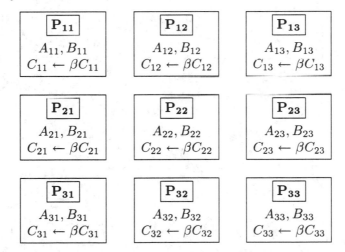

- Step 1:
  - (a) Shift my block of A i places to the West.
  - (b) Shift my block of B j places to the North.
  - (c) Accumulate the scaled product of the A and B blocks in the C block.

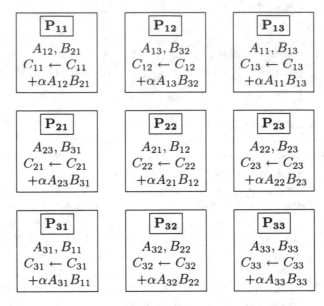

- Step 2:
  - (a) Shift my block of A one place to the East.
  - (b) Shift my block of B one place to the South.
  - (c) Accumulate the scaled product of the A and B blocks in the C block.

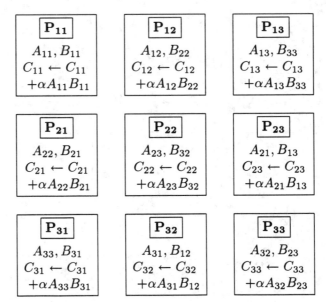

- Step 3:

  (a) Shift my block of A one place to the East.

  (b) Shift my block of B one place to the South.

  (c) Accumulate the scaled product of the A and B blocks in the C block.

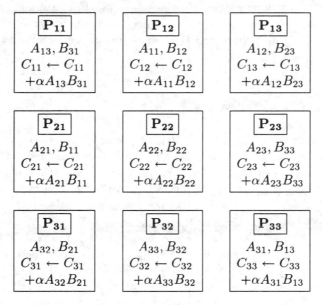

The net effect of the four steps is that each process P_{ij}, $i, j = 1, 2, 3$, computes

$$C_{ij} \leftarrow \alpha \sum_{k=1}^{3} A_{ik} B_{kj} + \beta C_{ij}.$$

The essential feature of this, our first example of a *block algorithm*, is that the matrices involved in the **sgemm** operation are subdivided into blocks (submatrices), and we found it convenient to map these blocks to processors arranged in a square lattice. Ignoring the shifting, the algorithm splits the problem into a number of subproblems which can be performed concurrently; each subproblem happens to be an **sgemm** operation but for much smaller matrices than the original **sgemm** operation. There remains a requirement, therefore, for an efficient serial implementation of **sgemm**.

Cannon's algorithm may also be implemented on a shared memory multiprocessor although, since the memory is a global resource, there is no need explicitly to shift blocks of matrices around. The algorithm is likely to be most efficient, in terms of minimising the movement of data to and from the caches of the processors, if the matrices are stored as block matrices in four-dimensional arrays of dimensions $b \times b \times p \times p$. A parallel implementation of **sgemm**, without Cannon's subdivision and involving simple loop spreading, as described in Section 3.3.3, is likely to be an even better approach.

3.4   Block algorithms

In the previous section we described Cannon's algorithm for matrix-matrix multiplication on a local memory multiprocessor. The study shows how a blocking strategy may be used to reduce a large linear algebra problem into a number of subproblems. We now describe a further block algorithm which more fully motivates the block approach, and then show how to express it in terms of Level 3 BLAS operations.

3.4.1   Block *LU* factorisation

One way of solving the system of linear equations $A\mathbf{x} = \mathbf{b}$, in which A is a (non-singular) $n \times n$ matrix and $\mathbf{x}$ and $\mathbf{b}$ are n-vectors, is first to express A as $A = LU$, the product of L, a unit diagonal, lower triangular matrix, and U, an upper triangular matrix. (Full details of how such a factorisation can be constructed and how the solution to the system may subsequently be obtained are given in the next chapter. For the moment we ask the reader to bear with us and accept that, under suitable conditions, such a factorisation exists and can be readily constructed.) We note that the computation can be arranged so that the factors L and U overwrite A, with the unit diagonal of L being implicit. We also note that the operations count for the LU factorisation is $O\left(n^3\right)$, compared with the $O\left(n^2\right)$ of the data to be manipulated, implying that we have the potentially favourable compute/communication ratio of $O\left(n\right)$.

We begin by expressing A as a block matrix of the form

$$
A = \begin{pmatrix}
A_{11} & A_{12} & \cdots & A_{1m} \\
A_{21} & A_{22} & \cdots & A_{2m} \\
\vdots & \vdots & \ddots & \vdots \\
A_{m1} & A_{m2} & \cdots & A_{mm}
\end{pmatrix},
$$

where each A_{ij} is a $b \times b$ matrix and $n = bm$. For simplicity we have assumed that the matrix A has a uniform block size, b, and that n is divisible by b. We aim to decompose A in the form $A = LU$, where L is a block unit lower triangular matrix and U is a block upper triangular matrix, that is

$$
L = \begin{pmatrix}
I_b & 0 & \cdots & 0 \\
L_{21} & I_b & \cdots & 0 \\
\vdots & \vdots & \ddots & \vdots \\
L_{m1} & L_{m2} & \cdots & I_b
\end{pmatrix}, \qquad
U = \begin{pmatrix}
U_{11} & U_{12} & \cdots & U_{1m} \\
0 & U_{22} & \cdots & U_{2m} \\
\vdots & \vdots & \ddots & \vdots \\
0 & 0 & \cdots & U_{mm}
\end{pmatrix},
$$

where each block is $b \times b$ and I_b is the $b \times b$ identity matrix.

Under the assumption that A_{11} is invertible (non-singular), we define the $n \times b$

matrix $L^{(1)}$ by

$$L^{(1)} = \begin{pmatrix} I_b \\ L_2^{(1)} \\ L_3^{(1)} \\ \vdots \\ L_m^{(1)} \end{pmatrix}, \tag{3.3}$$

where

$$L_i^{(1)} = A_{i1} A_{11}^{-1}, \qquad i = 2, 3, \ldots, m, \tag{3.4}$$

is a $b \times b$ matrix. We also introduce the $b \times n$ matrix $U^{(1)}$ as

$$U^{(1)} = \left(U_1^{(1)}, U_2^{(1)}, \ldots, U_m^{(1)} \right), \tag{3.5}$$

where $U_j^{(1)} = A_{1j}$ is a $b \times b$ matrix. Then

$$A = L^{(1)} U^{(1)} + \tilde{A}^{(1)},$$

where

$$\tilde{A}^{(1)} = \begin{pmatrix} 0 & 0 & \cdots & 0 \\ 0 & \tilde{A}_{22}^{(1)} & \cdots & \tilde{A}_{2m}^{(1)} \\ \vdots & \vdots & \ddots & \vdots \\ 0 & \tilde{A}_{m2}^{(1)} & \cdots & \tilde{A}_{mm}^{(1)} \end{pmatrix}, \tag{3.6}$$

and for $i, j = 2, 3, \ldots, m$,

$$\begin{aligned} \tilde{A}_{ij}^{(1)} &= A_{ij} - L_i^{(1)} U_j^{(1)}, \\ &= A_{ij} - A_{i1} A_{11}^{-1} A_{1j}, \end{aligned} \tag{3.7}$$

is also $b \times b$.

This can now be repeated on the non-zero blocks of $\tilde{A}^{(1)}$, and so on. After $k-1$ such stages of the block elimination procedure we have the partially eliminated matrix

$$\tilde{A}^{(k-1)} = \begin{pmatrix} 0 & 0 & \cdots & 0 & \cdots & \cdots & 0 \\ 0 & 0 & & \vdots & & & \vdots \\ \vdots & \vdots & \ddots & \vdots & & & \vdots \\ 0 & \cdots & \cdots & 0 & \cdots & \cdots & 0 \\ \vdots & & & \vdots & \tilde{A}_{kk}^{(k-1)} & \cdots & \tilde{A}_{km}^{(k-1)} \\ \vdots & & & \vdots & \vdots & & \vdots \\ 0 & \cdots & \cdots & 0 & \tilde{A}_{mk}^{(k-1)} & \cdots & \tilde{A}_{mm}^{(k-1)} \end{pmatrix}, \tag{3.8}$$

and the kth stage, which eliminates the next block of b rows and columns, is described by

$$\tilde{A}^{(k-1)} = L^{(k)} U^{(k)} + \tilde{A}^{(k)},$$

where $L^{(k)}$ and $U^{(k)}$ are $n \times b$ and $b \times n$ matrices, respectively, with blocks $L_i^{(k)}$ and $U_i^{(k)}$. Now, for $i \leq k - 1$ we let $L_i^{(k)} = U_i^{(k)} = 0$, and also choose $L_k^{(k)} = I_b$. The aim is then to choose the remaining blocks of $L^{(k)}$ and $U^{(k)}$ so that $\tilde{A}_{ij}^{(k)} = 0$, for $i \leq k$ or $j \leq k$. Assuming that $\tilde{A}_{kk}^{(k-1)}$ is non-singular, the desired effect is obtained by choosing

$$L^{(k)} = \begin{pmatrix} 0 \\ \vdots \\ 0 \\ I_b \\ L_{k+1}^{(k)} \\ \vdots \\ L_m^{(k)} \end{pmatrix}, \qquad U^{(k)} = \left(0, \ldots, 0, U_k^{(k)}, U_{k+1}^{(k)}, \ldots, U_m^{(k)}\right), \qquad (3.9)$$

where, for $i = k + 1, k + 2, \ldots, m$,

$$L_i^{(k)} = \tilde{A}_{ik}^{(k-1)} \left(\tilde{A}_{kk}^{(k-1)}\right)^{-1},$$

and for $j = k, k + 1, \ldots, m$,

$$U_j^{(k)} = \tilde{A}_{kj}^{(k-1)}.$$

Then

$$\tilde{A}^{(k)} = \begin{pmatrix} 0 & 0 & \ldots & 0 & \ldots & \ldots & 0 \\ 0 & 0 & & \vdots & & & \vdots \\ \vdots & & \ddots & \vdots & & & \vdots \\ 0 & \ldots & \ldots & 0 & \ldots & \ldots & 0 \\ \vdots & & & \vdots & \tilde{A}_{k+1,k+1}^{(k)} & \ldots & \tilde{A}_{k+1,m}^{(k)} \\ \vdots & & & \vdots & \vdots & & \vdots \\ 0 & \ldots & \ldots & 0 & \tilde{A}_{m,k+1}^{(k)} & \ldots & \tilde{A}_{mm}^{(k)} \end{pmatrix},$$

where, for $i, j = k + 1, k + 2, \ldots, m$,

$$\tilde{A}_{ij}^{(k)} = \tilde{A}_{ij}^{(k-1)} - L_i^{(k)} U_j^{(k)},$$
$$= \tilde{A}_{ij}^{(k-1)} - \tilde{A}_{ik}^{(k-1)} \left(\tilde{A}_{kk}^{(k-1)}\right)^{-1} \tilde{A}_{kj}^{(k-1)}.$$

After m stages of this elimination procedure we find that

$$A = L^{(1)}U^{(1)} + \tilde{A}^{(1)},$$
$$= L^{(1)}U^{(1)} + L^{(2)}U^{(2)} + \tilde{A}^{(2)},$$
$$\vdots$$
$$= L^{(1)}U^{(1)} + L^{(2)}U^{(2)} + \cdots + L^{(m)}U^{(m)} + \tilde{A}^{(m)},$$

where $\tilde{A}^{(m)}$ is the null matrix. Hence

$$A = \sum_{k=1}^{m} L^{(k)} U^{(k)} = LU, \tag{3.10}$$

where $L = \left(L^{(1)}, L^{(2)}, \ldots, L^{(m)}\right)$ is a block, unit diagonal, lower triangular matrix and

$$U = \begin{pmatrix} U^{(1)} \\ U^{(2)} \\ \vdots \\ U^{(m)} \end{pmatrix},$$

is block upper triangular. In terms of our initial notation, $L_{ij} = L_i^{(j)}$ and $U_{ij} = U_j^{(i)}$.

The block algorithm which we have described in this subsection might be termed a 'true' block LU factorisation. The algorithm produces block matrices; U is a block upper triangular matrix, but it is not upper triangular in an elementwise sense. We emphasise the argument since it contrasts with the *blocked* or *partitioned algorithms* of LAPACK which we consider in Section 3.4.3. It has recently been shown (Demmel and Higham, 1991) that, for a block size $b > 1$, stability of the true block LU factorisation is questionable.

3.4.2   Block LU factorisation in terms of BLAS

We now consider the computation of the LU factorisation of Section 3.4.1 in more detail and attempt to decompose it into calls to Level 3 BLAS. It is sufficient to consider the first stage of the elimination procedure, described by (3.3)–(3.7); subsequent stages involve similar computations.

The block matrices $L_i^{(1)}$ are defined in terms of A_{11}^{-1}. However, it is not necessary explicitly to invert A_{11}. Instead we form its LU factors; that is, we determine $\hat{L}_{11}$ and $\hat{U}_{11}$ such that

$$A_{11} = \hat{L}_{11}\hat{U}_{11}, \tag{3.11}$$

where $\hat{L}_{11}$ is a unit diagonal lower triangular matrix and $\hat{U}_{11}$ is upper triangular. Note that this computation (forming the LU factors of A_{11}) cannot be performed by making explicit calls to Level 3 BLAS but requires a call to a high-level linear algebra routine which computes the LU factors of a submatrix. Then, for $i = 2, 3, \ldots, m$,

$$\begin{aligned} L_i^{(1)} &= A_{i1}\hat{U}_{11}^{-1}\hat{L}_{11}^{-1}, \\ &= Z_i\hat{L}_{11}^{-1}, \end{aligned} \tag{3.12}$$

where

$$Z_i = A_{i1}\hat{U}_{11}^{-1}. \tag{3.13}$$

The computation of each Z_i from (3.13) is a particular instance of the Level 3 BLAS operation strsm, as is the computation of each $L_i^{(1)}$ using (3.12). Thus each of the block matrices $L_i^{(1)}$, $i = 2, 3, \ldots, m$, can be constructed using two calls to strsm. Further, the construction of each $L_i^{(1)}$ is independent of that of every other, and so all of the $L_i^{(1)}$ can be constructed concurrently.

Alternatively, if we let $L^{(1)} = \begin{pmatrix} I_b \\ \tilde{L}^{(1)} \end{pmatrix}$, where

$$\tilde{L}^{(1)} = \begin{pmatrix} L_2^{(1)} \\ L_3^{(1)} \\ \vdots \\ L_m^{(1)} \end{pmatrix},$$

is a $(m-1)b \times b$ matrix, then (3.4) can be expressed as

$$\tilde{L}^{(1)} = A^{(1)} A_{11}^{-1}, \tag{3.14}$$

where

$$A^{(1)} = \begin{pmatrix} A_{21} \\ A_{31} \\ \vdots \\ A_{m1} \end{pmatrix},$$

is also a $(m-1)b \times b$ matrix. Hence, $\tilde{L}^{(1)}$ may be computed using only two calls to strsm, where there are now $(m-1)b$ right-hand sides and where any parallelism must be exploited *within* the Level 3 BLAS strsm. The latter strategy is more closely related to the LAPACK approach which we describe in Section 3.4.3. Whichever strategy is adopted, the newly generated matrices overwrite the first block row and first block column of A.

The rest of the first stage of the algorithm forms the matrix $\tilde{A}^{(1)}$ defined by (3.6) and (3.7). From (3.7) the $(m-1) \times (m-1)$ non-zero blocks of $\tilde{A}^{(1)}$ can be computed (concurrently) using calls to the Level 3 BLAS operation sgemm. Alternatively, since

$$\tilde{A}^{(1)} = A - L^{(1)} U^{(1)},$$

we can compute the whole of $\tilde{A}^{(1)}$ using a single call of sgemm, with the parallelism again being exploited within the BLAS routine.

The study of the LU factorisation is a further illustration of how a large linear algebra problem can be divided into a number of subproblems. It also shows that if those subproblems make extensive use of BLAS operations, and, in particular, the coarse-grain Level 3 BLAS, then by providing efficient, parallel, implementations of the BLAS we can hope to obtain immediate improvements on a shared memory multiprocessor. Further, we will have achieved some portability, since the only non-portable components will be the BLAS implementations themselves. This is the motivation for the LAPACK project which we now consider.

3.4.3   LAPACK

We have already made it clear that parallelism at the outermost level of nested loops is likely to give the greatest speed-up. This necessarily means that when we are investigating parallel algorithms for high-level linear algebra operations, such as the solution of a system of simultaneous linear equations, it is necessary to look at the operation as a whole.

An alternative approach is to lower our sights a little and aim for an algorithm which gives acceptable speed-up, if not the optimum possible. Here we analyse the strategy of subdividing the problem using blocked or partitioned algorithms which maximise the use of BLAS, and particularly Level 3 BLAS. The expectation is that, by arranging for most of the computation to be in terms of Level 3 BLAS, good performance of the algorithm on vector and parallel computers will result simply by providing efficient (serial or parallel, as appropriate) versions of the BLAS. Such is the basis of LAPACK (Demmel, Dongarra, *et al.*, 1987, Anderson, Bai, *et al.*, 1992), an attempt to provide transarchitectural portability for both uniprocessors and shared memory multiprocessors.

The LAPACK software, which is available via NETLIB and has been implemented in the NAG library, is designed to cover the functionality of the LINPACK and EISPACK packages; that is, it is a collection of high-level linear algebra routines for the solution of systems of linear equations, linear least squares problems and eigenvalue problems. One of the design criteria for LAPACK was that the routines should make maximal use of Levels 1, 2 and 3 BLAS. Indeed, the success of LAPACK relies on the provision of an efficient set of BLAS on a given machine. For optimal performance it may be appropriate to have assembler-coded lower-level (Level 1 and possibly Level 2) BLAS, with the higher Level (2 and 3 BLAS making calls to these optimised lower-level routines. Further, if parallel versions of the BLAS (at, possibly, all levels) are provided, then LAPACK can be regarded as a portable parallel numerical library. The LAPACK routines themselves employ no parallelism directly and are written in standard Fortran 77. However, they can call parallel BLAS in exactly the same way that they would call serial BLAS.

The technique used in the LAPACK codes to maximise calls to BLAS, and, in particular, calls to Level 3 BLAS, relies on the use of blocked algorithms. We emphasise that LAPACK uses blocked, or partitioned, algorithms, rather than algorithms of the true block type, such as the block LU factorisation described in Sections 3.4.1 and 3.4.2. Parallelism is then applied not at the outermost level (by modifying blocks in parallel) but at lower levels, making use of parallel BLAS. Hence a collection of blocks is modified by making a call to a single Level 3 BLAS, as, for example, in (3.14), rather than by making separate calls to compute the individual blocks, as in (3.12) and (3.13).

As an example we consider the routine `sgetrf` with specification

```
SUBROUTINE SGETRF(M,N,A,LDA,IPIV,INFO)
```

which computes the LU decomposition of an $m \times n$ matrix A. As in the descriptions of the BLAS, LDA is the declared leading (first) dimension of A. IPIV is an integer array which is used to keep track of any reordering (*pivoting*) of the rows of A that may be necessary. The details of the possible need for such interchanges (a *pivotal strategy*) are discussed in the next chapter. INFO is an integer error flag.

To illustrate the way the algorithm works, consider the LU factorisation of the $n \times n$ matrix A, which we write as

$$A = \begin{pmatrix} A_{11} & A_{12} & A_{13} \\ A_{21} & A_{22} & A_{23} \\ A_{31} & A_{32} & A_{33} \end{pmatrix},$$

where the block matrices A_{ij} are each $b \times b$ with $n = 3b$. We introduce the block matrices L and U, written as

$$L = \begin{pmatrix} L_{11} & 0 & 0 \\ L_{21} & L_{22} & 0 \\ L_{31} & L_{32} & L_{33} \end{pmatrix}, \qquad U = \begin{pmatrix} U_{11} & U_{12} & U_{13} \\ 0 & U_{22} & U_{23} \\ 0 & 0 & U_{33} \end{pmatrix},$$

where all blocks are again $b \times b$, the diagonal blocks of L are unit diagonal, lower triangular, and the diagonal blocks of U are upper triangular. Multiplying out we find that we must have

$$A = \begin{pmatrix} L_{11}U_{11} & L_{11}U_{12} & L_{11}U_{13} \\ L_{21}U_{11} & L_{21}U_{12} + L_{22}U_{22} & L_{21}U_{13} + L_{22}U_{23} \\ L_{31}U_{11} & L_{31}U_{12} + L_{32}U_{22} & L_{31}U_{13} + L_{32}U_{23} + L_{33}U_{33} \end{pmatrix}.$$

We begin by forming the LU factors of the rectangular matrix consisting of the first n rows and b columns of A, that is, we find $\bar{L}_{11}$ and U_{11} such that

$$\bar{A}_{11} = \bar{L}_{11}U_{11},$$

where $\bar{A}_{11} = \begin{pmatrix} A_{11} \\ A_{21} \\ A_{31} \end{pmatrix}$ and $\bar{L}_{11} = \begin{pmatrix} L_{11} \\ L_{21} \\ L_{31} \end{pmatrix}$. This requires a single call to the routine `sgetf2`, with specification

```
SUBROUTINE SGETF2(M,N,A,LDA,IPIV,INFO)
```

`sgetf2`, the 'point' version of `sgetrf`, does not itself involve calls to Level 3 BLAS, but it has been written in such a way that it makes maximal use of Levels

2 and 1 BLAS. The integer array IPIV is present to take account of any reordering of the rows which may be necessary. In the following description we assume that no such reordering takes place. To factorise $\bar{A}_{11}$ we make the call

```
CALL SGETF2(M,NB,A(1,1),LDA,IPIV(1),INFO)
```

where NB is the block size.

We can now find U_{12} by solving the triangular system

$$L_{11}U_{12} = A_{12}.$$

This is an **strsm** operation which overwrites A_{12} with U_{12}. The desired result is achieved by

```
CALL STRSM('L','L','N','U',NB,NB,ONE,A,LDA,A(1,NB+1),LDA)
```

(The character parameters stand for *Left* multiplication, *Lower* triangular, *No* transpose, *Unit* diagonal.) The next stage is to update the blocks on and below the diagonal in the second block column. We form

$$\bar{A}_{22} \leftarrow -\bar{L}_{21}U_{12} + \bar{A}_{22},$$

where $\bar{A}_{22} = \begin{pmatrix} A_{22} \\ A_{32} \end{pmatrix}$ and $\bar{L}_{21} = \begin{pmatrix} L_{21} \\ L_{31} \end{pmatrix}$. This can be achieved by a single call to the Level 3 BLAS **sgemm** of the form

```
     CALL SGEMM('N','N',M-NB,NB,NB,-ONE,A(NB+1,1),LDA,
    +           A(1,NB+1),LDA,ONE,A(NB+1,NB+1),LDA)
```

(*No* matrix transpose, twice.) Having updated the rectangular matrix $\bar{A}_{22}$ we can form its LU factors according to

$$\bar{A}_{22} = \bar{L}_{22}U_{22},$$

where $\bar{L}_{22} = \begin{pmatrix} L_{22} \\ L_{32} \end{pmatrix}$. This is again achieved by a call to **sgetf2**, now of the form

```
CALL SGETF2(M-NB,NB,A(NB+1,NB+1),LDA,IPIV(NB+1),INFO)
```

with the factors overwriting A_{22} and A_{32}. We can also find U_{13} and U_{23} by solving the triangular system

$$\begin{pmatrix} L_{11} & 0 \\ L_{21} & L_{22} \end{pmatrix} \begin{pmatrix} U_{13} \\ U_{23} \end{pmatrix} = \begin{pmatrix} A_{13} \\ A_{23} \end{pmatrix},$$

using

```
CALL STRSM('L','L','N','U',2*NB,NB,ONE,A,LDA,A(1,2*NB+1),LDA)
```

with the solution overwriting A_{13} and A_{23}. We now must modify the $(3,3)$ diagonal block according to

$$\bar{A}_{33} \equiv A_{33} \leftarrow -(L_{31} \quad L_{32}) \begin{pmatrix} U_{13} \\ U_{23} \end{pmatrix} + A_{33},$$

which requires a single call to **sgemm** of the form

```
      CALL SGEMM('N','N',M-2*ND,NB,2*NB,-ONE,A(2*NB+1,1),LDA,
     +           A(1,2*NB+1),LDA,ONE,A(2*NB+1,2*NB+1),LDA)
```

and, finally, we calculate the LU factors of $\bar{A}_{33}$,

$$\bar{A}_{33} = L_{33}U_{33},$$

by using a third, and final, call to **sgetf2** of the form

```
      CALL SGETF2(M-2*NB,NB,A(2*NB+1,2*NB+1),LDA,IPIV(2*NB+1),INFO)
```

with the factors overwriting A_{33}.

It should be noted that this algorithm deals with the blocks of A by columns. At the first stage the three blocks in the first block column of A are factored. At the second stage the blocks in the second block column of A are modified using the blocks in the first block column of the factorisation, and then the appropriate blocks are factored. Finally, at the third stage the blocks in the third block column of A are modified by blocks in the first two block columns of the factorisation, and the appropriate block is factored. In a serial environment, arranging the computation in this way can be advantageous since, for a suitable block size, we can expect the block column of A which is being factored to be resident in the cache memory and/or in the main memory rather than on disk. We expect some similar benefit in a parallel environment also.

We repeat that the above is a blocked, rather than a block, algorithm. It results in an LU factorisation of A in which L is a unit diagonal, lower triangular matrix and U is an upper triangular matrix. (Contrast this with the block LU factorisation of Section 3.4.1 in which U is upper triangular in a block, rather than element, sense.) Both factorisations reduce to the point algorithm of Section 4.3.1 when $b = 1$, but only the blocked algorithm permits, in any sensible way, the full use of row interchanges that the point algorithm employs to maintain stability.

Shared memory parallel LAPACK implementation

As we have already indicated, LAPACK codes themselves employ standard Fortran 77 only and thus support no explicit parallelism. The codes will work in both serial and parallel environments and so portability is retained. Where parallelism can be employed, implicitly, is in the BLAS. Initially LAPACK was aimed at shared memory multiprocessors, particularly those with pipeline arithmetic units. We first consider the implementation of LAPACK in such an environment.

We recall that the LAPACK routine `sgetrf` computes the LU factors of a general $m \times n$ matrix A. The first stage is to form the LU factors of the initial block column of A using `sgetf2`. Precisely how this may be achieved is left for discussion in Section 4.3, but here we note that it is possible to derive an algorithm which makes use of Levels 2 and 1 BLAS operations. In Section 3.3 we considered possible parallel implementations of these operations and noted that, whilst we may not be able to achieve a major increase in performance because of the low compute/communication ratio, some gain is nevertheless possible. Since, in a shared memory environment, there is no direct association of memory with a particular processor, the block distribution of A amongst the memory modules is not critically important.

Having formed $\bar{L}_{11}$ and U_{11} we can proceed to modify the remaining blocks. The way that this is done is not unique and an alternative strategy to that employed earlier is outlined in Exercise 3.9. Whichever method is selected we are again faced with the problems of maintaining portability and maximising the use of data already in cache. When forming $\bar{A}_{22}$, we do not update the individual blocks separately, either in parallel or sequentially. Rather, the required result is achieved using a single call to a parallel version of `sgemm`.

We have still to address the question of how large the block size should be. A block size of unity is taken to mean no blocking; in the case of `sgetrf` this means that the factorisation is achieved using a single call to `sgetf2` with the values for m and n as supplied to `sgetrf`. If the block size is too large then the limitations of parallel Levels 1 and 2 BLAS will manifest themselves within the calls to `sgetf2` with large values for m and n. If the block size is too small then the overheads associated with a large number of routine calls will become significant. Unfortunately, it is not possible to lay down a universally valid set of hard and fast rules for any particular machine. However, experimentation has shown that for most machines and for most routines a block size of around 32 is close to optimal, with little variation in performance for values around this optimum.

Local memory parallel LAPACK implementation

More recently, attention has turned to the implementation of LAPACK in a local memory environment. Assuming that a routine is called from a host process, there are two major problems with simply adopting the approach as outlined for shared memory multiprocessors:

- We are unlikely to be able to achieve any speed-up for the Levels 1 and 2 BLAS (which means that the point algorithm may have to be executed sequentially).
- Each call to a Level 3 BLAS may require data to be loaded into, and the results returned from, the processor array.

The development of a local memory version of LAPACK, LAPACK 2, is still in its early stages. The blocking strategy remains at the core of LAPACK 2, but a decision has to be made as to what form of data distribution is appropriate. The use of wraparound block columns appears to be the most appropriate. Assuming that A has n columns, with n divisible by b, the block size, we write $A = (A_1, A_2, \ldots, A_{n/b})$, where each A_i is a $n \times b$ matrix, and allocate A_i to process $(i - 1) \bmod p$, where the processes are numbered in the range $[0, p - 1]$. The factorisation then consists of n/b stages, and at each stage a single process modifies its block column. Except on the last stage, it then distributes data to the other processes which use this information to update their own block columns. The algorithm is essentially a block version of the local memory scattered column decomposition we refer to at the end of the next chapter and so we curtail further discussion at this stage.

To assist in the writing of routines in LAPACK 2, a standard for a communications library is in the process of being established. This has led to the specification of what are variously known as the BLAMPS (*B*asic *L*inear *A*lgebra *M*essage *P*assing *S*ubprograms) or BLACS (*B*asic *L*inear *A*lgebra *C*ommunication *S*ubprograms); see Dongarra (1991b) and Dongarra and van Geijn (1991). Three basic types of communication routine are proposed:

- SGESD(M,N,A,LDA,IDEST,MSGID)

  sgesd sends a single precision M × N array A to a nominated process (IDEST). MSGID is an identifier for the message.

- SGERC(M,N,A,LDA,ISRC,MSGID)

  sgerc receives an array from a nominated process (ISRC).

- SGEBC(M,N,A,LDA,ISRC,MSGID)

  sgebc broadcasts an array from process ISRC to all other processes.

Variants of these routines are proposed for the case that A is trapezoidal (stored in the upper or lower triangle), and unit or non-unit diagonal; cf. strsm.

Also proposed are three global operations:

- SGMAX(N,X,Y,IY,IDEST,MSGID)

  sgmax returns in the vector Y the elementwise maxima (in absolute value) of the values in the distributed vectors X. The vector IY gives the identifiers of the processes which gave rise to each elementwise maximum value. The results vectors Y and IY are sent to the nominated process (IDEST) with the message identifier MSGID.

- SGMIN(N,X,Y,IY,IDEST,MSGID)

  sgmin returns the elementwise minimum values.

- SGSUM(N,X,Y,IDEST,MSGID)

  sgsum returns the elementwise sum of the values in the distributed vectors X.

3.5   Models for parallel libraries

The transition of existing codes from a sequential to a shared memory parallel environment has been greatly eased by the availability of parallelising compilers. This, and the provision of parallel BLAS, can be exploited to provide the basis of a parallel numerical software library. We have seen that this is the basis for LAPACK, a successful (for shared memory systems, at least) attempt at providing genuine transarchitectural portability.

The situation with regard to local memory systems is less clear. Parallelising compilers for such systems are in their relative infancy. Further, there are severe problems associated with exploiting high- or low-level routines if there is an accompanying need to redistribute data frequently. To illustrate the difficulties, we consider the situation in which a user writes a sequential code which makes several calls to parallel routines. One possible implementation model is that in which code and data are downloaded to the processor array each and every time a call to a parallel routine is made, and at the end of the call the results are returned to the host. The user may, for example, wish to compute the LU factors of a matrix A and then use those factors to solve a system of linear equations. Our implementation model requires code and data to be downloaded for the factorisation routine, and then for the solution routine, even though the data for the latter is the result of the former computation. Similarly, the repeated call of a routine with different data requires the code to be downloaded separately for each of the calls. Even if codes for predistributed data are provided some problems remain. In particular, a given data distribution may not match that expected by the routine. For example, for a system of linear equations, a distribution by columns may be appropriate for the factorisation phase, whilst a distribution by rows may be more suitable for the solution phase. For each phase the choice will be either

- to arrange the data (say, perform a matrix transpose) so that it fits the routine (this is likely to be expensive in terms of communications), or
- to use an algorithm which accepts a given data distribution (this is unlikely to result in an optimal implementation).

If a library is to be developed which assumes predistributed data, it is important to have available a number of efficient communication routines, such as those provided by the BLAMPS.

Exercises

3.1.   Loop unrolling can sometimes be used to speed up code on a uniprocessor; precisely how effective it is will depend on the amount of work per loop cycle. The idea is to reduce the loop overheads (incrementing the loop counter and testing it against the upper limit) by reducing the number of loop cycles. For example, the loop

```
do i = 1,n
  <code dependent on i>
end do
```

can be replaced by

```
do i = 1,n/2
  <code dependent on 2i-1>
  <code dependent on 2i>
end do
```

for n even. If n is odd some tidying up must be done after the loop. Employ this technique in a unit stride version of `sscal` for various levels of loop unrolling $(n/2, n/3, \text{etc.})$ and monitor the variation in performance. (You may need to take n to be very large, or call `sscal` many times, to obtain meaningful timings.)

3.2. Explore the possible loop orderings for (a) the Level 2 BLAS `ssyr`, and (b) the Level 3 BLAS `ssyrk`. In what way(s) can these operations be parallelised?

3.3. Write sequential Fortran code which uses (a) `saxpy`, and (b) `sdot` operations to implement the Level 2 BLAS `sgbmv`, with specification

```
subroutine sgbmv(trans,m,n,kl,ku,alpha,a,lda,x,incx,beta,y,incy)
```

in which `kl` and `ku` indicate the number of co-diagonals below and above the principal diagonal. It is recommended that initially you consider the non-transpose, unit-increment case only! How would you feel about a parallel algorithm that makes row accesses to a banded matrix stored in this way?

3.4. Let P be the matrix $P = I_n - 2\mathbf{v}\mathbf{v}^T$. Show that the transformation $A \leftarrow PA$ can be made by successive calls to the Level 2 BLAS `sgemv` and `sger`.

3.5. For each of the four permitted combinations of `TRANSA` and `TRANSB`, consider the possible implementations of the Level 3 BLAS `sgemm`, making use of calls to Level 2 and/or Level 1 BLAS as appropriate.

3.6. How may the Level 3 BLAS `strmm` be implemented in terms of the Level 2 BLAS `strsv`? Suggest a strategy for parallelising the `strmm` operation. How would you apportion the work/data in (a) a shared, and (b) a local memory environment?

3.7. Consider the block matrix-matrix multiplication

$$\begin{pmatrix} C_{11} & C_{12} \\ C_{21} & C_{22} \end{pmatrix} = \begin{pmatrix} A_{11} & A_{12} \\ A_{21} & A_{22} \end{pmatrix} \begin{pmatrix} B_{11} & B_{12} \\ B_{21} & B_{22} \end{pmatrix}.$$

Verify that, if the intermediate matrices P_i satisfy

$$P_1 = (A_{11} + A_{22})(B_{11} + B_{22}),$$
$$P_2 = (A_{21} + A_{22}) B_{11},$$
$$P_3 = A_{11}(B_{12} - B_{22}),$$
$$P_4 = A_{22}(B_{21} - B_{11}),$$
$$P_5 = (A_{11} + A_{12}) B_{22},$$
$$P_6 = (A_{21} - A_{11})(B_{11} + B_{12}),$$
$$P_7 = (A_{12} - A_{22})(B_{21} + B_{22}),$$

then

$$C_{11} = P_1 + P_4 - P_5 + P_7,$$
$$C_{12} = P_3 + P_5,$$
$$C_{21} = P_2 + P_4,$$
$$C_{22} = P_1 + P_3 - P_2 + P_6,$$

(Strassen, 1969). Indicate how this algorithm can be used recursively to implement the Level 3 BLAS (Higham, 1990).

3.8. Assuming that no pivoting is required, express the L and U factors of A as 3×3 block matrices. Hence express A in terms of 3×3 block $\bar{L}\bar{D}U$ factors, where $\bar{L}$ is a block, unit diagonal, lower triangular, matrix, and $\bar{D}$ is a block diagonal matrix. Use this result to relate the block LU factors of A as defined in Section 3.4.1 to the blocks of the LU factors of A as defined in Section 3.4.3.

3.9. An alternative (and simplified) implementation of the LAPACK routine sgetrf is given by

```
do j = 1,n,b
   call sgemm('n','n',n-j+1,b,j-1,-one,a(j,1),lda,
  +               a(1,j),lda,one,a(j,j),lda)
   call sgetf2(n-j+1,b,a(j,j),lda,ipiv(j),info)
   if(j+b.le.n)then
      call sgemm('n','n',b,n-j-jb+1,j-1,-one,a(j,1),lda,
  +               a(1,j+b),lda,one,a(j,j+b),lda)
      call strsm('l','l','n','u',b,n-j-b+1,one,a(j,j),lda,
  +               a(j,j+b),lda)
   end if
end do
```

By considering the case $n = 9$, $j = 3$ indicate how this algorithm differs from that outlined in Section 3.4.3.

3.10. If A is symmetric and *positive definite* (that is, $\mathbf{x}^T A \mathbf{x} > 0$ for all non-zero vectors $\mathbf{x}$), a decomposition of the form $A = LL^T$ (the *Cholesky decomposition*) exists, where L is lower triangular. By considering a 3×3 block matrix, derive a block algorithm for Cholesky decomposition which makes use of the Level 3 BLAS sgemm, strsm and ssyrk. The routine spotf2, with specification

```
subroutine spotf2(uplo,n,a,lda,info)
```

computes the Cholesky factorisation of a matrix A of order n stored in either upper or lower triangular form. Use this to construct a section of code implementing your algorithm (Dongarra, Du Croz, *et al.*, 1990a).

3.11.   What are the likely effects that variations in the block size will have on the performance of the LAPACK routine sgetrf on a uniprocessor and on a shared memory multiprocessor? What can you say about the LAPACK 2 equivalent?

Further reading

For a general bibliography of parallel numerical algorithms see Ortega, Voigt, *et al.* (1990).

A discussion of the issues involved in the parallelisation of the Level 2 BLAS sgemv on both shared and local memory systems is given by Golub and Van Loan (1989). Phillips (1991) considers various BLAS implementations on an Encore Multimax and their use in conjunction with LAPACK routines.

The Liverpool Parallel Library (Brown, Delves, *et al.*, 1990), is a suite of library routines specifically targeted at a Parsys/Telmat Supernode transputer array. The current contents of the library are limited and oriented towards linear algebra algorithms, making use, wherever possible, of single-processor low-level assembler BLAS. The library adopts a host/slave model with the slave codes written in occam, whilst occam and Fortran versions of the host codes are provided to interface with the user's program. In general the host codes do little more than check parameter values and pass data to the slaves. Collaborative work between NAG and the University of Liverpool, funded as part of the ESPRIT Supernode II (P2528) project, is extending the functionality of this library. A rival development is the TopExpress Occam Procedure Library (TopExpress, 1989) which operates in a similar way. Sadly, the company has ceased development of this library but continues to market it. Allan (1990) has developed a transputer library based on the Fortnet environment in which routines are available that, for example, assign local memory for storage, and fetch a block of data from storage. Whilst the data (arrays, etc.) are distributed throughout the processor array, the programmer's view is of a global memory resource apart from explicit gather/scatter, etc., operations.

CHAPTER 4

Gaussian elimination:
a case study

4.1 Linear algebra as a basic computational tool

In the previous chapter we discussed a number of fundamental linear algebra operations which collectively constitute the BLAS. From these we can construct higher-level routines which themselves will often be components of yet higher-level algorithms. As an example consider the problem, in optimisation, of finding a local minimum point, $\mathbf{x}^*$, of the real-valued function $f(\mathbf{x})$, $\mathbf{x} \in \mathcal{R}^n$. That is, we search for a point $\mathbf{x}^*$ for which $f(\mathbf{x}^*) \leq f(\mathbf{x})$ for all $\mathbf{x}$ in some neighbourhood of $\mathbf{x}^*$. One way of solving this problem is to use Newton's method, which we describe in detail in Section 6.4.2. Starting at some point $\mathbf{x}^{(1)}$, we compute iterates $\mathbf{x}^{(k+1)}$ according to the scheme

$$\mathbf{x}^{(k+1)} = \mathbf{x}^{(k)} + \alpha^{(k)}\mathbf{p}^{(k)}, \tag{4.1}$$

where

$$G\left(\mathbf{x}^{(k)}\right)\mathbf{p}^{(k)} = -\mathbf{g}\left(\mathbf{x}^{(k)}\right), \tag{4.2}$$

and $\alpha^{(k)}$ is calculated by searching from $\mathbf{x}^{(k)}$, in the direction $\mathbf{p}^{(k)}$, for a minimum value of f (a *line search*). In (4.2) $G(\mathbf{x})$ is the $n \times n$ symmetric *Hessian matrix* of $f(\mathbf{x})$, with components $G_{ij} = \partial^2 f(\mathbf{x})/\partial x_i \partial x_j$, and $\mathbf{g}(\mathbf{x})$ is the *gradient vector* of $f(\mathbf{x})$, with components $g_i = \partial f(\mathbf{x})/\partial x_i$. We note that (4.1) is a Level 1 BLAS operation (**saxpy**) and that the determination of $\mathbf{p}^{(k)}$ in (4.2) involves the solution of a system of simultaneous linear equations, an operation which, we later show, can be expressed in terms of Levels 2 and 1 BLAS (and Level 3 BLAS if we employ a blocked algorithm).

In linear algebra there are three classes of problem with which we can be faced:

1. The solution of the system of simultaneous linear equations

$$A\mathbf{x} = \mathbf{b}, \tag{4.3}$$

in which the coefficient matrix A is square (as, for example, in (4.2)). This is probably the most important class of problem. We study the general form in detail in the remainder of this and much of the next chapter. For the moment we remark that the system may, or may not, possess a unique solution.

2. The solution of a system of simultaneous linear equations (4.3) in which the coefficient matrix A is rectangular, say $m \times n$. We choose to distinguish between two distinct cases:

   - If $m < n$ there are fewer equations than unknowns and we refer to an *underdetermined system*. There is no unique solution to the system (the emphasis being on the non-uniqueness) and we can arbitrarily fix $n - m$ of the unknowns; the solution of the original system for the remaining m unknowns is then a problem of class 1.

   - If $m > n$ there are more equations than unknowns and we refer to an *overdetermined system*. Again, there is, in general, no unique solution (the emphasis now being on the non-existence of a solution). Since it is not possible to satisfy all of the equations simultaneously we compromise and attempt to compute a solution which approximately satisfies the equations, for example, by minimising $\|A\mathbf{x} - \mathbf{b}\|_2$, when we refer to the solution as a *least squares solution*. The computation of such a solution is discussed in Section 5.5.

3. For an $n \times n$ matrix A the determination of one or more scalars λ (*eigenvalues*) and corresponding vectors $\mathbf{x}$ (*eigenvectors*) such that $A\mathbf{x} = \lambda \mathbf{x}$. The solution of a problem of this form is discussed in Section 5.6.

We begin with problems of class 1.

4.2   Gaussian elimination

Consider the solution of the system of linear equations (4.3) where A is a known $n \times n$ square matrix (the *matrix of coefficients*) with (i, j) entry a_{ij}, $\mathbf{b}$ is a known n-vector (the *right-hand side vector*) with ith entry b_i, and the n-vector $\mathbf{x}$ (with ith entry x_i) is to be determined. We assume here that A is *dense*, or *full* (that is, has relatively few zeros). If A is *sparse* (possesses few non-zero entries, say, less that 10%) then it is still possible to apply the following techniques, but it is usual, for efficiency and storage considerations, to modify the basic methods, or to employ entirely different methods; see Sections 5.3, 5.4 and 7.4.

Written out in full, the system (4.3) is

$$a_{11}x_1 + a_{12}x_2 + \cdots + a_{1n}x_n = b_1,$$
$$a_{21}x_1 + a_{22}x_2 + \cdots + a_{2n}x_n = b_2,$$
$$\vdots \qquad\qquad\qquad\qquad \tag{4.4}$$
$$a_{n1}x_1 + a_{n2}x_2 + \cdots + a_{nn}x_n = b_n.$$

We observe that the solution is invariant under the following operations:

1. The addition of a multiple of one equation to another.
2. A reordering of the equations (and, hence, the rows of A, provided that we simultaneously reorder the corresponding components of **b**).

In terms of (4.3), the matrix form of the system, (1) is a `saxpy` operation involving two rows of A, whilst (2) is an `sswap` operation. In contrast to reordering the rows, we observe that a reordering of the columns of A is not simply an `sswap` operation since it involves a reordering of the unknowns.

The standard method of computing **x** is based on the use of *Gaussian elimination*. The basis of this method is to transform the original system (4.3) to

$$U\mathbf{x} = \mathbf{b}', \tag{4.5}$$

where U is an upper triangular matrix. The transformation of A to upper triangular form, known as *forward elimination*, employs only elementary operations of the forms (1) and (2) above. Hence the solution vector remains unchanged. Solving the upper triangular system (4.5) is easily achieved using *backward substitution*. We now consider in some detail these two steps.

4.2.1 Forward elimination

In its simplest form the forward elimination process (the transformation of (4.3) to (4.5)) relies on the use of operations of the form (1) above only. The need for operations of the form (2) will become apparent later. By subtracting suitable multiples of the first equation in (4.4) from all remaining equations we can eliminate the unknown x_1 from these equations to obtain a system in which the only non-zero entry in the first column of the coefficient matrix, $A^{(1)}$, is in the diagonal, $(1, 1)$, position. The first row of $A^{(1)}$ is equal to the first row of A and the remaining non-zero elements are denoted $a_{ij}^{(1)}$. Formally, assuming $a_{11} \neq 0$, we can eliminate x_1 from equation 2 by subtracting a_{21}/a_{11} times the first equation from the second. This transforms all the coefficients in equation 2, and the right-hand side entry b_2. Extending this to the general case, x_1 can be eliminated from equation i by the subtraction of a_{i1}/a_{11} times the first equation from the ith. The quantities a_{i1}/a_{11}, $i = 2, 3, \ldots, n$, used in this elimination process, are termed the *multipliers*. To summarise, after this, the first, stage of elimination, the original system (4.3) has been transformed to

$$A^{(1)}\mathbf{x} = \mathbf{b}^{(1)}, \tag{4.6}$$

where

$$A^{(1)} = \begin{pmatrix} a_{11} & a_{12} & \cdots & a_{1n} \\ 0 & a_{22}^{(1)} & \cdots & a_{2n}^{(1)} \\ \vdots & \vdots & \ddots & \vdots \\ 0 & a_{n2}^{(1)} & \cdots & a_{nn}^{(1)} \end{pmatrix}, \qquad \mathbf{b}^{(1)} = \begin{pmatrix} b_1 \\ b_2^{(1)} \\ \vdots \\ b_n^{(1)} \end{pmatrix},$$

and the new values are given by

$$a_{ij}^{(1)} = a_{ij} - a_{i1}a_{1j}/a_{11},$$

$$b_i^{(1)} = b_i - a_{i1}b_1/a_{11},$$

for $i, j = 2, 3, \ldots, n$.

To proceed we note that equations $2, 3, \ldots, n$ of (4.6) constitute a linear system of $n - 1$ equations in the $n - 1$ unknowns $x_2, x_3, \ldots, x_n$. Thus, the elimination process outlined above can be applied to this subsystem. Formally, assuming $a_{22}^{(1)} \neq 0$, the unknown x_2 can be eliminated from equation $i = 3, 4, \ldots, n$, by the subtraction from it of $a_{i2}^{(1)}/a_{22}^{(1)}$ times equation 2 in (4.6), yielding the system

$$A^{(2)}\mathbf{x} = \mathbf{b}^{(2)}, \tag{4.7}$$

where

$$A^{(2)} = \begin{pmatrix} a_{11} & a_{12} & a_{13} & \cdots & a_{1n} \\ 0 & a_{22}^{(1)} & a_{23}^{(1)} & \cdots & a_{2n}^{(1)} \\ 0 & 0 & a_{33}^{(2)} & \cdots & a_{3n}^{(2)} \\ \vdots & \vdots & \vdots & \ddots & \vdots \\ 0 & 0 & a_{n3}^{(2)} & \cdots & a_{nn}^{(2)} \end{pmatrix}, \qquad \mathbf{b}^{(2)} = \begin{pmatrix} b_1 \\ b_2^{(1)} \\ b_3^{(2)} \\ \vdots \\ b_n^{(2)} \end{pmatrix},$$

and the new values are given by

$$a_{ij}^{(2)} = a_{ij}^{(1)} - a_{i2}^{(1)}a_{2j}^{(1)}/a_{22}^{(1)},$$

$$b_i^{(2)} = b_i^{(1)} - a_{i2}^{(1)}b_2^{(1)}/a_{22}^{(1)},$$

for $i, j = 3, 4, \ldots, n$. This completes the second stage of the elimination.

In general, after the $(k-1)$th stage of elimination we are left with a system of linear equations

$$A^{(k-1)}\mathbf{x} = \mathbf{b}^{(k-1)}, \tag{4.8}$$

where

$$A^{(k-1)} = \begin{pmatrix} a_{11} & a_{12} & \cdots & a_{1,k-1} & a_{1k} & \cdots & a_{1n} \\ 0 & a_{22}^{(1)} & \cdots & a_{2,k-1}^{(1)} & a_{2k}^{(1)} & \cdots & a_{2n}^{(1)} \\ \vdots & \ddots & \ddots & \vdots & \vdots & & \vdots \\ \vdots & & \ddots & a_{k-1,k-1}^{(k-2)} & a_{k-1,k}^{(k-2)} & \cdots & a_{k-1,n}^{(k-2)} \\ \vdots & & & 0 & a_{kk}^{(k-1)} & \cdots & a_{kn}^{(k-1)} \\ \vdots & & & \vdots & \vdots & \ddots & \vdots \\ 0 & \cdots & \cdots & 0 & a_{nk}^{(k-1)} & \cdots & a_{nn}^{(k-1)} \end{pmatrix},$$

and

$$\mathbf{b}^{(k-1)} = \left(b_1, b_2^{(1)}, \dots, b_{k-1}^{(k-2)}, b_k^{(k-1)}, \dots, b_n^{(k-1)}\right)^T.$$

Note that only the first $k-1$ equations involve $x_1, x_2, \dots, x_{k-1}$, and that, of these, the ith involves x_j for $j \geq i$ only. At the kth stage x_k is eliminated from equations $i = k+1, k+2, \dots, n$, of (4.8) by the subtraction of $a_{ik}^{(k-1)}/a_{kk}^{(k-1)}$ times equation k from each of these equations. This yields the system

$$A^{(k)}\mathbf{x} = \mathbf{b}^{(k)},$$

in which the first k rows and $k-1$ columns of $A^{(k)}$ and the first k elements of $\mathbf{b}^{(k)}$ are the same as those of $A^{(k-1)}$ and $\mathbf{b}^{(k-1)}$, the remaining elements in column k of $A^{(k)}$ are zero, and the other elements are given by

$$a_{ij}^{(k)} = a_{ij}^{(k-1)} - a_{ik}^{(k-1)}a_{kj}^{(k-1)}/a_{kk}^{(k-1)},$$
$$b_i^{(k)} = b_i^{(k-1)} - a_{ik}^{(k-1)}b_k^{(k-1)}/a_{kk}^{(k-1)},$$

for $i, j = k+1, k+2, \dots, n$.

After $n-1$ such elimination stages, we have the linear system

$$A^{(n-1)}\mathbf{x} = \mathbf{b}^{(n-1)}, \tag{4.9}$$

where $A^{(n-1)}$ is an upper triangular matrix.

Code 4.1 is a listing of a Fortran subroutine which implements this algorithm directly (and serially). Here, each transformed matrix and right-hand side vector component overwrites its predecessor and the multipliers overwrite the lower triangular part of the matrix. Why we choose to store the multipliers will become apparent in Section 4.3. Because of the order in which the loop indices appear, Code 4.1 is said to be the *kij* variant of forward elimination. Other possible implementations result if we reorder the loops, and we consider these in Section 4.4.

When analysing a numerical algorithm, whether serial or parallel, there are, arguably, two principle questions that we should ask:

1. How accurate is it?
2. How much does it cost?

We consider (1) briefly in Section 4.2.2. Here we consider (2). Gaussian elimination may be classified as a *direct method* in the sense that the steps of the method, and therefore its total cost, are known a priori. Assume that each floating-point operation (addition, subtraction, multiplication or division) takes the same unit of time for execution. At the kth stage of the elimination process we have to compute $n-k$ multipliers (1 flop each) and modify $n-k$ right-hand side components (each component involves 2 flops, 1 for multiplication and 1 for subtraction),

```
      subroutine gelim(a,lda,b,n)
c Gaussian elimination (forward elimination)
c
c On entry a, with leading dimension lda, is the coefficient matrix,
c b is the right-hand side vector and n is the problem size
c
c On exit a and b are transformed
      integer lda,n
      real a(lda,*),b(*)
      integer i,j,k
      do k = 1,n-1
c Eliminate x(k) from equations k+1, k+2, ...,n
      do i = k+1,n
c Compute a multiplier for equation i and overwrite the (i,k) entry of a
      a(i,k) = a(i,k)/a(k,k)
c Modify row i of the submatrix
      do j = k+1,n
      a(i,j) = a(i,j)-a(i,k)*a(k,j)
      end do
c Modify the i'th component of the right-hand side
      b(i) = b(i)-a(i,k)*b(k)
      end do
      end do
      end
```

Code 4.1 *Forward elimination*

making a total of $3(n - k)$ flops. Added to this is the cost of modifying the $(n - k) \times (n - k)$ submatrix

$$
\begin{pmatrix}
a_{k+1,k+1}^{(k-1)} & \cdots & a_{k+1,n}^{(k-1)} \\
\vdots & \ddots & \vdots \\
a_{n,k+1}^{(k-1)} & \cdots & a_{nn}^{(k-1)}
\end{pmatrix},
$$

which requires $2(n-k)^2$ flops. (It is not necessary explicitly to reduce the elements $a_{ik}^{(k-1)}$, $i = k + 1, k + 2, \ldots, n$, to zero, even if we decide not to store the multipliers.) We note that this implies that the submatrix modification is the most costly component of each stage of forward elimination. The overall cost of the $n-1$ elimination steps can now be determined as $\sum_{k=1}^{n-1} \left(3(n - k) + 2(n - k)^2 \right) = \sum_{k=1}^{n-1} \left(3k + 2k^2 \right) = 2n^3/3 + O\left(n^2\right)$. Thus, forward elimination is an $O\left(n^3\right)$ process (on $O\left(n^2\right)$ data).

4.2.2   Partial pivoting

We observe that if any $a_{kk}^{(k-1)}$ (a *pivotal element*) is zero, $k = 1, 2, \ldots, n - 1$, the forward elimination process breaks down. This may mean that the matrix is *singular* (that is, no unique solution to the linear system exists), but this is not necessarily the case. As long as at least one of the elements, $a_{kk}^{(k-1)}, a_{k+1,k}^{(k-1)}, \ldots, a_{nk}^{(k-1)}$, is non-zero at the kth stage of elimination, the process may continue by interchanging rows. Even if $a_{kk}^{(k-1)}$ is not identically zero, it may be so small in relation to subsequent elements in the kth column that the multipliers generated are very large, with a consequent amplification of round-off errors, resulting in the process becoming unstable. The effect of this may be to render the solution to the final system (4.9) highly inaccurate. The solution to this problem also relies on row interchanges.

Considering again the kth elimination stage, we search, on and below the diagonal, the kth column of $A^{(k-1)}$ (the *pivotal column*) for the element of largest modulus. Suppose that it is $a_{rk}^{(k-1)}$ for some $r \geq k$. If this value is deemed to be close to 0 (for example, less than some prescribed tolerance) we conclude that the matrix is singular and abort the elimination process. Otherwise, if $r \neq k$, we interchange equations r and k and the elimination proceeds using the 'new' kth row (the *pivotal row*) to perform the eliminations. The process is known as *partial pivoting*. An alternative is *complete pivoting* in which we search the whole submatrix yet to be eliminated for the element of largest modulus and, using both row and column interchanges, make this the pivotal element. As we remarked earlier, column interchanges involve a reordering of the unknowns and thus require additional bookkeeping. Partial pivoting is usually considered to be sufficient. Henceforth any reference to pivoting implies partial pivoting unless otherwise qualified.

An implementation of forward elimination with partial pivoting is given in Code 4.2. On successful exit the integer error flag `ifail` contains the value zero. If singularity of the coefficient matrix is detected `ifail` contains the index of the elimination stage at which this occurred. The role of `ipiv` is explained at the end of Section 4.3.2.

4.2.3 Backward substitution

Having transformed the original system to the form

$$U\mathbf{x} = \mathbf{b}', \tag{4.10}$$

where

$$U = \begin{pmatrix} u_{11} & u_{12} & \cdots & u_{1,n-1} & u_{1n} \\ 0 & u_{22} & \cdots & u_{2,n-1} & u_{2n} \\ \vdots & & \ddots & \ddots & & \vdots \\ \vdots & & & \ddots & u_{n-1,n-1} & u_{n-1,n} \\ 0 & \cdots & \cdots & 0 & u_{nn} \end{pmatrix},$$

and the ith element of $\mathbf{b}'$ is b_i', we are now interested in recovering the solution vector $\mathbf{x}$. The final equation in the system (4.10) is

$$u_{nn}x_n = b_n',$$

which immediately gives

$$x_n = b_n'/u_{nn}.$$

```
      subroutine gelim(a,lda,b,n,tol,ipiv,ifail)
c Gaussian elimination with partial pivoting
c
c The parameters additional to Code 4.1 are:
c tol, an input tolerance used to test for singularity of a,
c ipiv, an integer vector to store pivoting information and
c ifail, an error flag whose value is 0 on successful exit,
c or the stage number at which a singularity was detected
      integer lda,n,ipiv(*),ifail
      real a(lda,*),b(*),tol
      integer i,j,k,piv
      real abspiv,absval,rdum
      intrinsic abs
c Assume no error to start with
      ifail = 0
c Perform elimination
      do k = 1,n-1
c Eliminate x(k) from equations k+1, k+2, ..., n
c
c Locate a pivot
      piv = k
      abspiv = abs(a(k,k))
      do i = k+1,n
        absval = abs(a(i,k))
        if (absval.gt.abspiv) then
          abspiv = absval
          piv = i
        end if
      end do
      ipiv(k) = piv
c Check for singularity
      if (abspiv.lt.tol) then
        ifail = k
        return
      else if (piv.ne.k) then
c Interchange the rows of a including the previously stored multipliers
        do j = 1,n
          rdum = a(k,j)
          a(k,j) = a(piv,j)
          a(piv,j) = rdum
        end do
c Interchange the corresponding components of b
        rdum = b(k)
        b(k) = b(piv)
        b(piv) = rdum
      end if
      do i = k+1,n
c Compute a multiplier for equation i
        a(i,k) = a(i,k)/a(k,k)
c Modify row i of the submatrix
        do j = k+1,n
          a(i,j) = a(i,j)-a(i,k)*a(k,j)
        end do
c Modify the i'th component of the right-hand side
        b(i) = b(i)-a(i,k)*b(k)
      end do
      end do
c Final check for singularity
      if (abs(a(n,n)).lt.tol) then
        ifail = n
      end if
      end
```

Code 4.2 *Forward elimination with partial pivoting*

Working backwards, the $(n-1)$th equation in (4.10) is

$$u_{n-1,n-1}x_{n-1} + u_{n-1,n}x_n = b'_{n-1},$$

which, given that x_n is known, can be solved for x_{n-1}. Formally,

$$x_{n-1} = \left(b'_{n-1} - u_{n-1,n}x_n\right)/u_{n-1,n-1}.$$

Generalising, the ith equation in (4.10) is

$$u_{ii}x_i + u_{i,i+1}x_{i+1} + \cdots + u_{in}x_n = b'_i,$$

so that, having determined $x_n, x_{n-1}, \ldots, x_{i+1}$, the unknown x_i is given by

$$x_i = \frac{1}{u_{ii}}\left(b'_i - \sum_{j=i+1}^{n} u_{ij}x_j\right).$$

We refer to this process as *backward substitution*.

A Fortran subroutine to implement backward substitution in the form described here is given in Code 4.3, in which we assume that U is stored in the upper triangle of A, so that a call to **bacsub** may follow a call to **gelim**. Again, other possibilities exist if we reorder the loops, and these are considered in Section 4.4.3. Note that the solution components could overwrite the corresponding right-hand side elements as they are computed.

```
      subroutine bacsub(a,lda,b,x,n)
c Backward substitution
c
c On entry the upper triangle of a, of leading dimension lda, contains
c the coefficient matrix and b the right-hand side vector of length n
c
c On exit x contains the solution vector
      integer lda,n
      real a(lda,*),b(*),x(*)
      integer i,j
      x(n) = b(n)/a(n,n)
      do i = n-1,1,-1
        x(i) = b(i)
        do j = i+1,n
          x(i) = x(i)-a(i,j)*x(j)
        end do
        x(i) = x(i)/a(i,i)
      end do
      end
```

Code 4.3 *Backward substitution*

In backward substitution we require $2(n-i)$ flops to accumulate the ith sum, leading to a total of $\sum_{i=1}^{n-1} 2(n-i) = n(n-1)$ flops. To this we add a further n

flops for computing the x_is, giving a total of n^2 flops. Hence, backward substitution is an $O(n^2)$ process (on $O(n^2)$ data). This operation count is an order of magnitude less than that for elimination and suggests a compute/communication ratio of $O(1)$ only. Thus, when attempting to derive a parallel algorithm for solving a system of linear equations based on Gaussian elimination followed by backward substitution, it is parallelisation of the elimination process which is likely to yield the most significant gains.

4.3   Matrix factorisation

Provided we know the right-hand side vector, the process of Gaussian elimination followed by backward substitution can be used to solve a system of linear equations. We saw in the previous section that the forward elimination process is applied to the coefficient matrix and the right-hand side vector simultaneously. If we have several right-hand sides then each stage of the forward elimination process must be applied to each right-hand side. An alternative is to separate the modification of the coefficient matrix from that of the right-hand side. That is, we first apply the whole elimination process to A, and only when this is complete do we then apply it to **b**. This has the advantage that each time we encounter a new right-hand side, whose value may not even be known at the start of the elimination process, only the right-hand side needs to be modified and the solution can then be recovered using backward substitution.

4.3.1   *LU* factorisation

We begin by showing that forward elimination (without pivoting) is equivalent to factorising A as

$$A = LU,$$

where L is a unit diagonal, lower triangular matrix and U is the upper triangular matrix $A^{(n-1)}$ of (4.9).

The first stage of forward elimination can be represented by

$$A = \begin{pmatrix} a_{11} & a_{12} & \cdots & a_{1n} \\ a_{21} & a_{22} & \cdots & a_{2n} \\ \vdots & \vdots & \ddots & \vdots \\ a_{n1} & a_{n2} & \cdots & a_{nn} \end{pmatrix} = \mathbf{l}^{(1)}\mathbf{u}^{(1)T} + \tilde{A}^{(1)} \tag{4.11}$$

where $l_1^{(1)} = 1$ and $\tilde{a}_{ij}^{(1)} = 0$ for $i = 1$ or $j = 1$, that is, the first row and column of $\tilde{A}^{(1)}$ have zero elements. Provided that $a_{11} \neq 0$, the unique solution to (4.11) is

$$\mathbf{l}^{(1)} = (1, a_{21}/a_{11}, a_{31}/a_{11}, \ldots, a_{n1}/a_{11})^T,$$
$$\mathbf{u}^{(1)} = (a_{11}, a_{12}, \ldots, a_{1n})^T,$$

(cf. the block matrices $L^{(1)}$ and $U^{(1)}$ of (3.3) and (3.5)) and then

$$\tilde{A}^{(1)} = A - \mathbf{l}^{(1)}\mathbf{u}^{(1)T},$$

$$= \begin{pmatrix} 0 & 0 & \cdots & 0 \\ 0 & a_{22} - a_{21}a_{12}/a_{11} & \cdots & a_{2n} - a_{21}a_{1n}/a_{11} \\ \vdots & \vdots & \ddots & \vdots \\ 0 & a_{n2} - a_{n1}a_{12}/a_{11} & \cdots & a_{nn} - a_{n1}a_{1n}/a_{11} \end{pmatrix}.$$

We observe that $\mathbf{l}^{(1)}$ consists of the multipliers of stage 1 of forward elimination and that the modified elements of $\tilde{A}^{(1)}$ are the same as the modified elements of $A^{(1)}$ in (4.6).

Clearly, a similar procedure can be used to eliminate the second row and column of $\tilde{A}^{(1)}$, and so on. After $k - 1$ stages of this elimination we have the partially eliminated matrix

$$\tilde{A}^{(k-1)} = \begin{pmatrix} 0 & 0 & \cdots & 0 & \cdots & \cdots & 0 \\ 0 & 0 & & \vdots & & & \vdots \\ \vdots & & \ddots & \vdots & & & \vdots \\ 0 & \cdots & \cdots & 0 & \cdots & \cdots & 0 \\ \vdots & & & \vdots & \tilde{a}_{kk}^{(k-1)} & \cdots & \tilde{a}_{kn}^{(k-1)} \\ \vdots & & & \vdots & \vdots & & \vdots \\ 0 & \cdots & \cdots & 0 & \tilde{a}_{nk}^{(k-1)} & \cdots & \tilde{a}_{nn}^{(k-1)} \end{pmatrix},$$

and the kth stage is described by

$$\tilde{A}^{(k-1)} = \mathbf{l}^{(k)}\mathbf{u}^{(k)T} + \tilde{A}^{(k)}, \tag{4.12}$$

where $l_i^{(k)} = u_i^{(k)} = 0$, $i \le k - 1$; $l_k^{(k)} = 1$; and $\tilde{a}_{ij}^{(k)} = 0$, for $i \le k$ or $j \le k$ (cf. the block matrices $L^{(k)}$ and $U^{(k)}$ of (3.9)). Provided that $\tilde{a}_{kk}^{(k-1)} \ne 0$, (4.12) is uniquely satisfied by

$$\mathbf{l}^{(k)} = \left(0, \ldots, 0, 1, \tilde{a}_{k+1,k}^{(k-1)}/\tilde{a}_{kk}^{(k-1)}, \ldots, \tilde{a}_{nk}^{(k-1)}/\tilde{a}_{kk}^{(k-1)}\right)^T, \tag{4.13}$$

$$\mathbf{u}^{(k)} = \left(0, \ldots, 0, \tilde{a}_{kk}^{(k-1)}, \tilde{a}_{k,k+1}^{(k-1)}, \ldots, \tilde{a}_{kn}^{(k-1)}\right)^T, \tag{4.14}$$

$$\tilde{A}^{(k)} = \begin{pmatrix} 0 & 0 & \cdots & 0 & \cdots & \cdots & 0 \\ 0 & 0 & & \vdots & & & \vdots \\ \vdots & & \ddots & \vdots & & & \vdots \\ 0 & \cdots & \cdots & 0 & \cdots & \cdots & 0 \\ \vdots & & & \vdots & \tilde{a}_{k+1,k+1}^{(k)} & \cdots & \tilde{a}_{k+1,n}^{(k)} \\ \vdots & & & \vdots & \vdots & & \vdots \\ 0 & \cdots & \cdots & 0 & \tilde{a}_{n,k+1}^{(k)} & \cdots & \tilde{a}_{nn}^{(k)} \end{pmatrix}, \tag{4.15}$$

where, for $i, j = k+1, k+2, \ldots, n,$

$$\tilde{a}_{ij}^{(k)} = \tilde{a}_{ij}^{(k-1)} - \tilde{a}_{ik}^{(k-1)} \tilde{a}_{kj}^{(k-1)} / \tilde{a}_{kk}^{(k-1)}. \tag{4.16}$$

After n stages of this elimination procedure we find that

$$
\begin{aligned}
A &= \mathbf{l}^{(1)} \mathbf{u}^{(1)T} + \tilde{A}^{(1)} \\
&= \mathbf{l}^{(1)} \mathbf{u}^{(1)T} + \mathbf{l}^{(2)} \mathbf{u}^{(2)T} + \tilde{A}^{(2)} \\
&\vdots \\
&= \mathbf{l}^{(1)} \mathbf{u}^{(1)T} + \mathbf{l}^{(2)} \mathbf{u}^{(2)T} + \cdots + \mathbf{l}^{(n)} \mathbf{u}^{(n)T} + \tilde{A}^{(n)},
\end{aligned}
$$

where $\tilde{A}^{(n)}$ is the null (or zero) matrix. We use n stages here, rather than the $n-1$ of Section 4.2.1, but the final stage simply defines $\mathbf{l}^{(n)} = (0, 0, \ldots, 0, 1)^T$ and $\mathbf{u}^{(n)} = \left(0, 0, \ldots, 0, \tilde{a}_{nn}^{(n-1)}\right)^T$. Hence

$$A = \sum_{k=1}^{n} \mathbf{l}^{(k)} \mathbf{u}^{(k)T}, \tag{4.17}$$

(cf. (3.10)) and the vectors $\mathbf{l}^{(k)}$ and $\mathbf{u}^{(k)}$ are given by (4.13) and (4.14). Equation (4.17) can be written as

$$A = LU, \tag{4.18}$$

where $L = \left(\mathbf{l}^{(1)}, \mathbf{l}^{(2)}, \ldots, \mathbf{l}^{(n)}\right)$ is a unit diagonal, lower triangular, matrix and $U = \left(\mathbf{u}^{(1)}, \mathbf{u}^{(2)}, \ldots, \mathbf{u}^{(n)}\right)^T$ is upper triangular.

Partial pivoting can be introduced at each stage of the above elimination procedure by interchanging rows as described in Section 4.2.2. A row interchange of the matrix A can be represented by PA, where P is a *permutation matrix*, which is the identity matrix I_n with the rows suitably reordered. Suppose that we wish to interchange rows r and s; then P is the identity matrix I_n with rows r and s interchanged. After n stages of the elimination we have matrices L and U such that $LU = PA$, where P is a permutation matrix which represents the accumulation of all of the interchanges performed.

4.3.2   Forward and backward substitution

Suppose that we have formed the LU factors of A and that we now wish to solve the system of equations $A\mathbf{x} = \mathbf{b}$. Since $LU\mathbf{x} = \mathbf{b}$, it follows that $U\mathbf{x} = L^{-1}\mathbf{b}$. We define $\mathbf{y}$, the *intermediate vector*, by $\mathbf{y} = L^{-1}\mathbf{b}$. Rather than invert L we find $\mathbf{y}$ by solving the lower triangular system

$$L\mathbf{y} = \mathbf{b}.$$

Here the components of **y** are computed in the order $y_1, y_2, \ldots, y_n$, using

$$y_i = b_i - \sum_{j=1}^{i-1} l_{ij} y_j,$$

remembering that L has a unit diagonal; this procedure is known as *forward substitution*. It is equivalent to modifying the right-hand side vector **b** as forward elimination proceeds in the way described in Section 4.2.1. The solution, **x**, may then be recovered from the upper triangular system

$$U\mathbf{x} = \mathbf{y},$$

using backward substitution in the normal manner.

The factorisation of A into the product of a unit diagonal, lower triangular matrix and an upper triangular matrix is usually called the *Doolittle reduction*. The *Crout reduction* is identical except that it factorises A into the product of a lower triangular matrix and a unit diagonal, upper triangular matrix.

When implementing the Doolittle reduction of A it is conventional to overwrite A with the LU factors as they are calculated, noting that it is not necessary to store the diagonal elements of L (or of U in the case of the Crout reduction) since they have unit value. The code to perform this reduction using explicit interchanges is essentially that given in Code 4.2, with the references to the right-hand side vector omitted. However, it is important that we keep a record of the pivoting undertaken so that the computation, by forward substitution, of the intermediate vector **y** is performed correctly. A simple strategy is to employ an integer vector `ipiv` and, at the kth stage, to set `ipiv(k)` = `piv`, the index of the pivotal row, as used in Code 4.2.

4.3.3   Rectangular matrices

Earlier in this section we described the LU factorisation of a square $n \times n$ matrix A. It is straightforward to extend the ideas to the LU factorisation of a rectangular $m \times n$ matrix (although, if $m \neq n$, we cannot then go on to solve a system of linear equations since A is no longer invertible.) The only changes from the description given earlier in this section are that the vectors $l^{(1)}, l^{(2)}, \ldots, l^{(n)}$, are now of length m instead of n, and the number of stages in the elimination procedure is $\min(m - 1, n)$. Thus Code 4.2 can be amended to cater for this situation simply by modifying the upper limits of the k and i loops appropriately.

A particular instance of this situation is when we choose $n = m + 1$ and set the final column of A equal to the right-hand side vector, **b**, of a system of m simultaneous linear equations, with the coefficient matrix equal to the first m columns of A; we refer to an *augmented matrix*. LU factorisation applied to A transforms the final column to the vector $\mathbf{b}'$ which corresponds to forward elimination applied to **b** (cf. (4.5)). The solution to the original system can now be recovered by backward substitution in the normal way.

4.4   BLAS variants

We now consider the implementation of forward elimination, pivoting, and forward and backward substitutions in terms of the BLAS introduced in Chapter 3. We do this for two reasons. The first is that we can analyse the various possible implementations of these computations using what is, effectively, a shorthand notation. The second is that, in the spirit of LAPACK, we will, by using parallel versions of the BLAS, immediately have a parallel algorithm for solving a system of linear equations, at least for a shared memory system. Because it is computationally the more expensive we first consider forward elimination.

4.4.1   BLAS versions of forward elimination

Assuming, for the moment, that pivoting is not necessary, forward elimination is represented by the three nested loops shown in Code 4.4 (taken directly from Code 4.1), in which k represents the stages and the i and j loops correspond to the computation of the multipliers and the elements of the matrix (4.15) given by (4.16).

```
do k = 1,n-1
  do i = k+1,n
    a(i,k) = a(i,k)/a(k,k)
    do j = k+1,n
      a(i,j) = a(i,j)-a(i,k)*a(k,j)
    end do
  end do
end do
```

Code 4.4 *Forward elimination – kij variant*

The innermost, j, loop is a (row) **saxpy** operation, and can be replaced by

```
call saxpy(n-k,-a(i,k),a(k,k+1),lda,a(i,k+1),lda)
```

in which, we note, both increments are equal to `lda` since we are dealing with rows of the matrix. An alternative is to rearrange the i and j loops of forward elimination to give the kji variant shown in Code 4.5. Here, we first have to compute the multipliers (an **sscal** operation) before subtracting a multiple of column k from column j (for $j = k+1, k+2, \ldots, n$) of the submatrix to be modified ($n - k$ **saxpy** operations). Hence, both variants can be expressed in terms of Level 1 BLAS.

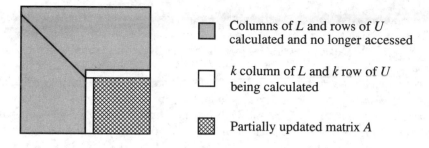

Columns of L and rows of U calculated and no longer accessed

k column of L and k row of U being calculated

Partially updated matrix A

Figure 4.1 *Submatrix forward elimination*

```
do k = 1,n-1
  do i = k+1,n
    a(i,k) = a(i,k)/a(k,k)
  end do
  do j=k+1,n
    do i = k+1,n
      a(i,j) = a(i,j)-a(i,k)*a(k,j)
    end do
  end do
end do
```

Code 4.5 *Forward elimination – kji variant*

In either of the two cases considered here each cycle of the k loop represents the rank-one modification

$$\tilde{A}^{(k)} \leftarrow \tilde{A}^{(k-1)} - \mathbf{l}^{(k)}\mathbf{u}^{(k)^T},$$

(cf. (4.12)), a Level 2 BLAS `sger` operation, and thus Codes 4.4 and 4.5 can be replaced by Code 4.6, with the ordering of the i and j loops specified within `sger`. For the reasons discussed in Section 3.3.2, the ji variant of `sger` is usually preferred in a serial Fortran environment.

```
do k = 1,n-1
  call sscal(n-k,one/a(k,k),a(k+1,k),1)
  call sger(n-k,n-k,-one,a(k+1,k),1,a(k,k+1),lda,a(k+1,k+1),lda)
end do
```

Code 4.6 *Forward elimination in terms of* `sger`

The kij and kji variants of forward elimination are often described as *forward looking algorithms*. Given a pivotal column, a submatrix is modified using that column, either in row order or in column order, as appropriate. In Figure 4.2 we indicate the situation at the kth stage. The white region indicates that portion of the matrix which is being used for modification, whilst the hatched region shows the portion to be modified. For obvious reasons we refer to the kij and kji variants as *submatrix forward elimination*.

In the submatrix forward elimination variants a cycle of the outer, k, loop results in a partial modification of the submatrix to the right of column k and below row k. An alternative approach is to modify completely a given column using all preceding columns, which now hold the multipliers. The starting point is to reorder the loops represented by Code 4.4 so that the j loop is outermost. Thus, the jki variant is as shown in Code 4.7. For a given value of j, the l loop denotes the computation of the multipliers in column $j-1$, an sscal operation, whilst the i loop denotes the subtraction of a multiple (a_{kj}) of a portion of column k from column j (a saxpy operation) for $k < j$.

```
do j = 2,n
  do l = j,n
    a(l,j-1) = a(l,j-1)/a(j-1,j-1)
  end do
  do k = 1,j-1
    do i = k+1,n
      a(i,j) = a(i,j)-a(i,k)*a(k,j)
    end do
  end do
end do
```

Code 4.7 *Forward elimination – jki variant*

We split the i loop of the jki variant into the portion for which $i < j$ and that for which $i \geq j$. For $i < j$ the two innermost, k and i, loops describe the following sequence of steps, which modify elements in column j (except for the first) above the diagonal.

$$a_{2j} \leftarrow \quad a_{2j} - \quad a_{21}a_{1j} \, ,$$
$$a_{3j} \leftarrow \quad a_{3j} - \quad a_{31}a_{1j} - \quad a_{32}a_{2j} \, ,$$
$$\vdots$$
$$a_{j-1,j} \leftarrow a_{j-1,j} - a_{j-1,1}a_{1j} - a_{j-1,2}a_{2j} - \cdots - a_{j-1,j-2}a_{j-2,j} \, .$$

Thus, the newly updated elements, which we denote $\hat{a}_{ij}$, satisfy the equations

$$\hat{a}_{1j} \qquad\qquad\qquad\qquad = a_{1j},$$
$$a_{21}\hat{a}_{1j} + \qquad \hat{a}_{2j} \qquad\qquad = a_{2j},$$
$$a_{31}\hat{a}_{1j} + \quad a_{32}\hat{a}_{2j} + \hat{a}_{3j} \qquad = a_{3j},$$
$$\vdots$$
$$a_{j-1,1}\hat{a}_{1j} + a_{j-1,2}\hat{a}_{2j} + \cdots + a_{j-1,j-2}\hat{a}_{j-2,j} + \hat{a}_{j-1,j} = a_{j-1,j},$$

and we recognise that these equations involve the solution of a lower triangular system, namely

$$\mathbf{a}^{(1)} \leftarrow \bar{A}^{(1)-1}\mathbf{a}^{(1)}, \tag{4.19}$$

an strsv operation, where $\mathbf{a}^{(1)} = (a_{1j}, a_{2j}, \ldots, a_{j-1,j})^T$ is of length $j-1$ and

$$\bar{A}^{(1)} = \begin{pmatrix} 1 & 0 & \cdots & & 0 \\ a_{21} & \ddots & & \ddots & \vdots \\ \vdots & \ddots & & \ddots & 0 \\ a_{j-1,1} & \cdots & a_{j-1,j-2} & & 1 \end{pmatrix}$$

is a lower triangular, unit diagonal, $(j-1) \times (j-1)$ matrix.

For $i \geq j$ the k and i loops of Code 4.7 modify the elements in column j on and below the diagonal. We have

$$
\begin{aligned}
a_{jj} &\leftarrow & a_{jj} - & a_{j,j-1}a_{j-1,j} - \cdots - & a_{j2}a_{2j} - & a_{j1}a_{1j}, \\
a_{j+1,j} &\leftarrow a_{j+1,j} - a_{j+1,j-1}a_{j-1,j} - & \cdots - a_{j+1,2}a_{2j} - & a_{j+1,1}a_{1j}, \\
&\vdots \\
a_{nj} &\leftarrow & a_{nj} - & a_{n,j-1}a_{j-1,j} - \cdots - & a_{n2}a_{2j} - & a_{n1}a_{1j},
\end{aligned}
$$

so that the newly updated values satisfy the equations

$$
\begin{aligned}
\hat{a}_{jj} &= & a_{jj} - & a_{j1}\hat{a}_{1j} - & a_{j2}\hat{a}_{2j} - \cdots - & a_{j,j-1}\hat{a}_{j-1,j}, \\
\hat{a}_{j+1,j} &= a_{j+1,j} - a_{j+1,1}\hat{a}_{1j} - & a_{j+1,2}\hat{a}_{2j} - & \cdots - a_{j+1,j-1}\hat{a}_{j-1,j}, \\
&\vdots \\
\hat{a}_{nj} &= & a_{nj} - & a_{n1}\hat{a}_{1j} - & a_{n2}\hat{a}_{2j} - \cdots - & a_{n,j-1}\hat{a}_{j-1,j}.
\end{aligned}
$$

These equations involve a matrix-vector product

$$
\mathbf{a}^{(2)} \leftarrow -\bar{A}^{(2)}\mathbf{a}^{(1)} + \mathbf{a}^{(2)},
$$

an **sgemv** operation, where $\mathbf{a}^{(1)}$ has been calculated by (4.19), $\mathbf{a}^{(2)}$ comprises the final $n - j + 1$ elements of column j of A, that is, $\mathbf{a}^{(2)} = (a_{jj}, a_{j+1,j}, \ldots, a_{nj})^T$, and

$$
\bar{A}^{(2)} = \begin{pmatrix}
a_{j1} & a_{j2} & \cdots & a_{j,j-1} \\
a_{j+1,1} & a_{j+1,2} & \cdots & a_{j+1,j-1} \\
\vdots & \vdots & \ddots & \vdots \\
a_{n1} & a_{n2} & \cdots & a_{n,j-1}
\end{pmatrix}.
$$

Thus, the jki (and jik) variants of forward elimination can be written as shown in Code 4.8.

```
do j = 2,n
   call sscal(n-j+1,one/a(j-1,j-1),a(j,j-1),1)
   call strsv('l','n','u',j-1,a(1,1),lda,a(1,j),1)
   call sgemv('n',n-j+1,j-1,-one,a(j,1),lda,a(1,j),1,one,a(j,j),1)
end do
```

Code 4.8 *Forward elimination in terms of* `strsv` *and* `sgemv`

In contrast to submatrix forward elimination, the jki and jik variants are termed *backward looking*, since, at the jth stage (which corresponds to j-1 in Code 4.8), the whole of the jth column is completely modified by all previous stages of the elimination procedure and is then available as the next pivot column. The situation is illustrated in Figure 4.2, and we refer to the variants as *column forward elimination* (cf. the implementation of **sgetrf** described in Section 3.4.3).

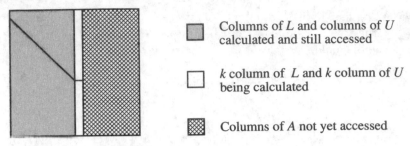

Figure **4.2** *Column forward elimination*

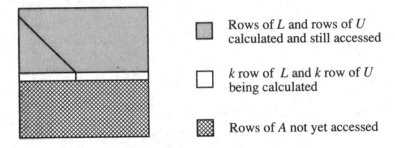

Figure **4.3** *Row forward elimination*

The ikj (ijk) variant of forward elimination applied to A is equivalent to the jki (jik) variant applied to A^T, the only substantial difference being that the former deals with rows whilst the latter deals with columns at the outermost level. We refer to these ikj and ijk variants collectively as *row forward elimination* (Figure 4.3).

This relationship between row and column forward elimination suggests an alternative implementation of the column and row variants of forward elimination which is, essentially, a hybrid of the two. At the jth stage, elements on and below the diagonal of the jth column are modified as in Code 4.8 but, instead of modifying the elements above the diagonal of the jth column using **strsv**, we can modify the elements to the right of the diagonal of the jth row using **sgemv**. We note that the second call to **sgemv** accesses the matrix A by rows. The required code is given in Code 4.9 and the situation is represented pictorially in Figure 4.4.

```
do j = 2,n
   call sscal(n-j+1,one/a(j-1,j-1),a(j,j-1),1)
   call sgemv('n',n-j+1,j-1,-one,a(j,1),lda,a(1,j),1,one,a(j,j),1)
   call sgemv('t',j-1,n-j,-one,a(1,j+1),lda,a(j,1),lda,one,
   +          a(j,j+1),lda)
end do
```

Code **4.9** *Forward elimination in terms of* sgemv *only*

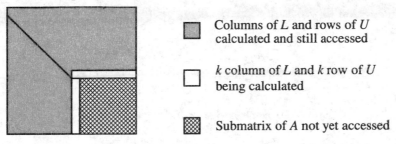

Columns of L and rows of U calculated and still accessed

k column of L and k row of U being calculated

Submatrix of A not yet accessed

Figure 4.4 *Row/column hybrid forward elimination*

4.4.2 Pivoting

We recall that, at each stage of elimination, explicit pivoting involves

- locating the element with largest modulus in a column, and
- interchanging two rows.

The first can be implemented using a call to `isamax`, whilst the second is an `sswap` operation. Considering the *jki* variant, at the beginning of each cycle of the j loop we set

```
piv = j-2+isamax(n-j+2,a(j-1,j-1),1)
```

(remembering that, in Code 4.8, the initial value of j is 2) to locate the pivotal row and then interchange rows using

```
call sswap(n,a(j-1,1),lda,a(piv,1),lda)
```

The interchanging of rows in a column-based language is unfortunate, and a better strategy might be to interchange only the elements in the column currently being considered. This means that, at the start of each cycle of the j loop, we must additionally implement all previous interchanges.

4.4.3 Forward and backward substitutions

Since forward and backward substitutions are, essentially, the same procedure (ignoring any interchanges that may need to be applied to the right-hand side vector before forward substitution begins), it is sufficient to consider one of them only. We arbitrarily choose backward substitution.

Backward substitution in which the solution vector overwrites the right-hand side vector is a Level 2 BLAS `strsv` operation. In the case of an upper triangular, non-transposed, non-unit diagonal, matrix A, Code 4.3 provides a possible implementation in which the inner, j, loop represents a dot product. Hence we may write the code as shown in Code 4.10.

```
do i = n,1,-1
  x(i) = (b(i)-sdot(n-i,a(i,i+1),lda,x(i+1),1))/a(i,i)
end do
```

Code 4.10 *Backward substitution in terms of* sdot

Assuming that x is initialised to b, an alternative implementation results if we reorder the i and j loops to give Code 4.11. The inner, i, loop is a **saxpy** operation, and hence the code may be rewritten as shown in Code 4.12.

```
do j = n,1,-1
  x(j) = x(j)/a(j,j)
  do i = 1,j-1
    x(i) = x(i)-a(i,j)*x(j)
  end do
end do
```

Code 4.11 *Backward substitution – ji variant*

```
do j = n,1,-1
  x(j) = x(j)/a(j,j)
  call saxpy(j-1,-x(j),a(1,j),1,x,1)
end do
```

Code 4.12 *Backward substitution in terms of* saxpy

4.4.4   The choice of variant

Each of the variants of forward elimination presented above involves the same arithmetic operations. On the face of it, therefore, there would appear to be nothing to choose between them, and the same can be said of the variants of forward and backward substitutions. However, even in a sequential environment the execution times for the different variants are unlikely to be the same. The main criterion likely to distinguish between the various approaches concerns the way two-dimensional arrays are stored. In Fortran matrices are stored by columns and thus we might expect a column-based algorithm to perform optimally if it avoids a significant amount of paging between a local cache and main memory, or between main memory and secondary storage. For languages such as Pascal, in which matrices are stored by rows, we would expect a row-based algorithm to fair better than the alternatives. For a further discussion on these matters, see Ortega (1988).

In the next section we consider forward elimination on parallel architectures, concentrating on the kij/kji variants. This choice is partly influenced by the fact that these are, arguably, the easiest to follow. In the context of a local memory system the choice is also influenced by the way the matrix is distributed; inevitably, the data distribution and the choice of algorithm go hand in hand. We also have a few words to say about other variants.

4.5 Parallel equation solution

We now consider ways of parallelising elimination and backward (and, implicitly, forward) substitution. In the case of a shared memory system the principal concern is to distribute the computational workload across the processors and we will, at this stage, be content simply to use parallel versions of the Levels 1 and 2 BLAS. A secondary consideration which we do not address here is the maximal use of data held in the local caches. In a local memory environment we have similar considerations. Indeed it is necessary to match the work sharing and data distribution aspects of our algorithm. The aim is to construct an algorithm which exhibits good load balancing properties and, at the same time, limits the amount of data communication.

4.5.1 Parallel forward elimination

Since it has the higher operation count we initially concentrate our efforts on speeding up forward elimination. We begin by taking Code 4.4, the kij variant, as the basis for our attempts at parallelisation. This code involves three nested loops and we analyse these in turn:

- *The outer, k, loop*

  The k loop represents the stages of Gaussian elimination. Unfortunately, the scope for parallelism appears to be somewhat limited. We recall that the first stage ($k = 1$) involves the subtraction of a multiple of the first equation from equations $2, 3, \ldots, n$, whilst stage 2 involves the subtraction of a multiple of the second equation from equations $3, 4, \ldots, n$, but only after these equations have been modified by the stage 1 procedure. There is an essential dependence of the computations of the kth stage on the computations of the earlier stages. When, in the absence of pivoting, a pivot row (or column for the kji variant) is available it can be used to modify all subsequent rows (columns), and, as soon as the first such row has been modified, it is available as the next pivot row. Thus, in principle, we can perform forward elimination using several pivot rows concurrently. However, this imposes synchronisation restrictions and thus reduces the parallelism of the inner loops. Further, partial pivoting requires that the modifications to the pivot column must be completed before the pivot can be determined. Thus, whilst in principle several pivots can be used in parallel, there are yet more synchronisation restrictions on the inner loops. We conclude that there is limited scope only for parallelism in this outer loop.

✓ • *The middle, i, loop*
   The i loop describes the modification of the ith equation by the kth equation, each of which is an independent computation. Hence we can parallelise this loop directly.

✓ • *The inner, j, loop*
   The j loop describes the modification of the elements of a row of the coefficient matrix. Since the modification of each element of the row is independent of the modification of every other element, this loop may also be parallelised directly.

Of the three loops, therefore, the inner two offer the most scope for parallelisation. The parallelisation of the outer of these two loops should give the greatest improvement as the amount of work that each process has to perform between synchronisation points is then maximised.

Reordering the loops as described in Section 4.4.1 yields a similar situation. For the kji variant the middle and inner loops again represent independent processes. For the jki (or jik) variant the outer, j, loop represents the modification of column j using all previous columns. It is possible to modify columns in parallel, but it is necessary to synchronise actions so that, say, column r is fully modified before it is used in the updating of column q, for $q > r$. The dependence is similar to that exhibited by the k loop above. A similar situation holds for the ikj and ijk variants.

4.5.2   Parallel backward substitution

We first consider the implementation of backward substitution represented by Codes 4.3 and 4.10. The outer, i, loop represents the computation of each component of the solution vector. Since, for $r < q$, x_q must be available before x_r can be determined, this loop must be performed sequentially. The inner, j, loop is an **sdot** operation which can be parallelised, but it has synchronisation restrictions (Section 3.3.1).

For the versions given in Codes 4.11 and 4.12, the outer, j, loop represents the subtraction from the solution vector of a multiple of the jth column of the upper triangle of A, and offers little scope for parallelism. In this case the inner, i, loop is a **saxpy** operation which describes independent processes and thus can be parallelised directly.

4.5.3 Equation solution in a shared memory environment

We consider here the parallelisation of forward elimination and backward substitution in the context of the Encore Multimax 520 using the parallel extensions to Fortran 77 supported by the EPF compiler.

First, we consider the *kij* (and, by implication, the *kji*) variant (Code 4.4), in which the innermost loop is a `saxpy` operation on the rows (or columns) of the submatrix to be modified whilst the middle loop is an `sger` operation. In the former case we can parallelise the `saxpy` operation using simple loop spreading, to give Code 4.13.

```
parallel
  integer j
  do all (j = k+1:n)
    a(i,j) = a(i,j)-a(i,k)*a(k,j)
  end do all
end parallel
```

Code 4.13 *Forward elimination – parallelisation of* saxpy

For the `sger` operation we can again use loop spreading to give Code 4.14, if we update the rows of **a**. We could equally have updated the columns of **a** by interchanging the *i* and *j* loops. It is important to recognise that, whilst we have expressed `sger` in terms of calls to `saxpy`, the `saxpy` operations must, themselves, be sequential as EPF does not support nested parallel blocks.

```
parallel
  integer i
  do all (i = k+1:n)
    a(i,k) = a(i,k)/a(k,k)
    call saxpy(n-k,-a(i,k),a(k,k+1),lda,a(i,k+1),lda)
  end do all
end parallel
```

Code 4.14 *Forward elimination – parallelisation of* sger

Referring to Code 4.6, an equivalent description of the above is to note that each cycle of the *k* loop of forward elimination can be expressed as sequential calls to routines which exploit parallelism: `sscal` and `sger`. Assuming that no parallelism is employed elsewhere in the program, the first `sscal` call ($k = 1$) involves starting up a number of parallel processes to which portions of the loop are allocated. We recall that the number of processors, and, hence, processes is determined statically before execution begins. The overhead associated with the start-up of a process is quite high, and these overheads accumulate as we add more processes. Thus, we would expect to experience slow-down rather than speed-up if the performance of the parallel `sscal` was timed in isolation; unless the vector length is extremely large, the cost of one `sscal` operation (which, we recall, requires $O(n)$ floating-point operations) is likely to be much lower than the process creation overheads. When the parallel block inside `sscal` is left all processes idle except one, which continues with the next, sequential, part of the code. (Of course, the processors on which these processes were executing do not idle; they are free to perform some other task, such as the partial execution of another job.) When the next parallel block is encountered (inside `sger`) there is no need to start up new parallel processes since they already exist. All that needs to be done is to apportion work to them. Thus we incur the process creation overheads once only. On leaving this second parallel block the processes idle again and are next given work by `sscal` on the second cycle of the k loop. Thus, the cost of the single set of overheads must be measured against the cost of the total amount of work which can take place in parallel. Since Gaussian elimination is an $O(n^3)$ process we can, therefore, expect reasonable speed-up for relatively modest values of n.

As we remarked earlier, parallelisation at the middle loop can be expected to give the most significant gains. To assess, experimentally, the two approaches (parallelisation of the inner or middle loop) outlined here we show, in Figure 4.5, speed-ups obtained using $n = 256$ and various numbers of processors. The matrix A was initialised with random numbers in the range $[0, 1]$, except for the diagonal entries, each of which was set to 256. We have, therefore, ensured that the coefficient matrix is *diagonally dominant* (that is, on each row the modulus of the element on the diagonal is greater that the sum of the moduli of all other elements on that row). It can be shown (Golub and Van Loan, 1989, p. 119) that Gaussian elimination without pivoting applied to a system of equations in which the coefficient matrix is diagonally dominant is numerically stable. Clearly, we cannot expect 100% efficiency, not least because, eventually, the length of the spread loops is less than the number of processes available. However, under the assumption that $n \gg p$ this is not a serious problem. We repeat the caveat mentioned in Section 1.3.2 that it is difficult to obtain accurate timings on a multi-user shared memory system, even when there is only one application running. The results summarised in Figure 4.5 represent the best times observed over several consecutive runs executed within the same program. The best time is unlikely to be the first as this, and only this, contains the process creation overheads. However, for $n = 256$ these overheads are not significant. The results from the two parallelisation strategies are virtually indistinguishable and hence only one set of results is plotted in Figure 4.5. Our conclusion is, therefore, that for this particular hardware/software combination and for this particular problem size there is little to choose between the approach of parallelising the middle loop (the outer loop of `sger`) and of parallelising the inner (`saxpy`) loop. For other shared memory systems and larger problem sizes this may not be the case and the advice would always be to employ parallelism at the outermost level.

In neither of the two cases considered above have we explicitly had to bother about the allocation of processes (the modification of a row, or the elements of a row) to processors. Further, the changes needed to accommodate partial pivoting

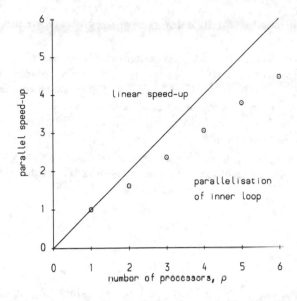

Figure 4.5 *Speed-ups for equation solution on an Encore Multimax 520*

are trivial; the derivation of a parallel `isamax` follows directly from the parallel `sdot` described in Section 2.4.5 (Exercise 2.5); each process deals with a portion of the pivot column and finds the local maximum of that portion, after which the maximum of these local maxima must be found, and this requires synchronisation of the processes. A parallel `sswap` employs simple loop spreading only. Parallel backward substitution follows by employing parallel versions of `sdot` or `saxpy` in Codes 4.10 or 4.12.

It is interesting to look briefly at the column and row/column hybrid variants of Gaussian elimination as expressed in Codes 4.8 and 4.9, in which the j loop no longer consists of independent processes. In the former case the calls to `strsv` and `sgemv` in each cycle of the j loop are not independent and must be performed in the given order. Hence it is necessary to parallelise at the BLAS level. The `sgemv` operation can be easily parallelised at the outermost loop if it is computed in terms of dot products (Section 3.3.2). The `strsv` operation can be parallelised easily at the innermost loop only; further, there is a sequential computation to be performed (division by a diagonal element of the coefficient matrix). A similar situation occurs in Code 4.9, except that the two calls to `sgemv` within each cycle of the j loop are independent and could be assigned to two separate processes to be obeyed concurrently. However, given the limitations of EPF, each of these processes could not itself spawn other processes. Hence we limit the potential speed-up. Again it is more sensible to parallelise at the BLAS level.

4.5.4   Equation solution in a local memory environment

We now consider the implementation of submatrix forward elimination followed by backward substitution within a local memory environment. For simplicity we assume that the right-hand side vector is modified as the forward elimination takes place. Breaking this up into *LU* factorisation followed by forward and backward substitutions is straightforward but adds to what is already destined to be a lengthy section of code.

Before proceeding an initial decision has to be made as to what is likely to be an appropriate data distribution. The obvious possibilities for the distribution of the matrix A are

- by rows,
- by columns, or
- by blocks.

We consider first the distribution of A by rows, since this is often seen as the natural way to describe the algorithm. We will find it convenient to distribute the components of the right-hand side vector, and accumulate the components of the solution vector, in a conforming manner. That is, if a process is allocated row i of A then it also has the ith component of **b** and is responsible for computing the ith component of **x**.

Restricting our attention to a row distribution still leaves a number of possible avenues open, the most obvious of which are

- block rows, and
- scattered rows.

In a block row distribution consecutive rows of A are grouped into blocks. Suppose that we have p processes/processors numbered $0, 1, \ldots, p-1$, with, for convenience, n divisible by p. Then, to the process labelled 0 we allocate rows $1, 2, \ldots, n/p$, of A, to the process labelled 1 we allocate rows $n/p + 1, n/p + 2, \ldots, 2n/p$, and so on. That is, process j gets rows $jn/p + 1, jn/p + 2, \ldots, (j + 1)n/p$. If n is not an integral multiple of p then we have an uneven share of rows to apportion. Let $m = \lfloor n/p \rfloor$ and $q = n \bmod p$. Then we allocate $m + 1$ rows to processes $j = 0, 1, \ldots, q - 1$, and m rows to processes $j = q, q + 1, \ldots, p - 1$.

In a scattered row distribution consecutive rows are not allocated to the same process but, usually, to processes which are, in some sense, adjacent. For example, if a wraparound approach is adopted we allocate row 1 to process 0, row 2 to process 1, and so on, up to row p which is allocated to process $p-1$. We then cycle round the processes and allocate row $p+1$ to process 0, row $p+2$ to process 1, and so on. In general, row j is allocated to process $(j - 1) \bmod p$. On the latest local memory parallel computers the adjacency of processors (and, hence, processes) is not so important, since the throughrouting communications capabilities (for example, the Direct-Connect routing system in the case of the Intel iPSC/2) of these machines means that all processors are *effectively adjacent*. Nevertheless, for clarity, a systematic allocation of rows to processes is to be recommended.

We now consider whether it is more efficient to use a block or scattered row (or, for that matter, column) distribution. We see from the two innermost loops of Code 4.4 that, at the kth stage of the elimination, row k of A is used to modify all remaining rows of an $(n-k) \times (n-k)$ submatrix. If a block row distribution is employed then we will soon reach a situation in which we have under-utilisation of some of the processes. Recall that process 0 has allocated to it the first $\lceil n/p \rceil$ rows. It follows that it will become idle as soon as the first $\lceil n/p \rceil$ stages of forward elimination have been completed. However, by employing a scattered row distribution we can ensure that processes begin to become idle only after the first $n - p$ stages have finished, and for $p \ll n$ the process idle time will be of little significance.

Having decided to adopt a scattered row distribution we now need to define the processes that are required. We choose to employ two types:

- *A host process*
  This allocates the processors in the cube and loads the slave processes on these node processors. It also distributes the scattered rows to the slave processes, but it does not perform any floating-point computations.
- *A slave process*
  The slave processes together, and concurrently, perform the floating-point computations of Gaussian elimination and backward substitution.

As with the iPSC/2 dot product of Section 2.4.5, we assume p slave processes executing on p node processors of the hypercube, and a single host process running on a separate processor (the hypercube SRM).

First we consider the host process, of which the subroutine of Code 4.15 represents a part, for a hypercube implementation and note that the slave processes, each with a process identifier of 1, the same as that of the host process, are assumed to run on the processors with node identifiers $0, 1, \ldots, p - 1$. The host process distributes the rows of the coefficient matrix A and corresponding elements of **b** to the appropriate slave processes and receives the solution vector **x** from one of the slave processes. For each row, i, of A the following information is gathered in the solution vector **x** which here is acting as temporary storage:

- **x(1)** – the row number,
- **x(2:n+1)** – the row of the matrix, which is copied from **a** using **scopy** (one processor version),
- **x(n+2)** – the corresponding component of the right-hand side vector.

A single start-up overhead is thus incurred when this data is sent to a slave process.

```
      subroutine gelim(n,a,lda,b,x,cube)
c gelim uses a distributed row-based Gaussian elimination algorithm
c to solve the n linear equations A x = b on the numslave processors
c of an Intel iPSC/2 Hypercube
c
c n        - an integer which is the number of equations in the system -
c             unchanged on exit
c a        - the real two-dimensional coefficient matrix - unchanged on exit
c lda      - an integer which is the leading dimension of a - unchanged
c             on exit
c b        - the real right-hand side vector - unchanged on exit
c x        - a real vector used to return the solution
c cube     - a character string which is the number and type of nodes of
c             the hypercube to be allocated - unchanged on exit
c
c Assume that an integer variable requires 4 bytes of storage and
c that a real variable also requires 4 bytes of storage
      integer intlen,reallen
      parameter(intlen = 4,reallen = 4)
      integer n,lda
      character*4 cube
      real a(lda,*),b(*),x(*)
      integer pid,itype,ibytes,numslave,numnodes,nextslave,i,j,numsent,
     +          nsent
      external getcube,setpid,load,numnodes,csend,scopy,crecv,relcube
c Get the cube, set the host pid and load the slave processes
      call getcube('gecube',cube,' ',0)
      call setpid(1)
      pid = 1
      call load('slave.out',-1,pid)
c The number of slave nodes, numslave, is passed on to the slave processes
      itype = 1
      ibytes = intlen
      numslave = numnodes()
      call csend(itype,numslave,ibytes,-1,pid)
c n is passed on to the slave processes
      itype = 2
      call csend(itype,n,ibytes,-1,pid)
c The next n messages send, one row at a time, the row index, the
c corresponding row of a and the corresponding elements of b to the
c slave processes
c
c The rows are distributed cyclically amongst the slave processes
      nextslave = 0
      itype = 3
      ibytes = (n+2)*reallen
      do i = 1,n
        x(1) = float(i)
        call scopy(n,a(i,1),lda,x(2),1)
        x(n+2) = b(i)
        call csend(itype,x,ibytes,nextslave,pid)
        if (nextslave.eq.numslave-1) then
          nextslave = 0
        else
          nextslave = nextslave + 1
        end if
      end do
c Receive the solution vector x from one of the slave processes
      itype = 4
      call crecv(itype,x,ibytes)
      call relcube('gecube')
      end
```

Code 4.15 *Host process*

The slave processes, Code 4.16, perform the forward elimination and backward substitution steps. Each slave process concurrently performs the following operations:

1. Store a list of the node numbers of all the other slave processes for use when broadcasting data.
2. Receive my own rows of A and corresponding elements of **b**. The components of **b** are stored as an additional column of A; effectively, we have a scattered row distribution of an augmented matrix.
3. Perform elimination steps:
   - Either broadcast the next pivot row, or receive the next pivot row from the broadcast of another slave process.
   - Call **sscal** and **sger** (one processor versions), as in Code 4.6, to perform the elimination steps.
4. Perform backward substitution:
   - Either calculate and broadcast the next element of the solution vector, or receive it from the broadcast of another slave process.
   - Call **saxpy** (one processor version), as in Code 4.12, to update the partial solution vector.

In addition, slave process 0 returns the solution vector to the host process. It should be noted that, on any of the slave processes, the ith row of A is not the ith row of the original matrix, but is the ith row of the matrix on this slave process. Thus the rows of A are always accessed via the pointers held in **index**.

```
      Program slave
c Slave code for Gaussian Elimination which uses a scattered row
c distribution of the coefficient matrix
c
c Assume that an integer variable requires 4 bytes of storage and
c that a real variable also requires 4 bytes of storage.
c Assume also a hypercube of at most 32 nodes
      integer lda,intlen,reallen,maxnodes
      real one
      parameter(lda = 512,intlen = 4,reallen = 4,one = 1.0e0,
     +         maxnodes = 32)   MAX-NODES = 8
      integer otherslave(maxnodes),index(lda),itype,ibytes,numslave,
     +        numother,thisslave,i,mynode,n,numrows,nextra,j,ipointer,
     +        nextindex,k,nsize,node,pid,myhost
      real a(lda,lda+1),x(lda+2)
      external crecv,mynode,scopy,gsendx,sscal,sger,saxpy,myhost,csend
c The first message received is the number of slave processors in use
      itype = 1
      ibytes = intlen
      call crecv(itype,numslave,ibytes)
c Given numslave, store a list of the node numbers of all the other
c slave processes in otherslave.
c Note that mynode() returns the node number of the current process,
c and that the slave processes are assumed to have node numbers in
c the range [0,numslave-1]
      numother = numslave-1
      thisslave = mynode()
      do i = 0,thisslave-1
        otherslave(i+1) = i
```

```
        end do
        do i = thisslave+1,numslave-1
          otherslave(i) = i
        end do
c The second message received is the dimension of the matrix a
        itype = 2
        call crecv(itype,n,ibytes)
c Each slave process calculates the number of rows which it will handle.
c Each handles either n/numslave or (n/numslave)+1 rows
        numrows = n/numslave
        nextra = n - numrows*numslave-1
        if (thisslave.le.nextra) then
            numrows = numrows+1
        end if
c The third message received contains the row index together
c the corresponding row of a and element of b
        do i = 1,numrows
          itype = 3
          ibytes = (n+2)*reallen
          call crecv(itype,x,ibytes)
          index(i) = int(x(1))
          call scopy(n+1,x(2),1,a(i,1),lda)
        end do
c ELIMINATION STEPS
c nextindex points to the next row on this slave process which is
c a potential elimination row.
c Each time a row is used in the elimination process nextindex is
c incremented to point to the next row on this slave process
        ipointer = 1
        nextindex = index(1)
        do k = 1,n-1
c Set itype and ibytes for passing of pivot rows.
c Note that itype and ibytes depend on k
          itype = 4+k
          ibytes = (n+2-k)*reallen
          if (k.eq.nextindex) then
c Pivot row is held on this slave process.
c Copy the pivot row to x and broadcast it to all the other slave
c processes.
c Also, provided ipointer<numrows, update nextindex and ipointer
            call scopy(n-k+2,a(ipointer,k),lda,x,1)
            call gsendx(itype,x,ibytes,otherslave,numother)
            if (ipointer.lt.numrows) then
              ipointer = ipointer+1
              nextindex = index(ipointer)
            end if
          else
c Pivot row is held on another slave process.
c The next message received is the pivot row which is stored in temp
            call crecv(itype,x,ibytes)
          end if
c Provided that k is less than the highest index held on this slave
c then perform eliminations (making calls to sscal and sger) using
c the pivot row held in temp
          if(k.lt.index(numrows))then
            nsize = numrows-ipointer+1
            call sscal(nsize,one/x(1),a(ipointer,k),1)
            call sger(nsize,n+1-k,-one,a(ipointer,k),1,x(2),1,
     +                a(ipointer,k+1),lda)
          end if
        end do
c BACKWARD SUBSTITUTION
c nextindex currently points to the highest index row on this
c slave process.
c Each time a row is used in the backward substitution process
c nextindex is decremented to point to the previous row on this
```

```
c slave process.
c Set itype and ibytes for passing the values of the solution vector.
c Again, itype depends on k
      ibytes = reallen
      do k = n,1,-1
         itype = 4+k
         if(k.eq.nextindex)then
c Backward substitution row is held on this slave process.
c Calculate the solution x(k) and broadcast it to all the other
c slave processes.
c Also, provided ipointer .gt. 1, update nextindex and ipointer
            x(k) = a(ipointer,n+1)/a(ipointer,k)
            call gsendx(itype,x(k),ibytes,otherslave,numother)
            if (ipointer.gt.1) then
               ipointer = ipointer-1
               nextindex = index(ipointer)
            end if
         else
c Backward substitution row is held on another slave process.
c The next message received is the solution value x(k)
            call crecv(itype,x(k),ibytes)
         end if
c Provided that k is greater than the lowest index held on this
c slave then update earlier rows using the value x(k) just calculated
         if (k.gt.index(1)) then
            call saxpy(ipointer,-x(k),a(1,k),1,a(1,n+1),1)
         end if
      end do
c Slave process running on node 0 returns the solution vector x to
c the host process
      if (thisslave.eq.0) then
         itype = 4
         ibytes = n*reallen
         node = myhost()
         pid = 1
         call csend(itype,x,ibytes,node,pid)
      end if
      end
```

$-x(k) * a(1,k) + a(1,n+1)$

Code 4.16 *Slave process*

In Figure 4.6 we give performance figures for this parallel implementation obtained on an Intel iPSC/2 Hypercube; the particular machine used has scalar accelerators on all the node processors. The linear system solved is the one used to time the corresponding EPF code in Section 4.5.3. The number of processors given is the number executing the slave processes; including an extra parameter in the subroutine **gelim** enables **numslave** to be set in the calling process, rather than set by a call to **numnodes** in **gelim**. The times used to calculate the speedups exclude the time taken to communicate data from and to the host process. We use this measure of time because it gives a better illustration of the degree of parallelisation of the code, and also because the host to node communication times on the iPSC/2 depend substantially on the specific hardware/software configuration in use. We also include in Figure 4.6 similar statistics for a larger system of linear equations of order 512. As might be expected, the speed-up figures (and consequently the efficiency) are better for the larger system.

The slave process in Code 4.16 could be made marginally more efficient in either of two ways:

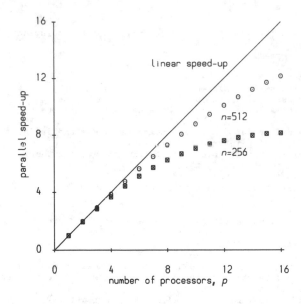

Figure 4.6 *Speed-ups for equation solution on an iPSC/2 Hypercube*

1. At each elimination step k the process which holds row $k+1$, the pivot row for the next step, could
   o modify row $k+1$,
   o broadcast, asynchronously, row $k+1$ as the pivot row to be employed at the next elimination step, and then
   o modify the remaining rows associated with this slave process.
2. A similar modification could be made to the backward substitution steps, that is, generate a component of the solution vector as early as possible and broadcast it, asynchronously, before modifying the remaining elements of the solution vector.

For our purposes the resulting complications to the code render these changes not worthwhile.

There are some difficulties associated with the introduction of partial pivoting into the scattered row-based code which we have just described. At the kth stage of the elimination the pivotal element is defined to be the maximum (in absolute value) element, on or below the diagonal, of the kth column of the partially eliminated matrix $A^{(k-1)}$. Since these elements are distributed amongst the various slave processes, identification of the pivotal element involves a maximisation operation applied across the slave processes. This is implemented by performing a local maximisation operation on each slave process and combining the results. It is inevitably less efficient than a similar maximisation operation on a single slave process. Further, the physical interchange of two rows incurs a communications overhead. If, instead, rows are interchanged only indirectly via an integer pointer array, load balancing may be distorted, since, for example, the first $\lceil n/p \rceil$ pivots may all be associated with process 0.

As an alternative, suppose that we decide to adopt a scattered column distribution of A. At each stage of the outer, k, loop, the process which holds the kth column has to broadcast the multipliers to every other process. If partial pivoting is employed, one process has the full pivot column and can thus easily determine the pivot element without communication (and asynchronous communication of pivot columns in the manner suggested by (1) above remains viable). The task of identifying the pivotal element could be distributed to processes which otherwise would be idle, but it would be inefficient to do this. When a pivot has been located rows can be interchanged, either physically or via pointers, without significant communications overhead, simply by broadcasting, with the multipliers, the indices of the rows to be interchanged.

The difficulty with a scattered column distribution concerns the backward substitution step. It is again natural to treat the right-hand side vector as just another column of A. At the end of the elimination only one process, therefore, has local access to the transformed vector, $\mathbf{b}'$, of (4.5). The **saxpy** variant (Code 4.12) of backward substitution is no longer appropriate; the i loop cannot be parallelised without incurring considerable communication overheads (a redistribution of a column of the upper triangular factor). More hopeful is the **sdot** variant (Code 4.10), but this necessitates a gather at each cycle of the i loop, followed by a broadcast of x_i; in a scattered row distribution it is only the latter form of communication that is required.

Exercises

4.1. To what (block) variants of LU do the blocked algorithms of Section 3.4.3 and Exercise 3.9 correspond?

4.2. Develop algorithms for block(ed) forward and backward substitutions.

4.3. A scattered blocked row distribution of a matrix A is similar to a scattered row distribution except that adjacent rows are grouped into blocks of width b, say, and then allocated to processes in a wraparound fashion. Compare and contrast the merits of scattered blocked row and column distributions of A for computing a solution to the linear system of equations $A\mathbf{x} = \mathbf{b}$. Does a block size of $b > 1$ offer any advantage over a block size of $b = 1$?

4.4. Gaussian elimination breaks down if a zero pivot is detected. Modify Code 4.16 so that it performs partial pivoting and returns an appropriate message to the host process and terminates if a singular coefficient matrix is detected. Modify Code 4.15 accordingly.

4.5. Modify Codes 4.15 and 4.16 so that the host process receives the *LU* factors of the matrix supplied to the slaves. How does the algorithm change if the Crout factorisation is required? Write host and slave processes which implement a parallel version of the Level 3 BLAS strsm for execution on a hypercube. How can this be combined with the code for *LU* factorisation to determine a matrix inverse? What are the drawbacks of splitting the factorisation and solution phases in this way?

4.6. If *A* is symmetric and positive definite a decomposition of the form $A = LL^T$ exists (the Cholesky decomposition), with *L* lower triangular (cf. Exercise 3.10). Verify that the components of *L* satisfy

$$l_{jj} = \left(a_{jj} - \sum_{k=1}^{j-1} l_{jk}^2 \right)^{\frac{1}{2}},$$

$$l_{ij} = \frac{1}{l_{ii}} \left(a_{ij} - \sum_{k=1}^{i-1} l_{ik} l_{jk} \right), \qquad i = j+1, j+2, \ldots, n,$$

for $j = 1, 2, \ldots, n$. (Positive definiteness ensures that all the square roots are of positive quantities.) Determine the operation count for Cholesky decomposition, assuming that the cost of a square root is 1 flop. Express the algorithm in terms of Levels 1 and 2 BLAS and explore the scope for parallelism. In the context of a local memory multiprocessor, what would be a suitable data distribution for *A*? How will this affect the way you implement forward and backward substitutions to solve the system $Ax = b$?

4.7. Gaussian elimination followed by backward substitution will yield an approximate solution only ($\mathbf{x}^{(1)}$, say) to the system $Ax = b$ because of the introduction of round-off error. This approximation may be improved by using a process known as *iterative refinement*. For $k = 1, 2, \ldots$, we

1. calculate the residual vector $\mathbf{r}^{(k)} = \mathbf{b} - A\mathbf{x}^{(k)}$,
2. solve the linear system $A\delta\mathbf{x}^{(k)} = \mathbf{r}^{(k)}$, and
3. calculate $\mathbf{x}^{(k+1)} = \mathbf{x}^{(k)} + \delta\mathbf{x}^{(k)}$,

until some suitable convergence criterion is satisfied (say, $\|\delta\mathbf{x}^{(k)}\| < \epsilon$). Forming $\mathbf{r}^{(k)}$ requires a matrix-vector multiplication involving *A*. Using the *LU* factors is not advisable and so it is necessary to retain a copy of the original matrix. However, the *LU* factors can be used to solve for $\delta\mathbf{x}^{(k)}$ using forward and backward substitutions. Indicate how parallelism can be introduced into each of the three stages individually. What is a suitable data distribution for *A* on a local memory multiprocessor? How does this influence the algorithms you employ for each of the stages (1), (2) and (3)? Indicate the communications traffic involved.

4.8. The following code forms the $U^T U$ decomposition of a symmetric positive definite banded matrix A, with semi-bandwidth (that is, the number of super-diagonals or subdiagonals) equal to m.

```
do j = 1,n
  info = j
  s = zero
  ik = m+1
  jk = max0(j-m,1)
  mu = max0(m+2-j,1)
  if (m.ge.mu) then
    do k = mu,m
      t = (a(k,j)-sdot(k-mu,a(ik,ik),1,a(mu,j),1))/a(m+1,jk)
      a(k,j) = t
      s = s+t*t
      ik = ik-1
      jk = jk+1
    end do
  end if
  s = a(m+1,j)-s
  if (s.le.zero) goto 10
  a(m+1,j) = sqrt(s)
end do
info = 0
10 continue
```

The code assumes that the upper triangle only of A is stored, by columns, with rows representing diagonals. The principal diagonal starts at position a(m+1,1), the first superdiagonal starts at a(m,2), and so on. The upper triangular factor U overwrites A. Work out what the code is doing and investigate the scope for parallelisation in a shared memory multiprocessor environment. Suggest a loop reordering which improves the situation.

4.9. An alternative to Gaussian elimination is the *Gauss–Jordan scheme*; at stage k, the variable x_k is eliminated from all equations other than the kth. After $n-1$ stages, A is reduced to diagonal form and, provided that the operations are simultaneously applied to the right-hand side vector, the solution can be immediately recovered. Determine the operation count of this scheme and compare it with that of Gaussian elimination. How well does Gauss–Jordan parallelise? (Cosnard, 1991.)

4.10. Consider the block representation of A employed in Section 3.4.3. Routines are available which

- form the LU factors of a diagonal, pivot, block,
- update a block in the block pivot row using the lower triangular matrix of the pivot block,
- form the multipliers in blocks in the block pivot column using the upper triangular matrix of the pivot block, and
- update a submatrix block using blocks in the block pivot row and column.

In each case the blocks are assumed to be square. Construct a graph (the *dependency graph*) which indicates the order in which these routines need to be called and with which actual arguments. What does this say about the potential for parallelism at the block level? What effect will partial pivoting have on the graph? What are the advantages and disadvantages of restricting the search for a pivot to a diagonal block?

4.11.   How can pivoting involving only column interchanges be incorporated into Gaussian elimination? What are the implications of this with respect to the storage pattern of the coefficient matrix for a parallel implementation on a local memory multiprocessor?

Further reading

Gallivan, Plemmons, *et al.* (1990) review the potential for parallelism in linear algebra computations. In particular, they consider the BLAS and various block implementations of *LU* decomposition. This work also contains an extensive bibliography.

A comparative analysis of the variants of *LU* decomposition, in the absence of pivoting, on local memory systems is given by Ortega and Romine (1988). Given appropriate characteristics of the system, they conclude that the *kij* variant using scattered row storage is optimal. The strict synchronisation points we have imposed are blurred in the way we suggest for *QR* factorisation in Section 5.2. Chu and George (1987) consider the *kij* variant with pivoting and explicit row interchanges and construct a timing model for a local memory machine. An alternative strategy for data distribution is considered by Fox, Johnson, *et al.* (1988). Assuming p^2 processors, they suggest that A should be subdivided into $p \times p$ square blocks A_{ij}, with the (r, s) element of each block allocated to processor $(r + s) \bmod p^2$. See also van de Vorst (1988) and Bisseling and van de Vorst (1989). Scattered column local memory implementations of Gaussian elimination are considered by Geist and Heath (1986), Moler (1986), Robert (1991) and Saad (1986).

Ortega and Romine (1988) also consider variants of Cholesky decomposition and conclude that the *kji* form with scattered columns is potentially the best. George, Heath, *et al.* (1986) suggest that for shared memory systems *jki* performs optimally.

CHAPTER 5

Further linear algebra

5.1 There is more to life than Gaussian elimination

In Chapter 4 we considered the solution of the system of linear equations

$$A\mathbf{x} = \mathbf{b}, \tag{5.1}$$

for the n-vector $\mathbf{x}$, with A an $n \times n$ matrix and $\mathbf{b}$ an n-vector. In particular, we developed methods which were, either directly or indirectly, based on the triangular (LU) factorisation of A, and investigated the scope for parallelism on both shared and local memory multiprocessors. There are a number of other ways of determining $\mathbf{x}$ and several of the alternatives are considered in the next three sections of this chapter. Some of these are direct; the number of operations required to compute $\mathbf{x}$ is known a priori. Others are indirect; they are iterative in nature. For such methods the number of operations involved will depend on the accuracy of an initial guess at a solution to (5.1), and on the convergence criteria employed. Again we look at the general, dense case in which it is assumed that A has few zero elements. If A is sparse (has a small number of non-zero elements, say less than 10%) such methods will involve a significant amount of unnecessary computation. We therefore look at methods specifically targeted at sparse systems, both for the structured case, in which the non-zero elements follow a particular, simple, pattern, and for the general unstructured case. Certain other methods are left to Chapter 7 where they are discussed in the context of the patterns produced by numerical methods for the solution of ordinary and partial differential equations.

In Section 5.5 we consider the solution of (5.1) for the case of an overdetermined system, that is, A is $m \times n$ with $m > n$, and $\mathbf{x}$ and $\mathbf{b}$ are conforming n- and m-vectors respectively, whilst in Section 5.6 we look at the eigenvalue/eigenvector problem of determining one or more pairs of values $(\lambda, \mathbf{x})$ such that

$$A\mathbf{x} = \lambda\mathbf{x}.$$

There is a connection with earlier material in that certain of the methods considered in Sections 5.5 and 5.6 rely on the orthogonal matrices which are introduced in the next section.

5.2 *QR* factorisation

Like *LU* decomposition, the aim of *QR* factorisation is to transform the system (5.1) into one which is easy to solve. Specifically, the aim is to determine a matrix Q and an upper triangular matrix R such that

$$A = QR,$$

with Q *orthogonal*, that is, $Q^T Q = I_n$. If we replace A in (5.1) by QR and premultiply by Q^T, then, using the orthogonality of Q, it follows that

$$R\mathbf{x} = Q^T\mathbf{b}. \tag{5.2}$$

Having determined the *QR* factorisation of A, the solution of (5.1) therefore requires

- multiplication of the right-hand side vector $\mathbf{b}$ by Q^T, and
- backward substitution to solve the upper triangular system (5.2),

and we deal with these aspects of the solution phase in Section 5.2.5. It should be noted that, as with the *LU* decomposition, the transformation of $\mathbf{b}$ (premultiplication by Q^T) can be performed as the factorisation is being constructed. Further, we show that it is not necessary to form Q explicitly; rather, we express it as a product of orthogonal matrices, each of which can be represented in a condensed form.

The orthogonal matrix Q can be computed in at least three ways,

1. as the product of *Householder reflection matrices*,
2. as the product of *Givens rotation matrices*, or
3. via the generation of a set of orthogonal vectors from the columns of A using *modified Gram–Schmidt orthogonalisation*.

For a general full matrix, factorisation using Householder reflections is the technique most widely used in practice (but, in a serial environment, limited to an overdetermined system), and we consider this approach in detail in the next two subsections. Factorisation using Givens rotations is, on the face of it, a more expensive process, but there are circumstances where the method may be of use and, since it parallelises well, attention to this approach is given in Sections 5.2.3 and 5.2.4. The modified Gram–Schmidt algorithm is, in many ways, similar to factorisation using Householder transformations, and we leave its consideration to an exercise (Exercise 5.2).

We later indicate that, whichever of (1) or (2) is employed, QR factorisation is inherently more expensive than Gaussian elimination/LU decomposition. The reader may therefore wonder about the suitability of the QR factorisation for the solution of a system of linear equations, and in a serial environment it must be admitted that the method has little to offer. However, in a local memory parallel environment the lack of need for pivoting in the QR factorisation could be very important; whether it will ever win will depend on the speed of interprocessor communications.

We have a further reason for considering QR factorisation here. It can readily be extended to the case of an overdetermined system of linear equations, and Householder and Givens matrices have important roles in the determination of eigenvalues and eigenvectors. The relevance of these last two points will become clear in Sections 5.5 and 5.6.

5.2.1   Householder reflection matrices

An $n \times n$ Householder reflection matrix P takes the form

$$P = I_n - 2\mathbf{v}\mathbf{v}^T, \tag{5.3}$$

where $\mathbf{v}$, normalised so that $\|\mathbf{v}\|_2 = 1$, is termed a *Householder vector*. (We apologise for the notation, which is fairly standard. Do not confuse P with a permutation matrix.) It is easy to show the following:

- P is orthogonal and symmetric, and hence $PP = I_n$.
- For any two distinct non-zero n-vectors $\mathbf{x}$ and $\mathbf{z}$ with $\|\mathbf{x}\|_2 = \|\mathbf{z}\|_2$ there exists a Householder matrix P such that $P\mathbf{x} = \mathbf{z}$. In particular, the choice $\mathbf{v} = \mathbf{y}/\|\mathbf{y}\|_2$, where $\mathbf{y} = \mathbf{x} \pm \|\mathbf{x}\|_2 \mathbf{e}_1$ and $\mathbf{e}_1$ is the first unit vector (the first column of I_n), leads to
$$P\mathbf{x} = \mp \|\mathbf{x}\|_2 \mathbf{e}_1.$$

The second result is fundamental to QR factorisation using Householder transformations since it allows the introduction of zeros below the diagonal of a matrix. Difficulties with large relative errors in the normalisation of $\mathbf{y}$ can be avoided if the arbitrary sign in its definition is chosen so that $\mathbf{y} = \mathbf{x} + \text{sign}(x_1) \|\mathbf{x}\|_2 \mathbf{e}_1$, where x_1 is the first element of $\mathbf{x}$ (Golub and Van Loan, 1989, p. 196). Then

$$P\mathbf{x} = -\text{sign}(x_1) \|\mathbf{x}\|_2 \mathbf{e}_1.$$

Let the columns of A be denoted $\mathbf{a}_i$ and introduce the $n \times (n-1)$ matrix $\hat{A}_1$ such that $A = (\mathbf{a}_1, \hat{A}_1)$. Then a Householder vector $\mathbf{v}_1$ (and hence an orthogonal matrix $P^{(1)}$) can be chosen so that the matrix $A^{(1)} = P^{(1)}A$ has zeros in its first column below the diagonal element. Specifically, $\mathbf{v}_1$ is chosen so that

$$P^{(1)}\mathbf{a}_1 = \left(I_n - 2\mathbf{v}_1\mathbf{v}_1^T\right)\mathbf{a}_1 = \mp \|\mathbf{a}_1\|_2 \mathbf{e}_1.$$

Hence

$$A^{(1)} = \begin{pmatrix} \mp \|\mathbf{a}_1\|_2 & & & \\ 0 & & & \\ \vdots & P^{(1)}\mathbf{a}_2 & \cdots & P^{(1)}\mathbf{a}_n \\ 0 & & & \end{pmatrix} = \begin{pmatrix} \mp \|\mathbf{a}_1\|_2 & & \\ 0 & & \\ \vdots & P^{(1)}\hat{A}_1 \\ 0 & \end{pmatrix}.$$

It should be clear that this, the first, stage of orthogonal reduction is similar to the first stage of Gaussian elimination:

- We construct the Householder vector $\mathbf{v}_1 = \mathbf{y}_1 / \|\mathbf{y}_1\|_2$, where $\mathbf{y}_1 = \mathbf{a}_1 + \text{sign}(a_{11}) \|\mathbf{a}_1\|_2 \mathbf{e}_1$ is chosen to introduce zeros into the first column of $A^{(1)}$. (Compare this with determining the multipliers in the first stage of Gaussian elimination.) This calculation can be expressed in terms of Level 1 BLAS:
  - The determination of $\|\mathbf{a}_1\|_2$ is an snrm2 operation.
  - The normalisation of $\mathbf{v}_1$ is an sscal operation. It also involves the computation of $\|\mathbf{y}_1\|_2$, but this can be achieved without recourse to a further call to snrm2; $\mathbf{y}_1$ is defined in terms of $\mathbf{a}_1$ and hence $\|\mathbf{y}_1\|_2$ can be deduced from $\|\mathbf{a}_1\|_2$ and vice versa.
- We then modify the remaining entries in A appropriately. We have

$$P^{(1)}\hat{A}_1 = \left(I_n - 2\mathbf{v}_1\mathbf{v}_1^T\right)\hat{A}_1 = \hat{A}_1 - 2\mathbf{v}_1\mathbf{w}_1^T,$$

where $\mathbf{w}_1 = \hat{A}_1^T \mathbf{v}_1$. Here the required result may be achieved using calls to Level 2 BLAS (rather than the Level 3 BLAS ssymm):
  - sgemv can be used to form $\mathbf{w}_1$.
  - sger performs a rank-one modification to the matrix $\hat{A}_1$.

Having introduced zeros into the first column, attention can be turned to the second column where a second Householder vector is chosen to introduce zeros below the diagonal element, and so on.

After $k - 1$ stages of this orthogonal reduction we have

$$A^{(k-1)} = P^{(k-1)}P^{(k-2)} \cdots P^{(1)}A,$$

$$= \begin{pmatrix} R^{(k-1)} & S^{(k-1)} \\ 0 & \tilde{A}^{(k-1)} \end{pmatrix},$$

where $R^{(k-1)}$ is a $(k-1) \times (k-1)$ upper triangular matrix, $S^{(k-1)}$ is $(k-1) \times (n-k+1)$ and $\tilde{A}^{(k-1)}$ is $(n-k+1) \times (n-k+1)$. The kth stage eliminates the subdiagonal entries of the kth column of $A^{(k-1)}$ or, equivalently, the subdiagonal entries of the first column of $\tilde{A}^{(k-1)}$. If the columns of $\tilde{A}^{(k-1)}$ are denoted by $\tilde{\mathbf{a}}_k^{(k-1)}, \tilde{\mathbf{a}}_{k+1}^{(k-1)}, \ldots, \tilde{\mathbf{a}}_n^{(k-1)}$, then we can choose a Householder vector $\tilde{\mathbf{v}}_k$, of length $n - k + 1$, so that

$$\tilde{P}^{(k)}\tilde{\mathbf{a}}_k^{(k-1)} = \left(I_{n-k+1} - 2\tilde{\mathbf{v}}_k\tilde{\mathbf{v}}_k^T\right)\tilde{\mathbf{a}}_k^{(k-1)} = \mp \left\|\tilde{\mathbf{a}}_k^{(k-1)}\right\|_2 \mathbf{e}_1,$$

where $\mathbf{e}_1$ is the first column of I_{n-k+1}. Thus

$$\tilde{P}^{(k)}\tilde{A}^{(k-1)} = \begin{pmatrix} \mp\left\|\tilde{\mathbf{a}}_k^{(k-1)}\right\|_2 \\ 0 \\ \vdots \\ 0 \end{pmatrix} \quad P^{(k)}\tilde{\mathbf{a}}_{k+1}^{(k-1)} \quad \cdots \quad \tilde{P}^{(k)}\mathbf{a}_n^{(k-1)} \end{pmatrix},$$

$$= \begin{pmatrix} \mp\left\|\tilde{\mathbf{a}}_k^{(k-1)}\right\|_2 & \hat{\mathbf{a}}_k^T \\ 0 & \tilde{A}^{(k)} \end{pmatrix},$$

where $\hat{\mathbf{a}}_k$ is an $(n-k)$-vector and $\tilde{A}^{(k)}$ is $(n-k) \times (n-k)$. Note that

$$\tilde{P}^{(k)}\tilde{\mathbf{a}}_j^{(k-1)} = \left(I_{n-k+1} - 2\tilde{\mathbf{v}}_k\tilde{\mathbf{v}}_k^T\right)\tilde{\mathbf{a}}_j^{(k-1)} = \tilde{\mathbf{a}}_j^{(k-1)} - 2\left(\tilde{\mathbf{v}}_k^T\tilde{\mathbf{a}}_j^{(k-1)}\right)\tilde{\mathbf{v}}_k,$$

$$j = k+1, k+2, \ldots, n,$$

so that each modified column of $\tilde{P}^{(k)}\tilde{A}^{(k-1)}$ depends only on the Householder vector $\tilde{\mathbf{v}}_k$ and on the column $\tilde{\mathbf{a}}_j^{(k-1)}$ being modified. This kth stage of the reduction is equivalent to the premultiplication of $A^{(k-1)}$ by the $n \times n$ Householder matrix $P^{(k)} = I_n - 2\mathbf{v}_k\mathbf{v}_k^T$, where $\mathbf{v}_k = (0, 0, \ldots, 0, \tilde{\mathbf{v}}_k^T)^T$. The zeros in $\mathbf{v}_k$ ensure

$$\underbrace{}_{k-1}$$

that the first $k-1$ rows and columns of $A^{(k-1)}$ remain unchanged.

After $n-1$ stages of this orthogonal reduction we have

$$A^{(n-1)} = P^{(n-1)}P^{(n-2)}\cdots P^{(1)}A = R^{(n-1)},$$

where $R^{(n-1)}$ is an $n \times n$ upper triangular matrix. Because each of the transformations $P^{(k)}$, $k = 1, 2, \ldots, n-1$, is symmetric and orthogonal, we can rewrite this as

$$A = P^{(1)}P^{(2)}\cdots P^{(n-1)}R^{(n-1)} = QR,$$

where $Q = P^{(1)}P^{(2)}\cdots P^{(n-1)}$ is an $n \times n$ orthogonal matrix and $R = R^{(n-1)}$.

A sequential Fortran implementation of QR factorisation is given in Code 5.1, in which it is assumed that A is nonsingular. It should be noted that at no time do we explicitly form any of the Householder matrices. Rather, the non-zero components of the Householder vectors are stored in the lower triangular part of A, with R stored in the upper triangle. On the face of it there appears to be one element of each Householder vector missing; $\mathbf{v}_1$ is of length n but there are only $n-1$ subdiagonal elements in the first column of A. There are various ways of overcoming this, the usual technique, employed in Code 5.1, being to scale each Householder vector $\mathbf{v}$ so that the first non-zero component is unity (rather than scale $\mathbf{v}$ to satisfy $\|\mathbf{v}\|_2 = 1$). Hence, for a general vector $\mathbf{x}$, we again define $\mathbf{y} = \mathbf{x} + \text{sign}(\alpha)\|\mathbf{x}\|_2\,\mathbf{e}_1$, where $\alpha = x_1$, but now set $\mathbf{v} = \mathbf{y}/(\alpha - \beta)$, where

$\beta = -\text{sign}(\alpha) \, \|\mathbf{x}\|_2$. The Householder transformation may then be written as $P = I_n - \tau \mathbf{v}\mathbf{v}^T$, where $\tau = (\beta - \alpha)/\beta$. When applying Householder transformations to A we can therefore store the Householder vectors below the diagonal of A, with the unit component being implicitly defined. It is, however, necessary to store τ and in Code 5.1 the one-dimensional array `tau` is used for this purpose. Before calls to `sgemv` and `sger` are made we explicitly insert the unit values for the Householder vectors, restoring the diagonal values of R after the calls have been made. Besides using `tau` to store the required constants, we can also employ it as temporary storage for the results of the matrix-vector products.

```
      subroutine qrfact(a,lda,tau,n)
c qrfact reduces the real n*n matrix A to upper triangular form using
c Householder transformations of the form P = I - tau v v^T
c
c The Householder vectors overwrite the lower triangular part of
c A and the upper triangular part of R overwrites the
c strictly upper triangular part of A
c
c The constants tau are returned in the n-vector tau which is also
c used for workspace
      real one,zero
      parameter(one = 1.0e0,zero = 0.0e0)
      integer lda,n
      real a(lda,*),tau(*)
      integer k
      real norm,alpha,beta,snrm2
      intrinsic sqrt,sign
      external snrm2,sgemv,sger
      do k = 1,n-1
c n-1 stages of elimination
c Compute the two-norm of the subdiagonal elements of column k of A
         norm = snrm2(n-k,a(k+1,k),1)
c Compute beta, the diagonal component of R
         alpha = a(k,k)
         beta = -sign(sqrt(alpha**2+norm**2),alpha)
c Compute the constant factor in the Householder transformation
         tau(k) = (beta-alpha)/beta
c Form the Householder vector
         call sscal(n-k,one/(alpha-beta),a(k+1,k),1)
c Set the unit element of the Householder vector
         a(k,k) = one
c Form the vector w in the unused entries in tau
         call sgemv('T',n-k+1,n-k,one,a(k,k+1),lda,a(k,k),1,zero,
     +               tau(k+1),1)
c Perform the rank-one update
         call sger(n-k+1,n-k,-tau(k),a(k,k),1,tau(k+1),1,a(k,k+1),lda)
c Restore the diagonal element of R
         a(k,k) = beta
      end do
      end
```

Code 5.1 *Householder QR factorisation*

An operation count reveals that the QR factorisation via Householder transformations requires $4n^3/3 + O\left(n^2\right)$ flops. Thus the computational cost is approximately twice that of calculating the LU factorisation and so, in a sequential environment at least, QR factorisation is not usually employed to determine a solution to (5.1) when the coefficient matrix is square. However, it should be noted that the QR factorisation is numerically stable without the need for pivoting. From the point of view of parallelisation this is a useful property. In a shared memory environment it removes what would otherwise be a potential synchronisation point. In a local memory environment it simplifies the communication that might otherwise be necessary.

5.2.2 Parallelisation of Householder QR factorisation

Many of the issues which affect the parallelisation of the forward elimination phase of Gaussian elimination (or, equivalently, LU decomposition) also arise when we consider parallelisation of the Householder QR factorisation. We again need to work with a code which involves three nested loops. In terms of the notation used in Chapter 4, Code 5.1 represents the kij or kji form of factorisation, depending on how the matrix-vector multiplication sgemv and the rank-one update sger are implemented. Looking at the three loops expressed in this order we make the following observations:

- The outer, k, loop represents the stages of the orthogonal reduction and presents restricted scope for parallelism. The difficulty is that the kth Householder vector can only be determined after the kth column of A has been transformed by the previous $k-1$ Householder matrices.
- The inner, i and j, loops, which are hidden in sgemv and sger, can readily be parallelised (see Section 3.3.2 for a discussion of parallel implementations of sgemv, and Section 4.5.3 for an EPF parallel implementation of sger).

On a shared memory machine the parallelism of the inner loops could be exploited simply by using calls to parallel versions of sgemv and sger (and to snrm2 and sscal). When considering Gaussian elimination we decided to look no further than this, but, now that we have some experience under our belt, it is worth considering whether there is any other potential parallelism that can be exploited.

Consider the first cycle of the outer loop ($k = 1$). There are three stages. We need to

1. construct a Householder vector $\mathbf{v}$,
2. construct an intermediate vector $\mathbf{w}$, and
3. perform a rank-one update on a matrix.

The use of parallel versions of the BLAS imposes synchronisation points between each of these stages. There is also a synchronisation point within the first stage, since a vector norm needs to be computed before a vector scale can be performed. Considering the kji variant and stage (3) in more detail, the sger operation may be parallelised by spreading the outer (with respect to sger), j, loop (using EPF's DO ALL construct, say). This will, inevitably, lead to a small amount of processor idle time unless the number of columns to be modified happens to be a multiple of the number of processes available. We observe that once the second column has been modified the second Householder vector can be computed whilst the remainder of the matrix is still being modified. This suggests a rather more complex programming model than that employed for Gaussian elimination. It consists of a number of tasks (the modification of a column, the computation of a Householder vector, etc.) which are allocated to a pool as and when they are ready to be performed. A set of processes is started, each of which selects a task from the pool whenever it would otherwise be idle. The additional programming effort is not insignificant and the clear and concise representation of the algorithm offered by the use of the BLAS is lost, but to get the most out of a parallel computing system and achieve optimal results the full potential of an algorithm must be realised. Note that, in this sense, the kji variant is to be preferred to the kij variant as, strictly, in the latter case, the kth cycle would have to be completed fully before the $(k + 1)$th Householder vector could be determined.

On a local memory machine the data distribution of A is, as usual, an important consideration in the design of an efficient algorithm. The very nature of the kji variant suggests that A should be distributed by scattered columns. At the kth cycle the first stage is to calculate the Householder vector $\tilde{\mathbf{v}}_k$. This requires a call to snrm2, followed by a call to sscal, and these operations can be performed by a single process without requiring the communication of data from any other process. The second and third stages are to calculate the intermediate vector $\mathbf{w}$ and to modify the entries in the trailing $(n-k+1) \times (n-k)$ block of A. The process which has calculated the kth Householder vector $\tilde{\mathbf{v}}_k$ needs to broadcast it to all other processes. Then each process can modify its own columns of the trailing block of A without any further communications. It is possible to maintain a copy of all the Householder vectors on any one processor (or, even, on each processor) since they have been globally broadcast during the factorisation. Given the initial data distribution it follows that the upper triangular matrix R will be stored by scattered columns.

As soon as a process has completed modification of its columns it must check to see whether it has column $k + 1$. If it has then it computes the next Householder vector (computes a norm and scales a vector). Better still, if a process has column $k + 1$ then it should calculate and broadcast (asynchronously) the next Householder vector as soon as possible, before modifying its remaining columns. The other processes, meanwhile, will be completing their column modifications, or held waiting for the next Householder vector to arrive. This local memory implementation is equivalent to the 'pool of tasks' shared memory implementation outlined above, and many of these detailed implementation issues are equally valid for the Gaussian elimination algorithm of Chapter 4.

It is clearly possible to reorder the loops to produce other variants of QR factorisation. A kij version suggests a distribution by scattered rows on a local memory system, with data communication (a distributed snrm2) when constructing the Householder vector (cf. the selection of a pivot in LU factorisation, Section 4.5.4).

The jki version would involve a sweep across the columns of the matrix, with each column being modified by all previous Householder transformations. Because of the need to compute a Householder vector from an entire column of A, variants with the i loop outermost are not feasible.

5.2.3 Givens rotation matrices

A second way of calculating a QR factorisation is to premultiply A by a sequence of Givens rotation matrices. The (r, s) element of the $n \times n$ Givens matrix G_{ij} is given by

$$(G_{ij})_{rs} = \begin{cases} \cos \theta, & r = s = i \text{ or } j, \\ \sin \theta, & r = i, s = j, \\ -\sin \theta, & r = j, s = i, \\ \delta_{rs}, & \text{otherwise,} \end{cases} \tag{5.4}$$

and $\delta_{rs} = \begin{cases} 1, & r = s, \\ 0, & r \neq s, \end{cases}$ is the usual Kronecker delta function. That is, G_{ij} is identical to the identity matrix I_n except for rows and columns i and j. Writing G_{ij} out in full, we have

$$
G_{ij} = \begin{array}{c} \\ \\ \\ \\ i \\ \\ \\ \\ j \\ \\ \\ \\ \end{array}
\left(\begin{array}{cccccccccccc}
1 & 0 & \cdots & \cdots & \cdots & \cdots & \cdots & \cdots & \cdots & \cdots & \cdots & 0 \\
0 & \ddots & & & & & & & & & & \vdots \\
\vdots & & 1 & & & & & & & & & \\
\vdots & & & \cos\theta & & & \sin\theta & & & & & \vdots \\
\vdots & & & & 1 & & & & & & & \\
\vdots & & & & & \ddots & & & & & & \vdots \\
\vdots & & & & & & 1 & & & & & \\
\vdots & & & -\sin\theta & & & \cos\theta & & & & \vdots \\
\vdots & & & & & & & & 1 & & & \vdots \\
\vdots & & & & & & & & & \ddots & & 0 \\
0 & \cdots & \cdots & \cdots & \cdots & \cdots & \cdots & \cdots & \cdots & & 0 & 1
\end{array} \right),
$$

for some θ as yet to be determined. Premultiplication of a matrix by G_{ij} is referred to as a *Givens rotation in the (i, j) plane*. It is straightforward to prove the following results:

- G_{ij} is orthogonal (the result follows from the identity $\sin^2 \theta + \cos^2 \theta = 1$).
- For any n-vector $\mathbf{x}$, the n-vector $\mathbf{y} = G_{ij}\mathbf{x}$ satisfies

$$y_k = \begin{cases} x_k, & k \neq i \text{ or } j, \\ x_i \cos \theta + x_j \sin \theta, & k = i, \\ -x_i \sin \theta + x_j \cos \theta, & k = j. \end{cases}$$

It follows that, for given indices i and j, we can choose a value for θ, and thus values for $\cos\theta$ and $\sin\theta$, such that $y_j = 0$. A little manipulation shows that the choice $\cos\theta = x_i/r$ and $\sin\theta = x_j/r$, where $r = \pm\sqrt{x_i^2 + x_j^2}$, has the required effect, and then $y_i = r$. Note that there is no need explicitly to compute the angle θ. It is even possible to avoid the calculation of the square root; see Gentleman (1973) for details.

- Premultiplication of A by G_{ij} affects rows i and j only.

In a serial environment the most common use of Givens rotations is in introducing zeros in specific locations in a matrix, for example, as is required by Jacobi's method which we describe in Section 5.6.3. If it is required to introduce zeros into all the subdiagonal elements of a column of a full matrix then the lower operation count of Householder reflection matrices makes them the preferred choice. Nevertheless, Givens rotations can be used to form a QR factorisation and this may have some advantages in a parallel environment, as we show in Section 5.2.4.

Starting with A, we can eliminate the $(n, 1)$ element using premultiplication by the Givens rotation matrix $G_{n-1,n}^{(1)}$ for a suitable choice of θ (or, more precisely, for a suitable choice of $\sin\theta$ and $\cos\theta$). If $r = \sqrt{a_{n-1,n}^2 + a_{nn}^2}$, then we define $\sin\theta = a_{n1}/r$ and $\cos\theta = a_{n-1,1}/r$, and then $A^{(1)} = G_{n-1,n}^{(1)}A$ is such that $a_{n1}^{(1)} = 0$ and $a_{n-1,1}^{(1)} = r$. Further, all the elements in $A^{(1)}$, except those in rows $n-1$ and n, are identical to the elements of A. The $(n-1, 1)$ element of $A^{(1)}$ can now be eliminated by premultiplication of $A^{(1)}$ by the Givens rotation matrix $G_{n-2,n-1}^{(1)}$ for a suitable choice of θ. If $A^{(2)} = G_{n-2,n-1}^{(1)}A^{(1)}$, then only rows $n-2$ and $n-1$ of $A^{(1)}$ are modified. Thus the zero entry for $a_{n1}^{(1)}$ generated by the previous step remains unchanged. Continuing in this way and, finally, using the Givens rotation matrix $G_{12}^{(1)}$ to eliminate the $(2, 1)$ element of $A^{(n-2)}$, we have

$$A^{(n-1)} = G_{12}^{(1)}G_{23}^{(1)}\cdots G_{n-2,n-1}^{(1)}G_{n-1,n}^{(1)}A,$$

where $A^{(n-1)}$ has the form

$$A^{(n-1)} = \begin{pmatrix} * & * & \cdots & * \\ 0 & * & \cdots & * \\ \vdots & \vdots & \ddots & \vdots \\ 0 & * & \cdots & * \end{pmatrix},$$

in which $*$ indicates a possibly non-zero element. That is, $n-1$ Givens transformations are employed to give the same result as a single Householder transformation.

The subdiagonal elements of the second column of $A^{(n-1)}$ can now be eliminated in turn by premultiplication by a new set of Givens rotation matrices $G_{n-1,n}^{(2)}, G_{n-2,n-1}^{(2)}, \ldots, G_{23}^{(2)}$, with θ chosen appropriately in each case. The effect is to leave the zeros introduced into the first column of $A^{(n-1)}$ unchanged. Continuing in this way, subdiagonal elements in succeeding columns of (transformed versions of) A are eliminated until it is reduced to upper triangular form; in general, this requires premultiplication of A by $n(n-1)/2$ Givens rotation matrices and has an operation count of $2n^3 + O\left(n^2\right)$ flops. Thus the Givens QR factorisation is approximately three times as expensive (in terms of computational cost) as the LU factorisation of A, and about 50% more expensive than the Householder QR factorisation.

Code 5.2 calculates a Givens QR factorisation, making use of the Level 1 BLAS srotg and srot. In Chapter 3 reference to these routines was omitted, partly because we would at that stage have needed to define a plane rotation. We now define the specifications and effects of these two routines:

- SUBROUTINE SROTG(A,B,C,S)
  srotg computes the cosine (c) and the sine (s) of the plane rotation involving a and b such that
  $$\begin{pmatrix} c & s \\ -s & c \end{pmatrix}\begin{pmatrix} a \\ b \end{pmatrix} = \begin{pmatrix} r \\ 0 \end{pmatrix}.$$

  The input values A and B are overwritten by r and z, respectively, where

  $$r = \sigma\sqrt{a^2 + b^2},$$
  $$\sigma = \begin{cases} \operatorname{sign} a, & |a| > |b|, \\ \operatorname{sign} b, & |a| \le |b|, \end{cases}$$
  $$z = \begin{cases} s, & |s| < c \text{ or } c = 0, \\ c^{-1}, & 0 < |c| \le s. \end{cases} \tag{5.5}$$

  Given a value for z, the quantities c and s which define the corresponding Givens rotation may be recovered using

  $$c = \begin{cases} 0, & z = 1, \\ \sqrt{1 - z^2}, & |z| < 1, \\ z^{-1}, & |z| > 1, \end{cases}$$
  $$s = \begin{cases} 1, & z = 1, \\ z, & |z| < 1, \\ \sqrt{1 - c^2}, & |z| > 1, \end{cases}$$

  (Dodson and Grimes, 1982).

- SUBROUTINE SROT(N,X,INCX,Y,INCY,C,S)
  srot applies the plane rotation defined by c and s to consecutive pairs of the components of the vectors $\mathbf{x}$ and $\mathbf{y}$, that is,

  $$\begin{aligned} x_i &\leftarrow cx_i + sy_i, \\ y_i &\leftarrow -sx_i + cy_i, \end{aligned} \qquad i = 1, 2, \ldots, n,$$

with the assignments taking place simultaneously. In line with the other Level 1 BLAS, N denotes the number of pairs of elements to which the rotation is to be applied, and INCX and INCY denote strides for X and Y.

Of these two routines, only srot offers any scope for parallelism, but it possesses a low operation count unless n is very large. In Code 5.2 for each i and j we use srotg to generate the Givens rotation in the $(i-1, i)$ plane. srot is then used to modify the corresponding rows of A to the right of column j. Note that Code 5.2 includes implicit storage of the Givens rotation matrices $G_{i-1,i}$, which is required if the right-hand side vector is not modified as the factorisation proceeds. Each Givens rotation matrix is represented by a single real number, z, as given by (5.5), which overwrites the matrix element which has been zeroed.

```
      do j = 1,n-1
        do i = n,j+1,-1
c Calculate the Givens transformation G(i-1,i) which eliminates the
c (i,j) entry of A
          call srotg(a(i-1,j),a(i,j),c,s)
c Pre-multiply A by G(i-1,i), modifying only rows i-1 and i
          call srot(n-j,a(i-1,j+1),lda,a(i,j+1),lda,c,s)
        end do
      end do
```

Code 5.2 *Givens QR factorisation*

5.2.4   Parallelisation of Givens *QR* factorisation

A simple way of parallelising the QR factorisation of Section 5.2.3 would be to concentrate on the Level 1 BLAS srot, but such an approach is unlikely to give any substantial improvement, particularly on a local memory machine. We observe that the granularity is increased if several of the Givens rotations can be performed simultaneously. Suppose that the first step uses a rotation in the $(n-1, n)$ plane to zero the $(n, 1)$ element of A and the second step uses a rotation in the $(n-2, n-1)$ plane to zero the $(n-1, 1)$ element. On the next step simultaneous rotations in the $(n-3, n-2)$ and $(n-1, n)$ planes to zero the elements in the $(n-2, 1)$ and $(n, 2)$ positions, respectively, could be performed. On the fourth step simultaneous rotations in the $(n-4, n-3)$ and $(n-2, n-1)$ planes to zero the elements in positions $(n-3, 1)$ and $(n-1, 2)$ could be performed, whilst on the fifth step three simultaneous rotations in the $(n-5, n-4)$, $(n-3, n-2)$ and $(n-1, n)$ planes could be used to zero the elements in the $(n-4, 1)$, $(n-2, 2)$ and $(n, 3)$ positions, respectively. Clearly, this process of simultaneous elimination can be continued until all the lower triangular entries have been eliminated. For example, if we consider a 10×10 matrix A then the elimination pattern is as shown in Figure 5.1, where the integers indicate the steps at which the given elements are zeroed. Instead of requiring $n(n-1)/2 = 45$

$$\begin{pmatrix} * & * & * & * & * & * & * & * & * & * \\ 9 & * & * & * & * & * & * & * & * & * \\ 8 & 10 & * & * & * & * & * & * & * & * \\ 7 & 9 & 11 & * & * & * & * & * & * & * \\ 6 & 8 & 10 & 12 & * & * & * & * & * & * \\ 5 & 7 & 9 & 11 & 13 & * & * & * & * & * \\ 4 & 6 & 8 & 10 & 12 & 14 & * & * & * & * \\ 3 & 5 & 7 & 9 & 11 & 13 & 15 & * & * & * \\ 2 & 4 & 6 & 8 & 10 & 12 & 14 & 16 & * & * \\ 1 & 3 & 5 & 7 & 9 & 11 & 13 & 15 & 17 & * \end{pmatrix}$$

Figure 5.1 *Annihilation pattern for parallel Givens rotations*

steps to complete the elimination, each consisting of a single Givens rotation, the algorithm now requires just 17 steps, most of which consist of several concurrent Givens rotations, the maximum level of concurrency being 5 at step 9. In general, using this so-called 'standard' ordering of the Givens rotations, an $n \times n$ matrix can be reduced to upper triangular form in $2n - 3$ compound steps.

The order in which the elements of a matrix are eliminated is termed the *annihilation pattern*. Using the fact that the (i, j) element can be eliminated by a Givens rotation in the (k, i) plane for any k such that $a_{kj} \neq 0$, other, more efficient, but less straightforward, annihilation patterns can be developed. Details are given by Modi and Clarke (1984).

The parallelisation of QR factorisation using Givens rotations stems from the fact that each compound step consists of a number of independent rotations which can be performed concurrently. In addition, there is the possibility of some overlap between the steps. We restrict the discussion to the standard annihilation pattern without loss of generality. Then, considering Figure 5.1 again, there is little that we can do about the first three steps which must be performed in sequence. Recall that at step 4 we employ rotations in the $(n - 4, n - 3)$ and $(n - 2, n - 1)$ planes to zero the elements in positions $(n - 3, 1)$ and $(n - 1, 2)$. Once the latter rotation is complete we can begin the step 5 rotation in the $(n - 1, n)$ plane to zero the element in position $(n, 3)$. In addition, once element $(n - 3, 1)$ has been reduced to zero we can begin the step 5 rotations to zero elements $(n - 4, 1)$ and $(n - 2, 2)$. Hence the completion of each step need not be a synchronisation point, and we should attempt to exploit this fact using some appropriate dynamic task allocation strategy.

On a local memory machine we must, as usual, consider the storage pattern for A. We assume a one-to-one correspondence between processors and processes. Given that elimination of the (i, j) element requires modification to the $i-1$ and i rows of A, a column-based storage strategy would require a global broadcast from the process holding column j of A since this holds the two elements which define the Givens rotation. Row-based storage is much more attractive. As with LU decomposition there are several possibilities and we again consider two extremes:

- *Scattered rows*

    The advantage of scattered rows is that it is relatively easy to maintain good load balance properties since the rows at the beginning and at the end of the factorisation are uniformly distributed across the processors. The disadvantage is that every Givens rotation involves two (but only two) processes and there is, therefore, some communication involved in the execution of each rotation.

- *Block rows*

    The properties of a block rows strategy are essentially the converse of those for a scattered row distribution. Because neighbouring rows are likely to be on the same processor there is considerably less communications traffic involved in the execution of the rotations. However, the load balance properties of block rows are poor; the process which has the last rows of the matrix is the first to start its rotations and the last to finish.

If QR factorisation is to be followed by the solution of a system of equations then an appropriate storage strategy should be adopted which suits both numerical schemes (cf. matching LU factorisation with backward substitution). This again suggests a row-based algorithm; see Section 5.2.5.

As we have already remarked, Givens QR factorisation is about 50% more expensive than Householder factorisation (which, we recall, is approximately twice as expensive as LU factorisation) and hence, in a serial environment, it is an unattractive proposition. It is, perhaps, less clear which is likely to be the more efficient in a parallel environment, although if we can maintain a high process utilisation figure the Householder reduction will always tend to win.

5.2.5 QR factorisation and linear systems of equations

Having formed the QR factorisation of a matrix the solution of the system of equations (5.1) can be achieved in a straightforward manner. We consider first the case of Householder factorisation.

Suppose that we have been able to determine Householder transformations $P^{(k)}$, $k = 1, 2, \ldots, n-1$, such that

$$P^{(n-1)}P^{(n-2)}\cdots P^{(1)}A = R,$$

where R is upper triangular. Using the orthogonality of the matrices $P^{(k)}$, we can write (5.1) as

$$R\mathbf{x} = P^{(n-1)}P^{(n-2)}\cdots P^{(1)}\mathbf{b}.$$

Now, for each Householder matrix P we need to form $\mathbf{b}' = P\mathbf{b}$, where P is of the form $I_n - 2\mathbf{v}\mathbf{v}^T$. Hence, $\mathbf{b}' = \mathbf{b} - 2(\mathbf{v}^T\mathbf{b})\mathbf{v}$, and to achieve the required result we can

- form $\alpha = 2\mathbf{v}^T\mathbf{b}$ (an **sdot** operation), and

- form $\mathbf{b}' = \mathbf{b} - \alpha\mathbf{v}$ (a **saxpy** operation).

Code 5.3 implements these operations using the QR factorisation produced by Code 5.1.

```
      subroutine qrmodb(a,lda,tau,b,n)
c qrmodb modifies a right-hand side vector by the Householder
c transformations employed by qrfact to form the QR factors of an
c n*n matrix A
      real one
      parameter(one = 1.0e0)
      integer lda,n
      real a(lda,*),tau(*),b(*)
      integer i
      real alpha,beta,sdot
      external sdot,saxpy
c n-1 Householder transformations
      do i = 1,n-1
c Store the diagonal element of R and
c set the unit element of the Householder vector
         beta = a(i,i)
         a(i,i) = one
c Form the dot product of the Householder vector with the right-hand side
         alpha = sdot(n-i+1,a(i,i),1,b(i),1)
c Update the right-hand side
         call saxpy(n-i+1,-alpha*tau(i),a(i,i),1,b(i),1)
c Restore the diagonal element of R
         a(i,i) = beta
      end do
      end
```

Code 5.3 *Modification of the right-hand side using Householder transformations*

Unfortunately, the only possibility for exploiting parallelism would appear to be within the Level 1 BLAS, and this is not an attractive proposition, particularly on a local memory machine. An alternative approach is to treat $\mathbf{b}$ as the $(n+1)$th column of A (so that we form an augmented matrix, cf. Section 4.3.3) and modify it as the QR factors of A are being calculated. It will then not be necessary to store the Householder vectors unless they are required elsewhere. In either case the calculation of the solution vector $\mathbf{x}$ eventually involves the solution of a system of equations in which the coefficient matrix is upper triangular. This topic was considered in detail in Section 4.2.3 and, in a serial environment, Code 4.3 is again applicable. Parallelisation of the operation was reviewed in Section 4.5.2 and in Section 4.5.4 we indicated that, for a local memory system, a storage strategy based on scattered rows provided the best potential. Clearly, we may have some difficulty in matching the optimal data distributions for the QR factorisation and backward substitution phases. We recall that for Householder factorisation a scattered column distribution is the more natural.

With Givens rotations the modification of the right-hand side vector can be performed in parallel using non-overlapping transformations. We recall that storage by rows is more natural for Givens rotations, and hence we have a better match with the backward substitution phase.

5.3   Iterative methods for linear equations

All the methods considered so far for determining the solution of the system of linear equations (5.1) in which the coefficient matrix is square can be categorised as direct. Given exact arithmetic they compute the exact solution in a determinable number of arithmetic operations. For moderately sized dense systems this is the usual way of proceeding. For a large sparse system such an approach may be impractical for two reasons:

- The methods require $O\left(n^3\right)$ operations and, for n large, the computational cost may be prohibitively high.
- Even with sufficient storage capacity available, it is clearly wasteful to store large numbers of zeros explicitly, or to perform arithmetic operations involving zeros.

It is possible to modify factorisation methods to produce schemes which attempt to recognise and maintain the sparsity of a matrix, and this approach is considered in Section 5.4. First we investigate methods which are iterative in nature. Whilst it is possible to employ these methods for a dense system, direct methods are usually preferred. Hence, in the following, there is an implicit assumption that the coefficient matrix is sparse and that all operations exploit this sparsity.

5.3.1   Jacobi and Gauss–Seidel iterations

The simplest scheme is the *Jacobi iteration.* For a matrix A with non-zero diagonal elements the system (5.1) can be written as

$$x_i = \frac{1}{a_{ii}} \left(b_i - \sum_{j=1}^{i-1} a_{ij}x_j - \sum_{j=i+1}^{n} a_{ij}x_j \right), \qquad i = 1, 2, \ldots, n,$$

and this suggests the iteration

$$x_i^{(k+1)} = \frac{1}{a_{ii}} \left(b_i - \sum_{j=1}^{i-1} a_{ij}x_j^{(k)} - \sum_{j=i+1}^{n} a_{ij}x_j^{(k)} \right), \qquad i = 1, 2, \ldots, n, \qquad (5.6)$$

for $k = 1, 2, \ldots$, where $x_i^{(1)}$, $i = 1, 2, \ldots, n$, are given initial approximations.

To express the iteration in matrix form we observe that A may be written $A = L + D + U$, where L and U consist of the strictly lower and strictly upper triangular components of A respectively, that is

$$L = \begin{pmatrix} 0 & \cdots & & \cdots & 0 \\ a_{21} & \ddots & & & \vdots \\ \vdots & \ddots & \ddots & & \vdots \\ a_{n1} & \cdots & a_{n,n-1} & & 0 \end{pmatrix},$$

$$U = \begin{pmatrix} 0 & a_{12} & \cdots & & a_{1n} \\ \vdots & \ddots & \ddots & & \vdots \\ \vdots & & & \ddots & a_{n-1,n} \\ 0 & \cdots & \cdots & & 0 \end{pmatrix},$$

and $D = \mathrm{diag}(a_{11}, a_{22}, \ldots, a_{nn})$ is formed from the diagonal elements. Then $D\mathbf{x} = -(L + U)\mathbf{x} + \mathbf{b}$, and the Jacobi iteration is

$$\mathbf{x}^{(k+1)} = -D^{-1}(L + U)\mathbf{x}^{(k)} + D^{-1}\mathbf{b}. \tag{5.7}$$

A sufficient condition for convergence of the iteration (5.7) is that A be *strictly diagonally dominant*, that is

$$|a_{ii}| > \sum_{\substack{j=1 \\ j \neq i}}^{n} |a_{ij}|, \qquad i = 1, 2, \ldots, n,$$

(Varga, 1962, p. 73).

On a serial machine it is natural to implement the iterations so that the next set of iterates $\mathbf{x}^{(k+1)}$ is computed in increasing order of the subscript. Whichever order is adopted it is necessary to keep a record of the old set of values $\mathbf{x}^{(k)}$ until all the new values $\mathbf{x}^{(k+1)}$ have been calculated. Further, both are required to test for convergence (see below).

A related scheme is the *Gauss–Seidel iteration*, in which the updated values $x_i^{(k+1)}$ are used as soon as they are available. Assuming the components are computed in increasing order of the subscript, the iterations of this scheme are given by

$$x_i^{(k+1)} = \frac{1}{a_{ii}} \left(b_i - \sum_{j=1}^{i-1} a_{ij} x_j^{(k+1)} - \sum_{j=i+1}^{n} a_{ij} x_j^{(k)} \right), \qquad i = 1, 2, \ldots, n. \tag{5.8}$$

In matrix form we have $(L + D)\mathbf{x} = -U\mathbf{x} + \mathbf{b}$, so that the iteration (5.8) may be written

$$\mathbf{x}^{(k+1)} = -(L + D)^{-1}U\mathbf{x}^{(k)} + (L + D)^{-1}\mathbf{b}. \tag{5.9}$$

The Gauss–Seidel iteration will again converge if A is strictly diagonally dominant. Further, if A is symmetric then a condition which is both necessary and sufficient for convergence is that A be positive definite (Varga, 1962, p. 78). Fortunately a number of important problems satisfy these conditions.

For both the Jacobi and Gauss–Seidel iterations any practical implementation requires a test for convergence of the successive iterates. Two commonly used tests are

- $\left\| \mathbf{x}^{(k+1)} - \mathbf{x}^{(k)} \right\| \leq \epsilon$, or
- $\left\| \mathbf{x}^{(k+1)} - \mathbf{x}^{(k)} \right\| \leq \epsilon \left\| \mathbf{x}^{(k)} \right\|$,

where ϵ is the convergence parameter and either the two-norm or the infinity-norm is usually employed. The first is an absolute error test whilst the second is a relative error test which takes into account the magnitudes of the components of the solution vector. Because of the linear rates of convergence of the iterations, neither of these tests guarantees an accuracy of ϵ in the measure used.

Whether a direct or iterative method is the more efficient in a serial environment will depend on the convergence rate of the iteration and the number of terms in the summations (sparse dot products) which define the next set of iterates. We comment here that it is possible to improve the rate of convergence of the Gauss–Seidel iteration by using successive overrelaxation. We return to the subject in the context of a specific form for A in Section 7.4.1.

5.3.2   Parallel implementations of Jacobi and Gauss–Seidel iterations

It is immediately clear how to parallelise the Jacobi iteration. From (5.6) we see that the calculation of each component $x_i^{(k+1)}$, $i = 1, 2, \ldots, n$, is an independent computation. Another way of expressing this observation is to note, from (5.7), that the dominant computation of the Jacobi iteration is a (sparse) matrix-vector product. On a shared memory machine we simply spread the i loop which calculates the updated components $x_i^{(k+1)}$ of the approximate solution vector. On a local memory machine we need to arrange for the matrix A to be distributed by rows; whether the rows are blocked or scattered makes no real difference for a general A. Then the computation of the components $x_i^{(k+1)}$ can proceed concurrently in separate processes and this should be a well-balanced computation. At the end of each iteration every process broadcasts the components of $\mathbf{x}^{(k+1)}$ which it has computed to all other processes. If A possesses some structure it is worthwhile looking to see whether the equations can be reordered so that they may be decoupled in some way. That is, we attempt to group the equations so that the computation of $x_i^{(k+1)}$ depends mainly on components from within that group, and on few components which lie outside the group. A typical situation which may arise is that in which A is banded. If the groups are reasonably well balanced and each is allocated to a separate process, then the amount of interprocess communication at the end of each iteration will be reduced.

Gauss–Seidel iteration is rather more awkward to parallelise. The very nature of the iteration expressed by (5.8) is sequential unless we are content simply to parallelise the individual summations. Decoupling of the equations would again help; an individual process could perform the Gauss–Seidel iteration on its own block of $\mathbf{x}^{(k+1)}$, synchronising with other processes either to receive or to broadcast updated values at the end of each iteration. Unless the equations can be completely decoupled, this suggests that we have what is, in effect, a block Gauss–Seidel iteration, with all values obtained from other blocks being the values at the previous iteration, rather than values updated by the current iteration. An alternative is to split the variables into blocks of variables which are independent of each other and then update the blocks one at a time, with parallelism being employed within each block update. Such an approach is particularly viable if the equations result from the finite difference replacement to either a boundary value problem in ordinary differential equations, or a partial differential equation. We return to the matter in Section 7.4.1

Of the convergence tests, those which involve the two-norm require communication of data to accumulate a sum. The absolute error test with the infinity-norm can be applied independently by the separate processes; if each process indicates satisfactory convergence of its approximations then the overall iteration has converged. The relative error test with the infinity-norm can be evaluated in a similar way, although it also requires the calculation and broadcast of $\left\|\mathbf{x}^{(k)}\right\|_\infty$ from the previous iteration.

5.4   Direct methods for sparse matrices

When deriving algorithms to solve a system of equations in which the coefficient matrix is sparse, attempts to exploit the sparsity pattern should result in considerable savings in both storage and computations; failure so to do will waste these resources. For example, if we form either the LU factors (Chapter 4) or the QR factors (Section 5.2) of a sparse matrix A using an algorithm which does not exploit the sparsity of A then there is no reason to expect the factors themselves to be sparse. The effect is to incur a higher operation count, and higher storage costs, than might otherwise be necessary. Thus we need special algorithms to deal with these problems.

Sparsity of A can take one of two forms:

- Structured sparsity refers to a situation in which the non-zero elements of A occur in a regular pattern.
- General sparsity applies when the non-zero elements of A exhibit no particular pattern.

We have already remarked that sparse matrices with some form of structure frequently occur in the solution of ordinary and partial differential equations. This structure often takes the form of a band of non-zero elements about the principal diagonal. In Section 5.4.1 we consider the particular case of a tridiagonal matrix. Matrices which exhibit no pattern in their sparsity are, fortunately, less common. We consider a general sparse symmetric positive definite matrix in Section 5.4.2.

5.4.1 Tridiagonal matrices and cyclic reduction

We consider here the solution of a system of linear equations in which the coefficient matrix is (unsymmetric) tridiagonal. We write

$$
A = \begin{pmatrix}
b_1 & c_1 & & & \\
a_2 & b_2 & c_2 & & \\
& a_3 & \ddots & \ddots & \\
& & \ddots & \ddots & c_{n-1} \\
& & & a_n & b_n
\end{pmatrix}.
$$

It is straightforward to show that LU factorisation applied to A leads to lower and upper triangular matrices of the form

$$
L = \begin{pmatrix}
1 & & & & \\
l_2 & 1 & & & \\
& l_3 & \ddots & & \\
& & \ddots & \ddots & \\
& & & l_n & 1
\end{pmatrix}, \qquad
U = \begin{pmatrix}
u_1 & c_1 & & & \\
& u_2 & c_2 & & \\
& & \ddots & \ddots & \\
& & & \ddots & c_{n-1} \\
& & & & u_n
\end{pmatrix}.
$$

Rather than formally go through the elimination process we can form the product LU and equate coefficients to show that $u_1 = b_1$ and

$$
\begin{aligned}
l_i &= a_i/u_{i-1}, \\
u_i &= b_i - l_i c_{i-1},
\end{aligned} \qquad i = 2, 3, \ldots, n. \tag{5.10}
$$

Note that the $O\left(n^3\right)$ operation count of Gaussian elimination applied to a full matrix reduces to $O\left(n\right)$. Computing l_i and u_i from (5.10) yields an inherently sequential algorithm, since l_i depends on u_{i-1} and u_i depends on l_i. In this form, therefore, the factorisation is not particularly appropriate for implementation in a parallel environment.

An alternative approach is provided by the *cyclic reduction*, or *odd-even reduction*, *algorithm*, which we apply to the system $A\mathbf{x} = \mathbf{y}$. For $i = 3, 4, \ldots, n-2$, equations $i-1, i, i+1$ of this system are

$$
\begin{aligned}
a_{i-1}x_{i-2} + b_{i-1}x_{i-1} + c_{i-1}x_i & & &= y_{i-1}, \\
a_i x_{i-1} + & b_i x_i + & c_i x_{i+1} & &= y_i, \\
& a_{i+1}x_i + & b_{i+1}x_{i+1} + c_{i+1}x_{i+2} &= y_{i+1}.
\end{aligned}
$$

Treating this triple of equations in isolation, equation $i-1$ can be used to eliminate x_{i-1} from the ith equation. Similarly, equation $i+1$ can be used to eliminate x_{i+1} from the ith equation. The result is that the ith equation transforms to

$$
a_i^{(1)} x_{i-2} + b_i^{(1)} x_i + c_i^{(1)} x_{i+2} = y_i^{(1)},
$$

where

$$a_i^{(1)} = \alpha_i^{(1)} a_{i-1},$$

$$b_i^{(1)} = b_i + \alpha_i^{(1)} c_{i-1} + \beta_i^{(1)} a_{i+1},$$

$$c_i^{(1)} = \beta_i^{(1)} c_{i+1},$$

$$y_i^{(1)} = y_i + \alpha_i^{(1)} y_{i-1} + \beta_i^{(1)} y_{i+1}, \tag{5.11}$$

$$\alpha_i^{(1)} = -a_i/b_{i-1},$$

$$\beta_i^{(1)} = -c_i/b_{i+1}.$$

Performing these eliminations on all consecutive triples, noting that the first two and last two equations must be treated as special cases, the original system transforms to

$$A^{(1)}\mathbf{x} = \mathbf{y}^{(1)}, \tag{5.12}$$

where

$$A^{(1)} = \begin{pmatrix} b_1^{(1)} & 0 & c_1^{(1)} & & & & & \\ 0 & b_2^{(1)} & 0 & c_2^{(1)} & & & & \\ a_3^{(1)} & 0 & b_3^{(1)} & \ddots & & \ddots & & \\ & a_4^{(1)} & \ddots & \ddots & \ddots & & \ddots & \\ & & \ddots & \ddots & \ddots & \ddots & & \ddots \\ & & & & & & & c_{n-2}^{(1)} \\ & & & \ddots & \ddots & b_{n-1}^{(1)} & & 0 \\ & & & & a_n^{(1)} & & 0 & b_n^{(1)} \end{pmatrix},$$

and the elements of $A^{(1)}$ are given by (5.11) (the definitions also hold for $a_{n-1}^{(1)}$, $a_n^{(1)}$, $b_2^{(1)}$, $b_{n-1}^{(1)}$, $c_1^{(1)}$ and $c_2^{(1)}$), except that $b_1^{(1)} = b_1 + \beta_1^{(1)} a_2$ and $b_n^{(1)} = b_n + \alpha_n^{(1)} c_{n-1}$.

Clearly, we can form triples in (5.12) in a similar way, except that now, instead of consecutive equations, we consider alternate equations. Thus, for $i = 5, 6, \ldots, n-4$, equations $i-2, i, i+2$ of (5.12) are given by

$$a_{i-2}^{(1)} x_{i-4} + b_{i-2}^{(1)} x_{i-2} + c_{i-2}^{(1)} x_i = y_{i-2}^{(1)},$$

$$a_i^{(1)} x_{i-2} + b_i^{(1)} x_i + c_i^{(1)} x_{i+2} = y_i^{(1)},$$

$$a_{i+2}^{(1)} x_i + b_{i+2}^{(1)} x_{i+2} + c_{i+2}^{(1)} x_{i+4} = y_{i+2}^{(1)},$$

and we can combine these to eliminate both x_{i-2} and x_{i+2} from the ith equation to give

$$a_i^{(2)} x_{i-4} + b_i^{(2)} x_i + c_i^{(2)} x_{i+4} = y_i^{(2)},$$

where

$$a_i^{(2)} = \alpha_i^{(2)} a_{i-2}^{(1)},$$

$$b_i^{(2)} = b_i^{(1)} + \alpha_i^{(2)} c_{i-2}^{(1)} + \beta_i^{(2)} a_{i+2}^{(1)},$$

$$c_i^{(2)} = \beta_i^{(2)} c_{i+2}^{(1)},$$

$$y_i^{(2)} = y_i^{(1)} + \alpha_i^{(2)} y_{i-2}^{(1)} + \beta_i^{(2)} y_{i+2}^{(1)},$$ (5.13)

$$\alpha_i^{(2)} = -a_i^{(1)}/b_{i-2}^{(1)},$$

$$\beta_i^{(2)} = -c_i^{(1)}/b_{i+2}^{(1)}.$$

If we perform these eliminations on all triples of this form in the system (5.12), noting that this time the first four and last four equations must be treated as special cases, we obtain the system

$$A^{(2)}\mathbf{x} = \mathbf{y}^{(2)},$$ (5.14)

where

$$A^{(2)} = \begin{pmatrix} b_1^{(2)} & 0 & 0 & 0 & c_1^{(2)} & 0 & & & & & \\ 0 & b_2^{(2)} & 0 & 0 & 0 & c_2^{(2)} & & & & & \\ 0 & 0 & b_3^{(2)} & 0 & & & \ddots & & & & \\ 0 & 0 & 0 & b_4^{(2)} & & & & \ddots & & & \\ a_5^{(2)} & 0 & & & \ddots & & & & \ddots & & \\ 0 & a_6^{(2)} & & & & \ddots & & & & c_{n-5}^{(2)} & 0 \\ & & \ddots & & & & \ddots & & & 0 & c_{n-4}^{(2)} \\ & & & \ddots & & & & b_{n-3}^{(2)} & 0 & 0 & 0 \\ & & & & \ddots & & & 0 & b_{n-2}^{(2)} & 0 & 0 \\ & & & & & a_{n-1}^{(2)} & 0 & 0 & 0 & b_{n-1}^{(2)} & 0 \\ & & & & & 0 & a_n^{(2)} & 0 & 0 & 0 & b_n^{(2)} \end{pmatrix}$$

The elements of $A^{(2)}$ are defined by (5.13), with slightly modified definitions for $b_i^{(2)}, b_{n-i+1}^{(2)}, \ i = 1, 2, 3$.

We observe that the kth stage $(k \geq 1)$ of the method moves the off-diagonal bands 2^{k-1} positions further away from the diagonal. It follows that if $n = 2^N$, N stages of this cyclic reduction algorithm yield the system

$$A^{(N)}\mathbf{x} = \mathbf{y}^{(N)},$$ (5.15)

where $A^{(N)} = \text{diag}\left(b_1^{(N)}, b_2^{(N)}, \ldots, b_n^{(N)}\right)$. Hence, the solution of (5.15) may immediately be computed as

$$x_i = y_i^{(N)}/b_i^{(N)}, \qquad i = 1, 2, \ldots, n.$$

On a serial machine the cyclic reduction algorithm is uncompetitive with the Gaussian elimination algorithm, since it requires many more operations. We have $\log_2 n$ stages, each of which involves $O(n)$ operations, giving a total operation count of $O(n \log_2 n)$. This compares with the $O(n)$ operations count of Gaussian elimination. However, the technique has considerable potential for parallelisation. If we consider the first stage then the operations (5.11) which define the components of the cyclically reduced matrix $A^{(1)}$ can be performed concurrently. For example, the determination of $a_j^{(1)}$ and $a_k^{(1)}$ for $j \neq k$ are independent calculations, although they may depend on the same data. One difficulty of implementing the algorithm on a parallel machine is that, as the algorithm proceeds, the amount of computation involved in each stage of the reduction process decreases. A second difficulty, particularly for a local memory machine, is that the computation is fine-grain.

An alternative view of cyclic reduction is as a *divide and conquer algorithm*. Recall that the first stage of the reduction process transforms the original system to (5.12). If we assume that n is even and renumber the equations in the order $1, 3, \ldots, n-1, 2, 4, \ldots, n$, and we renumber the components of the vector $\mathbf{x}$ in a similar way, then (5.12) becomes

$$\tilde{A}^{(1)}\tilde{\mathbf{x}} = \tilde{\mathbf{y}}^{(1)}, \tag{5.16}$$

where

$$\tilde{A}^{(1)} = \begin{pmatrix} b_1^{(1)} & c_1^{(1)} & & & & & & & & \\ a_3^{(1)} & b_3^{(1)} & c_3^{(1)} & & & & & & & \\ & a_5^{(1)} & \ddots & & \ddots & & & & & \\ & & \ddots & \ddots & & c_{n-3}^{(1)} & & & & \\ & & & a_{n-1}^{(1)} & b_{n-1}^{(1)} & 0 & & & & \\ & & & & 0 & b_2^{(1)} & c_2^{(1)} & & & \\ & & & & & a_4^{(1)} & b_4^{(1)} & c_4^{(1)} & & \\ & & & & & & a_6^{(1)} & \ddots & \ddots & \\ & & & & & & & \ddots & \ddots & c_{n-2}^{(1)} \\ & & & & & & & & a_n^{(1)} & b_n^{(1)} \end{pmatrix},$$

and the relabelling gives $\tilde{\mathbf{y}}^{(1)} = \left(y_1^{(1)}, y_3^{(1)}, \ldots, y_{n-1}^{(1)}, y_2^{(1)}, y_4^{(1)}, \ldots, y_n^{(1)}\right)^T$, and $\tilde{\mathbf{x}} = (x_1, x_3, \ldots, x_{n-1}, x_2, x_4, \ldots, x_n)^T$. We observe that (5.16) consists of two independent $n/2 \times n/2$ tridiagonal systems of equations, which can be decoupled and treated concurrently. Further, if n is divisible by 4, the same approach can be applied to each subsystem to subdivide the problem into four independent $n/4 \times n/4$ tridiagonal systems, and so on. If the parallel system we employ is flexible enough, parallelism can be applied across and within the decoupled systems.

5.4.2   General sparse matrices

General sparse matrices are much more difficult to handle efficiently than those possessing structure. There is, therefore, a considerable literature devoted to numerical methods designed specifically for the solution of problems involving general sparse matrices. For an overview of the subject, from a serial perspective, see Duff, Erisman, *et al.* (1986), or George and Liu (1981). In this section we consider the particular case of a symmetric positive definite matrix A, although much of the discussion is equally applicable to a more general matrix.

Faced with the problem of solving a system of linear equations $A\mathbf{x} = \mathbf{b}$ in which A is symmetric and positive definite, we may choose first to form the Cholesky factorisation $A = LL^T$, where L is the lower triangular matrix of Exercise 4.6. The solution to the system may then be recovered using a combination of forward and backward substitutions; that is, we solve $L\mathbf{y} = \mathbf{b}$ for the intermediate vector $\mathbf{y}$, followed by $L^T\mathbf{x} = \mathbf{y}$ for the solution vector $\mathbf{x}$. The code for calculating the Cholesky factors is similar to the forward elimination codes of Chapter 4, but takes advantage of the symmetry of A. The kji variant (column-oriented submatrix forward elimination) is given in Code 5.4; it requires storage of the lower triangle of A only. The columns of L overwrite those of A as they are computed.

```
c n stages to calculate L (each calculates a column of L)
      do k = 1,n
c Calculate the k'th diagonal entry of L (overwrite a(k,k))
      a(k,k) = sqrt(a(k,k))
c Calculate the remainder of the k'th column of L (overwrite the k'th
c column of A)
         call sscal(n-k,one/a(k,k),a(k+1,k),1)
c Modify the columns of the remaining submatrix of A by the column of L
c just calculated
         do j = k+1,n
           call saxpy(n-j+1,-a(j,k),a(j,k),1,a(j,j),1)
         end do
      end do
```

Code 5.4 *Cholesky factorisation – kji variant*

Note that the $n - k$ calls to the Level 1 BLAS **saxpy** could be replaced by a single call to the Level 2 BLAS **ssyr**, but we have left the code in terms of calls to **saxpy** so that the implications of sparsity may more clearly be seen. Parallelisation of a code of this form has been considered in detail in Section 4.5. A limiting factor is that the kth column of A, $\mathbf{a}_k$, must be fully modified by columns $1, 2, \ldots, k - 1$ of L before it can be used to calculate $\mathbf{l}_k$, the kth column of L. If A contains a large number of zeros it may be possible to relax this restriction and it is this observation which justifies much of the following, in which we represent sparse matrices as full, $n \times n$, two-dimensional entities which contain a large number of zero entries. Any program which implements Code 5.4 for a sparse matrix should make use of the condensed storage strategies described in

$$A = \begin{pmatrix} * & * & * & * \\ * & * & & \\ * & & * & \\ * & & & * \end{pmatrix}$$

Figure 5.2 *4 × 4 sparse matrix*

the literature. See Duff, Erisman, *et al.* (1986) for details of the possibilities and Dodson, Grimes, *et al.* (1991a; 1991b) for definitions and model implementations of sparse versions of the Level 1 BLAS.

Consider the computations involved in Code 5.4 in the case that the matrix A is sparse. We would like to arrange that L be similarly sparse for two reasons:

- The introduction of non-zero elements will increase the total amount of computation required to form the Cholesky factors, and the intermediate and solution vectors using forward and backward substitutions.
- Any additional non-zero elements will need to be incorporated into the storage representation.

Unfortunately all we can say is that L has at least as many non-zeros as A; we cannot even guarantee that L and A have roughly the same number of non-zeros. For example, if A has the sparsity pattern shown in Figure 5.2, where $*$ denotes a non-zero element, then, for $k = 1$, the **saxpy** calls in Code 5.4 will make all zero elements in the lower triangle of A non-zero. This introduction of non-zero values is referred to as *fill-in*. We return to this later; for the moment we assume that there is only a limited amount of fill-in, so that the sparsity structures of A and L are, at least, similar.

In the kji variant of Cholesky factorisation, as expressed by Code 5.4, the calculation of the pivot column $\mathbf{l}_k$ is followed by the modification of $\mathbf{a}_j$, $j = k + 1, k + 2, \ldots, n$. The modification of each $\mathbf{a}_j$ is a **saxpy** operation in which the multiplying constant is a_{jk}. If this multiplier is zero then the corresponding **saxpy** call may be omitted. In an analogy with the discussion of QR factorisation in Section 5.2.2, once a column has been modified by all previous columns the determination of the next pivot column can take place concurrently with the updating of the remaining submatrix. However, the sparsity of A provides even greater potential for parallelisation, as we illustrate with reference to the symmetric 6×6 matrix of Figure 5.3. For reasons of clarity, we indicate the pattern of non-zeros in the lower triangle only and have deliberately chosen an example which involves no fill-in, so that the sparsity pattern of L is the same as that of A. In the general case, for $k = 1$, once the second column has been modified by $\mathbf{l}_1$, the next pivotal column, $\mathbf{l}_2$, can be computed (by forming a square root and scaling a vector) whilst $\mathbf{l}_1$ is being employed to modify $\mathbf{a}_3, \mathbf{a}_4, \mathbf{a}_5$ and $\mathbf{a}_6$. Here, since $a_{21} = 0$ (and hence $l_{21} = 0$), $\mathbf{a}_2$ is not modified by $\mathbf{l}_1$ and the pivot columns $\mathbf{l}_1$ and $\mathbf{l}_2$ can thus be calculated concurrently. Further, since $a_{31} = a_{32} = 0$ (and

$$A = \begin{pmatrix} * & & & & & \\ & * & & & & \\ & & * & & & \\ * & * & & * & & \\ & & & & * & \\ & * & * & * & * & * \end{pmatrix}$$

Figure 5.3 *6 × 6 sparse matrix*

hence $l_{31} = l_{32} = 0$), it follows that $\mathbf{a}_3$ will be modified by neither $\mathbf{l}_1$ nor $\mathbf{l}_2$, and thus $\mathbf{l}_3$ can be calculated concurrently with both $\mathbf{l}_1$ and $\mathbf{l}_2$. $\mathbf{l}_4$ cannot be determined until $\mathbf{a}_4$ has been fully modified by $\mathbf{l}_1$ and $\mathbf{l}_2$ since both a_{41} and a_{42} (and l_{41} and l_{42}) are non-zero. Because $a_{5j} = l_{5j} = 0$, $j = 1, 2, 3, 4$, the formation of $\mathbf{l}_5$ is independent of $\mathbf{l}_1$, $\mathbf{l}_2$, $\mathbf{l}_3$ and $\mathbf{l}_4$, and so can take place concurrently with any of the operations that have so far been described. Finally, $\mathbf{l}_6$ cannot be formed until $\mathbf{a}_6$ has been modified by $\mathbf{l}_2$, $\mathbf{l}_3$, $\mathbf{l}_4$ and $\mathbf{l}_5$. (Modification by $\mathbf{l}_1$ is not necessary since $a_{61} = l_{61} = 0$.) Note that it is the sparsity pattern of L, rather than of A, which dictates the potential for parallelism.

We describe the solution of a system of linear equations in the case that the coefficient matrix is symmetric, positive definite and sparse in terms of four phases:

- *Ordering*
  For a suitable reordering of the equations it may be possible to reduce the amount of fill-in that occurs. The less the fill-in, the less the additional space required to store L, the fewer the number of operations that will, in general, need to be performed to compute the Cholesky factorisation and in any subsequent forward and backward substitutions, and the greater the scope for determining pivot columns concurrently with each other and with submatrix modifications. (The qualification 'in general' is necessary; an ordering which minimises fill-in does not necessarily minimise the number of arithmetic operations involved in factorisation, and vice versa. A non-zero introduced early in the factorisation is likely to increase the operation count more than one which is introduced later.)
- *Symbolic factorisation*
  In order to be able to determine the potential for overlapping operations it is necessary to know the sparsity pattern of L, rather than of A. This may be achieved by performing the factorisation symbolically, recording only the amount of fill-in which will take place.
- *Numerical factorisation*
  Once the sparsity pattern of L has been found its entries can be computed.
- *Triangular equations solution*
  Having formed the Cholesky factors the original system of equations can be solved using forward and backward substitutions.

$$\bar{A} = PAP^T = \begin{pmatrix} * & & & * \\ & * & & * \\ & & * & * \\ * & * & * & * \end{pmatrix}$$

Figure 5.4 *Reordered 4 × 4 sparse matrix*

Whilst we have drawn a distinct line between each of these phases, in practice the dividing lines may be less clear, particularly between the symbolic and numeric factorisation phases. We consider each phase in turn.

Ordering

The aim of the ordering phase is, as we have commented, to limit the amount of fill-in that takes place in constructing L. We know that the solution to a system of linear equations is invariant under a reordering of the equations, and that this corresponds to premultiplication of the coefficient matrix A by a permutation matrix (Sections 4.2 and 4.3). Postmultiplication by a permutation matrix also leaves the solution unchanged except that the order of the unknowns is permuted. If P is such a permutation matrix and A is a symmetric positive definite matrix then $\bar{A} = PAP^T$ is also symmetric and positive definite. It has the same number of non-zero elements as A, but its sparsity pattern may be quite different. For example, consider the matrix of Figure 5.2. If we choose P to be I_4, but with the first and last rows interchanged, then the transformed matrix $\bar{A} = PAP^T$ has the form shown in Figure 5.4. We noted earlier that the lower triangular factor of A is full. In contrast, the lower triangular factor of $\bar{A}$ retains the sparsity pattern of $\bar{A}$.

Determining the permutation matrix which limits fill-in is a complex optimisation problem which is expensive computationally. Alternative approaches have been proposed which produce suboptimal solutions, which are nevertheless very effective. In particular we make reference to the *nested dissection* and *minimum degree algorithms* (Exercise 5.9). A full discussion of these algorithms is outside the scope of this book; we refer the reader to George and Liu (1981) for details. Henceforth we assume that A has been subjected to an appropriate reordering.

Symbolic factorisation

To illustrate how this may be achieved we return to the matrix of Figure 5.3. For this particular example the sparsity pattern of L is the same as that of the lower triangle of A. The first non-zero subdiagonal element in each column of L has an important bearing on the fill-in which occurs during the factorisation, and on the way the factorisation parallelises. For example, the first non-zero subdiagonal element in the first column is that in the fourth row, and so no fill-in can occur in columns 2 and 3 as a result of modification by l_1. In addition, the first non-zero subdiagonal in column 2 is also in the fourth row, whilst in column 3 it is in the sixth row. It follows that the calculation of l_1, l_2 and l_3 can

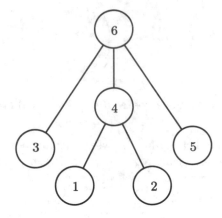

Figure 5.5 *Elimination tree for L corresponding to Figure 5.3*

be performed concurrently, which then enables the modifications of the trailing 3×3 submatrix of A by each of l_1, l_2 and l_3 to be performed concurrently (but properly synchronised).

A convenient way of expressing the structure of L is by its *elimination forest*. For each column, j, of L we define p_j (the *parent* of j) by

$$p_j = \min\{i : l_{ij} \neq 0, \ i > j\}.$$

Thus p_j is the row subscript of the first non-zero subdiagonal element in column j of L. If there are no such non-zero subdiagonal elements, then $p_j = j$. The elimination forest corresponding to the sparsity pattern of L has n vertices, labelled $1, 2, \ldots, n$, each corresponding to a column of L. If $p_j > j$, then vertex p_j is the parent of vertex j, and vertex j is one of several possible child vertices of p_j. In general the elimination forest consists of several disjoint trees, but we make the simplifying assumption that A is *irreducible* (that is, n is the only column of L with $p_j = j$), so that there is only one tree, with root at vertex n, in the elimination forest.

The elimination tree for the triangular factor of the matrix given in Figure 5.3 is displayed in Figure 5.5. It can be summarised by the vector (the *parent vector*) $\mathbf{p} = (4, 4, 6, 6, 6, 6)^T$. We observed earlier that the columns l_1 and l_2 can be formed in parallel. In terms of Figure 5.5, the vectors correspond to leaves of the tree. l_4 can only be formed once l_1 and l_2 are both available, as indicated by the dependence in the tree of node 4 on nodes 1 and 2. In this particular case, columns l_3 and l_5, which also correspond to leaf nodes, can be formed in parallel with l_1 and l_2, or with l_4. Finally, l_6 can be formed only when l_3, l_4 (and, by implication, l_1 and l_2) and l_5 are all available.

As a further example we consider the triangular factor L whose sparsity pattern is as shown in Figure 5.6. For this particular form of L the parent vector is $\mathbf{p} = (9, 9, 10, 10, 11, 11, 12, 12, 13, 13, 14, 14, 15, 15, 15)^T$, and the elimination tree is as given in Figure 5.7.

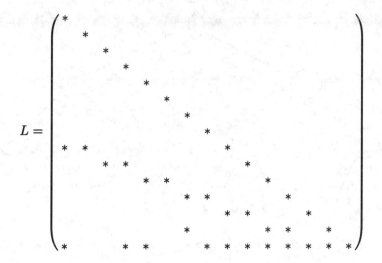

$$L =$$

Figure 5.6 *15 × 15 triangular factor*

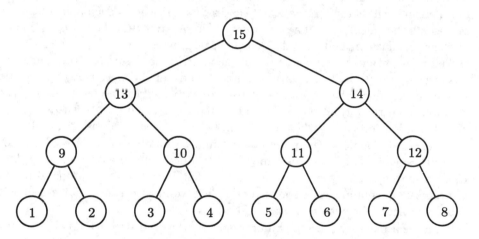

Figure 5.7 *Elimination tree for L of Figure 5.6*

Our description of elimination trees has been based on deriving the tree from the structure of the, as yet unknown, Cholesky factor L. Fortunately there are efficient algorithms which define the elimination tree of L directly from the structure of A; see, for example, Liu (1986), or Zmijewski and Gilbert (1988) for a parallel algorithm. It should be clear that, as far as parallelisation is concerned, the objective of the ordering step is to produce a reordered matrix for which the elimination tree is short and wide. An elimination tree with such a structure is also good in the sense of limiting the fill-in of the Cholesky factors.

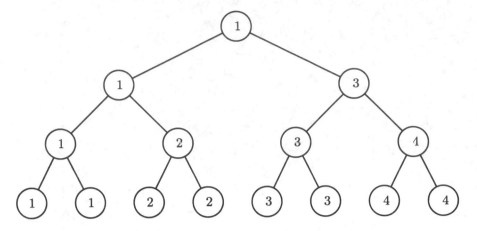

Figure 5.8 *Elimination tree showing process allocations*

Numerical factorisation

As has already been observed, the elimination tree provides the key to exploiting parallelism in the actual, numerical, Cholesky factorisation. Each node of the tree represents the calculation of a column of the lower triangular factor L, and nodes at the same level of the tree indicate calculations which are independent and can therefore be performed concurrently. Thus, referring to Figure 5.7, the calculation of each column $l_1, l_2, \ldots, l_8$ is independent and can be performed in parallel. Having determined these columns of L and having fully modified the remaining columns of A, the formation of the columns l_9, l_{10}, l_{11} and l_{12} constitutes a series of independent processes which can again be performed concurrently, and so on.

In terms of Code 5.4, the ability to calculate a number of pivot columns concurrently means that parallelism may be introduced into the outermost, k, loop. If, on a shared memory system, we permit a number of pivot columns to modify the remaining columns of A concurrently then we must arrange matters so as to avoid any problems with memory contention.

On a local memory machine an appropriate matrix distribution is by columns according to the structure of the elimination tree. Consider again the 15×15 example of Figure 5.6, whose elimination tree is as given in Figure 5.7, and suppose that four processes/processors are available. Then a suitable allocation of columns to processes is as shown in Figure 5.8. It can be seen, for example, that columns 1, 2, 9, 13, and 15 of A (and hence of L) are all allocated to the same process.

Triangular equations solution

It is again necessary fully to exploit the sparsity of L in forward and backward substitutions. Apart from this difference, the parallelisation of this phase is as summarised in Section 4.5.2. If, on a local memory machine, L is stored by scattered columns, calculation of the intermediate vector, **y**, by **saxpy** operations requires a redistribution of each column. Determining **y** by accumulating local dot products requires less communication and is, therefore, likely to be more appropriate. In contrast, L^T will be stored, implicitly, by scattered rows, which suggests the use of a local `saxpy`-based algorithm for backward substitution, and the only communication then is of the components of the solution as they are computed.

5.5   The linear least squares problem

Attention is now turned to the problem of determining a solution to the system (5.1) in which A is $m \times n$, with $m > n$. In the following we assume that A has full rank (that is, has n linearly independent columns). There is, in general, no solution which satisfies all equations in the system simultaneously and hence we need to redefine what we mean by a 'solution' in order that we may compute it. There are various options, but the criterion most commonly used is to determine the vector which minimises

$$S(\mathbf{x}) = \|A\mathbf{x} - \mathbf{b}\|_2^2. \tag{5.17}$$

We refer to S as the *sum of the squares of the residuals*, with the residuals being $r_i = (A\mathbf{x})_i - b_i$. The minimising vector $\mathbf{x}^*$ of (5.17) is then referred to as the *least squares solution* of the original overdetermined system. The techniques we describe can be applied to the rank deficient case, but we omit the details; see Golub and Van Loan (1989, p. 241).

5.5.1   The normal equations

From (5.17) we have

$$S(\mathbf{x}) = \mathbf{x}^T A^T A\mathbf{x} - 2\mathbf{b}^T A\mathbf{x} + \mathbf{b}^T\mathbf{b},$$

so that,

$$\nabla S(\mathbf{x}) = 2A^T A\mathbf{x} - 2A^T\mathbf{b},$$

where $\nabla S(\mathbf{x})$ is the n-vector with ith component $\partial S(\mathbf{x})/\partial x_i$. At a stationary point of S we must have $\nabla S(\mathbf{x}) = \mathbf{0}$ (and, since S is a sum of squares function, this must be a minimum point). Hence, the least squares solution $\mathbf{x}^*$ satisfies

$$A^T A\mathbf{x}^* = A^T\mathbf{b}, \tag{5.18}$$

and we refer to this system as the *normal equations*.

There are three stages to the determination of the least squares solution via the normal equations:

- Form the symmetric positive definite coefficient matrix $C = A^T A$. (C is symmetric, clearly. That it is positive definite follows from the observations that $\mathbf{x}^T A^T A \mathbf{x} = \|A\mathbf{x}\|_2^2$, and that A has full rank.)
- Form the right-hand side vector $\mathbf{d} = A^T \mathbf{b}$.
- Solve a system of linear equations with C as the coefficient matrix and $\mathbf{d}$ as the right-hand side vector.

On any machine (either serial or parallel) an efficient solution of (5.18) depends on exploiting the fact that C is symmetric and positive definite. We need only compute the lower (or upper) triangular part of C; the upper (lower) triangular part can be determined by symmetry. Further, C can be factored as $C = LL^T$, where L is lower triangular (the Cholesky factorisation), in a numerically stable way without the need for pivoting (Section 5.4.2).

Expressed in this way, the determination of the least squares solution involves synchronisation points between each of the stages, but this hides some of the potential parallelism of the problem. For example, as soon as the first column of C has been calculated the elimination steps which determine the first column of L can begin. Hence, throughout the solution, the elimination steps can proceed concurrently with the calculation of the columns of C, and of $\mathbf{d}$. In this way all the parallelism in the solution of the normal equations is exploited. On a local memory machine it is important that the elements of C and $\mathbf{d}$ should, as far as possible, be determined by the process which requires them for the factorisation. Recall that, for backward substitution, a scattered row distribution of a coefficient matrix is better than scattered columns, with the right-hand side vector allocated to processes in a conforming manner. The formation of $\mathbf{d}$ involves a matrix-vector multiplication which can be implemented in terms of **sdot** or **saxpy** operations. In either case, the process which calculates d_i requires knowledge of the ith row of A^T (the ith column of A) and the whole of $\mathbf{b}$. The formation of C involves a matrix-matrix product. Here, the process which calculates the ith row of C requires knowledge of not only the ith row of A^T but also the whole of A. If space is at a premium, this suggests that initially the columns of A be allocated to those processes which are to compute the corresponding columns of C. This will be followed by $\lceil n/p \rceil$ stages, with p the number of processes/processors, in which each process retains its own columns of A, and acquires a new set which allows it to determine further elements of the rows of C.

5.5.2　Solution using orthogonal transformations

An alternative approach to the solution of the linear least squares problem, which is preferable on the grounds of numerical stability, is based on the use of an orthogonal factorisation. Suppose that we can determine an $m \times m$ orthogonal matrix P such that

$$PA = \begin{pmatrix} R \\ 0 \end{pmatrix}, \tag{5.19}$$

where R is an $n \times n$, full rank, upper triangular matrix. If we write

$$Pb = \begin{pmatrix} \mathbf{b}_1 \\ \mathbf{b}_2 \end{pmatrix}, \tag{5.20}$$

where $\mathbf{b}_1$ is an n-vector and $\mathbf{b}_2$ is an $(m-n)$-vector, then, using the orthogonality of P,

$$
\begin{aligned}
S(\mathbf{x}) &= \|A\mathbf{x} - \mathbf{b}\|_2^2, \\
&= (A\mathbf{x} - \mathbf{b})^T (A\mathbf{x} - \mathbf{b}), \\
&= (A\mathbf{x} - \mathbf{b})^T P^T P (A\mathbf{x} - \mathbf{b}), \\
&= \|PA\mathbf{x} - P\mathbf{b}\|_2^2, \\
&= \left\| \begin{pmatrix} R \\ 0 \end{pmatrix} \mathbf{x} - \begin{pmatrix} \mathbf{b}_1 \\ \mathbf{b}_2 \end{pmatrix} \right\|_2^2, \\
&= \|R\mathbf{x} - \mathbf{b}_1\|_2^2 + \|\mathbf{b}_2\|_2^2.
\end{aligned}
$$

It follows that the minimum point $\mathbf{x}^*$ of S satisfies

$$R\mathbf{x}^* = \mathbf{b}_1, \tag{5.21}$$

and that $\|\mathbf{b}_2\|_2^2$ is the least squares error.

This approach to the linear least squares problem therefore involves three stages:

- Determine P such that (5.19) holds.
- Calculate $\mathbf{b}_1$ (and $\mathbf{b}_2$) from (5.20).
- Solve the triangular system (5.21).

Note that it is not necessary to compute $\mathbf{b}_2$, although this is useful since its norm is a measure of how well the least squares solution fits the original system of equations.

To determine P we employ Householder, or Givens, transformations. In the former case we determine n Householder matrices $P^{(k)}$, $k = 1, 2, \ldots, n$, (or, rather, the Householder vectors which define these transformations), each of size $m \times m$, so that

$$P^{(n)} P^{(n-1)} \cdots P^{(1)} A = \begin{pmatrix} R \\ 0 \end{pmatrix}.$$

The only modifications we need to make to Code 5.1 to deal with this situation are to

- include m as a parameter to the subroutine,
- change the upper limit of the outer k loop to n, and
- change n-k to m-k in the actual parameter lists of the calls to snrm2 and sscal, and n-k+1 to m-k+1 in the calls to sgemv and sger.

Parallelisation of this (and Givens) QR factorisation follows as for the case that A is square.

5.6　Eigenvalue/eigenvector problems

We next consider the eigenvalue/eigenvector problem, but it should be made clear, at the outset, that this description covers a variety of problems with which we may be faced. For a given matrix A we may wish to

1. find all the eigenvalues and eigenvectors, that is, n scalars λ_i, and corresponding vectors $\mathbf{x}_i$, such that

$$A\mathbf{x}_i = \lambda_i \mathbf{x}_i, \qquad i = 1, 2, \ldots, n, \tag{5.22}$$

2. find the eigenvalues only,
3. find the r largest (smallest) eigenvalues for some r, or
4. find the eigenvalue closest to some number μ.

The approach adopted will depend on the particular problem we are trying to solve.

To keep things simple we would like to ensure that the eigenvalues sought are real, as opposed to complex, numbers (although many of the methods we discuss can equally be applied to the complex case). Restricting A to be real does not, alone, guarantee this to be so. We require the additional condition that A is symmetric. We further assume the eigenvalues to be distinct (that is, $\lambda_i \neq \lambda_j$ for any $i \neq j$); then it is a simple matter to show that the eigenvectors are orthogonal, that is,

$$\mathbf{x}_i^T \mathbf{x}_j = 0, \qquad i \neq j, \tag{5.23}$$

and can be assumed real. It should be noted that eigenvectors are not unique in the sense that if $\mathbf{x}$ is an eigenvector then so is $\alpha \mathbf{x}$ for any (complex) scalar α. In the following we assume that eigenvectors have been normalised so that $\mathbf{x}^T \mathbf{x} = 1$.

Let X be the matrix with ith column $\mathbf{x}_i$ and $\Lambda = \mathrm{diag}(\lambda_1, \lambda_2, \ldots, \lambda_n)$. Then from (5.22)

$$AX = X\Lambda.$$

Further, (5.23) indicates that X is orthogonal, so that

$$X^T A X = \Lambda. \tag{5.24}$$

Clearly, the reduction of A to diagonal form offers a mechanism for determining all the eigenvalues of a matrix. All we need do is determine an orthogonal matrix X which satisfies (5.24). Casting our minds back to earlier sections of this chapter, we question whether Householder transformations have a role to play here. The answer is 'yes', but it is not possible to reduce A to diagonal form by pre- and postmultiplication by Householder matrices only. For example, suppose that P is chosen to zero the elements in the first column of A below the diagonal. Then

$$(PA)P^T = \left(P(PA)^T\right)^T,$$

and P cannot be guaranteed also to introduce zeros in the first column of $(PA)^T$ below the diagonal. Reduction to diagonal form is, therefore, a two-stage process. We

- use Householder transformations to reduce A to tridiagonal form, and then
- employ a different algorithm to determine the eigenvalues and eigenvectors of the resulting tridiagonal matrix.

We consider the reduction to tridiagonal form in the section which follows. In a serial environment a commonly employed approach to finding the eigenvalues of a tridiagonal matrix is the QR iteration which is based on the use of Givens rotation matrices. Unfortunately, the technique is not amenable to parallelisation and so, in Section 5.6.2, we consider alternative approaches. However, the use of Givens rotations iteratively to reduce A directly to diagonal form does have some attractions in a parallel environment. We discuss this approach in Section 5.6.3. Finally, in Section 5.6.4 we consider a method for determining a single eigenvalue/eigenvector pair.

5.6.1 Orthogonal reduction to tridiagonal form

Similar matrices

If P is an $n \times n$ non-singular matrix then the matrices A and $B = P^{-1}AP$ are said to be *similar*, and P is termed a *similarity transformation*. If the eigenvalues and eigenvectors of A are $(\lambda_i, \mathbf{x}_i)$, $i = 1, 2, \ldots, n$, that is, $A\mathbf{x}_i = \lambda_i \mathbf{x}_i$, then

$$APP^{-1}\mathbf{x}_i = \lambda_i \mathbf{x}_i,$$

which implies that

$$P^{-1}AP\mathbf{y}_i = \lambda_i \mathbf{y}_i,$$

where $\mathbf{y}_i = P^{-1}\mathbf{x}_i$. Hence, $(\lambda_i, P^{-1}\mathbf{x}_i)$, $i = 1, 2, \ldots, n$, are the eigenvalues and eigenvectors of B, that is, a similarity transformation leaves the eigenvalues unchanged, but the eigenvectors are multiplied by P^{-1}. This suggests that we employ similarity transformations to reduce A to a matrix B which has a simpler form, determine the eigenvalues and eigenvectors of B, and then recover the eigenvectors of A from those of B using

$$\mathbf{x}_i = P\mathbf{y}_i. \tag{5.25}$$

For stability reasons it is desirable that P be orthogonal, in which case we refer to an orthogonal similarity transformation.

Householder transformations

One way of reducing the symmetric matrix A to tridiagonal form is to employ $n - 2$ similarity transformations involving Householder reflection matrices. We write A as

$$A = \begin{pmatrix} a_{11} & \hat{\mathbf{a}}_1^T \\ \hat{\mathbf{a}}_1 & A_1 \end{pmatrix},$$

where $\hat{\mathbf{a}}_1$ is an $(n-1)$-vector and A_1 is $(n-1) \times (n-1)$. The first stage involves constructing the Householder matrix $P_1 = I_n - 2\mathbf{v}_1\mathbf{v}_1^T$, where $\mathbf{v}_1 = \begin{pmatrix} 0 \\ \mathbf{y}_1 / \|\mathbf{y}_1\|_2 \end{pmatrix}$ is of length n, $\mathbf{y}_1 = \hat{\mathbf{a}}_1 \pm \|\hat{\mathbf{a}}_1\|_2 \, \mathbf{e}_1$ is of length $n - 1$, and $\mathbf{e}_1$ is the first column of I_{n-1}. By construction, premultiplication of A by P_1 has the effect of reducing to zero all elements in the first column from position 3 downwards. Further, and importantly, the first row of A remains unchanged. Hence, postmultiplication of $P_1 A$ by P_1 leaves the zeros introduced into the first column unchanged and, at the same time, reduces to zero all entries in the first row from position 3 onwards. In detail, we have

$$P_1 A P_1 = \begin{pmatrix} a_{11} & \mp \|\hat{\mathbf{a}}_1\|_2 & 0 & \cdots & 0 \\ \mp \|\hat{\mathbf{a}}_1\|_2 & & & & \\ 0 & & & \tilde{A}_2 & \\ \vdots & & & & \\ 0 & & & & \end{pmatrix}.$$

We observe that the $(n-1) \times (n-1)$ matrix $\tilde{A}_2$ is given by

$$\tilde{A}_2 = \tilde{P}_1 A_1 \tilde{P}_1, \tag{5.26}$$

where $\tilde{P}_1$ is the $(n-1) \times (n-1)$ Householder matrix $\tilde{P}_1 = I_{n-1} - 2\tilde{\mathbf{v}}_1\tilde{\mathbf{v}}_1^T$, and $\tilde{\mathbf{v}}_1 = \mathbf{y}_1 / \|\mathbf{y}_1\|_2$. If we introduce the $(n-1)$-vectors $\mathbf{w}_1 = A_1\tilde{\mathbf{v}}_1$ and $\mathbf{u}_1 = \mathbf{w}_1 - \left(\mathbf{w}_1^T\tilde{\mathbf{v}}_1\right)\tilde{\mathbf{v}}_1$, then (5.26) becomes

$$\tilde{A}_2 = A_1 - 2\tilde{\mathbf{v}}_1\mathbf{u}_1^T - 2\mathbf{u}_1\tilde{\mathbf{v}}_1^T, \tag{5.27}$$

which is a symmetric rank-two update.

In the second stage of the process we attempt to reduce to zero entries in the first column of $\tilde{A}_2$ from row 3 onwards, and in the first row from column 3 onwards (or, equivalently, in the second column of $P_1 A P_1$ from row 4 onwards and in the second row from column 4 onwards), whilst retaining the zeros introduced at the first stage, and so on. After $k - 1$ such stages we know that A is orthogonally similar to the partially reduced matrix

$$P_{k-1}P_{k-2}\cdots P_1 A P_1 \cdots P_{k-2}P_{k-1}$$

$$= \begin{pmatrix} a_{11} & \tilde{a}_{12} & 0 & \cdots & \cdots & \cdots & \cdots & 0 \\ \tilde{a}_{12} & \tilde{a}_{22} & \ddots & \ddots & & & & \vdots \\ 0 & \ddots & \ddots & \ddots & \ddots & & & \vdots \\ \vdots & \ddots & \ddots & \ddots & \tilde{a}_{k-1,k} & 0 & \cdots & 0 \\ \vdots & & \ddots & \tilde{a}_{k-1,k} & \tilde{a}_{kk} & \cdots & \cdots & \tilde{a}_{kn} \\ \vdots & & & 0 & \vdots & \ddots & & \vdots \\ \vdots & & & \vdots & \vdots & & \ddots & \vdots \\ 0 & \cdots & \cdots & 0 & \tilde{a}_{kn} & \cdots & \cdots & \tilde{a}_{nn} \end{pmatrix}.$$

We introduce the $(n-k+1) \times (n-k+1)$ matrix $\tilde{A}_k = \begin{pmatrix} \tilde{a}_{kk} & \cdots & \tilde{a}_{kn} \\ \vdots & \ddots & \vdots \\ \tilde{a}_{kn} & \cdots & \tilde{a}_{nn} \end{pmatrix}$.

Then, if we write

$$\tilde{A}_k = \begin{pmatrix} \tilde{a}_{kk} & \hat{\mathbf{a}}_k^T \\ \hat{\mathbf{a}}_k & A_k \end{pmatrix},$$

where $\hat{\mathbf{a}}_k$ is of length $n-k$, and construct the $(n-k+1) \times (n-k+1)$ Householder matrix $\tilde{P}_k = I_{n-k+1} - 2\mathbf{v}_k\mathbf{v}_k^T$, where $\mathbf{v}_k = \begin{pmatrix} 0 \\ \mathbf{y}_k/\|\mathbf{y}_k\|_2 \end{pmatrix}$, $\mathbf{y}_k = \hat{\mathbf{a}}_k \pm \|\hat{\mathbf{a}}_k\|_2\,\mathbf{e}_1$, and $\mathbf{e}_1$ is the first column of I_{n-k}, then

$$\tilde{P}_k\tilde{A}_k\tilde{P}_k = \begin{pmatrix} \tilde{a}_{kk} & \mp\|\hat{\mathbf{a}}_k\|_2 & 0 & \cdots & 0 \\ \mp\|\hat{\mathbf{a}}_k\|_2 & & & & \\ 0 & & & \tilde{A}_{k+1} & \\ \vdots & & & & \\ 0 & & & & \end{pmatrix}, \tag{5.28}$$

where $\tilde{A}_{k+1}$ is $(n-k) \times (n-k)$. As in the first stage, it is straightforward to show that the modified elements in (5.28) are given by

$$\tilde{A}_{k+1} = A_k - 2\tilde{\mathbf{v}}_k\mathbf{u}_k^T - 2\mathbf{u}_k\tilde{\mathbf{v}}_k^T, \tag{5.29}$$

where

$$\tilde{\mathbf{v}}_k = \mathbf{y}_k/\|\mathbf{y}_k\|_2,$$
$$\mathbf{w}_k = A_k\tilde{\mathbf{v}}_k,$$
$$\mathbf{u}_k = \mathbf{w}_k - \left(\mathbf{w}_k^T\tilde{\mathbf{v}}_k\right)\tilde{\mathbf{v}}_k.$$

Hence, this, the kth, stage of the reduction process consists of two steps:

- Calculate the $n - k$ Householder vector $\tilde{\mathbf{v}}_k$. As in Section 5.2.1 this involves two Level 1 BLAS operations: snrm2 to compute a vector two-norm, and sscal to scale a vector.
- Calculate the modified elements of $\tilde{A}_{k+1}$ of (5.29). This involves the Level 2 BLAS operation ssymv (to determine $\mathbf{w}_k$), the Level 1 BLAS sdot and saxpy to form $\mathbf{u}_k$, and, finally, the Level 2 BLAS ssyr2 to form the symmetric rank-two update to A_k given by (5.29).

Since we begin with a symmetric matrix it is sufficient to store only the lower or upper triangle. If we decide to store the lower triangle then the space into which zeros are implicitly introduced can be used to store the Householder vectors by columns; these will be required if we subsequently wish to determine the eigenvectors of A.

To summarise this phase of the operation, if we write P_k as the $n \times n$ Householder matrix $P_k = I_n - 2\hat{\mathbf{v}}_k \hat{\mathbf{v}}_k^T$, where $\hat{\mathbf{v}}_k = \underbrace{(0, 0, \ldots, 0,}_{k} \tilde{v}_k^T)^T$, then after $n - 2$ stages of the reduction process A is orthogonally similar to

$$T = P_{n-2}P_{n-3}\cdots P_1 A P_1 \cdots P_{n-3}P_{n-2},$$

where T is symmetric and tridiagonal.

An implementation of the reduction, by Householder matrices, of the $n \times n$ symmetric matrix A to tridiagonal form is given in Code 5.5, which is a straightforward modification of Code 5.1. The original matrix is stored in the lower triangular part of A. The Householder vectors are again chosen so that their initial component is unity, and hence the Householder transformations take the form $P = I_n - \tau \mathbf{v}\mathbf{v}^T$. These vectors overwrite the lower triangle of A, with the unit elements implicitly defined. The constants τ are stored in the vector tau. For each transformation the vectors $\mathbf{w} = A\mathbf{v}$ and $\mathbf{u} = \mathbf{w} - (1/2)\tau \left(\mathbf{w}^T \mathbf{v} \right) \mathbf{v}$ are formed in the workspace array work, and $PAP = A - \tau \mathbf{v}\mathbf{u}^T - \tau \mathbf{u}\mathbf{v}^T$.

Parallelisation of Householder tridiagonalisation

We are again faced with the problem of parallelising a code which involves three nested loops. In terms of Code 5.5, the kji/kij variant, we distinguish between the outer, k, loop and the inner, i and j, loops:

- The k loop performs the $n - 2$ stages of the reduction to tridiagonal form. On the face of it this loop is inherently sequential since the kth Householder vector cannot be determined until the kth column of A has been fully modified by both pre- and postmultiplication by the preceding $k - 1$ Householder matrices. Taking our lead from Section 5.2.2, it is possible to construct the next Householder vector while the remaining rows and columns of A are being updated. However, the fact that A is modified by both pre- and postmultiplication means that we have to be even more careful with synchronisation. At the kth stage the $(k+1)$th column of A cannot be modified

```
      subroutine tridiag(a,lda,tau,n,work)
c tridiag reduces the real n*n symmetric matrix A to tridiagonal
c form using Householder transformations of the form P = I - tau v v^T
c
c Only the lower triangle of A is required
c
c The Householder vectors overwrite the lower triangular part of A
c whilst the diagonal and co-diagonal define the tridiagonal matrix
c
c The constants tau are returned in the n-vector tau and
c the vector work is used for workspace
      real one,zero,half
      parameter(one = 1.0e0,zero = 0.0e0,half = 0.5e0)
      integer lda,n
      real a(lda,*),tau(*),work(*)
      integer k
      real norm,alpha,beta,snrm2,sdot
      intrinsic sqrt,sign
      external snrm2,sscal,ssymv,sdot,saxpy,ssyr2
      do k = 1,n-2
c n-2 similarity transformations
c
c Compute the two-norm of the elements of column k of A from row
c k+2 downwards
      norm = snrm2(n-k-1,a(k+2,k),1)
c Compute beta, the co-diagonal element of column k
      alpha = a(k+1,k)
      beta = -sign(sqrt(alpha**2+norm**2),alpha)
c Compute the constant factor in the Householder transformation
      tau(k) = (beta-alpha)/beta
c Form the Householder vector
      call sscal(n-k-1,one/(alpha-beta),a(k+2,k),1)
c Set the unit element of the Householder vector
      a(k+1,k) = one
c Form the vector w in the work array
      call ssymv('l',n-k,tau(k),a(k+1,k+1),lda,a(k+1,k),1,zero,
     +           work(k+1),1)
c Compute the intermediate vector u, overwriting w
      alpha = -half*tau(k)*sdot(n-k,work(k+1),1,a(k+1,k),1)
      call saxpy(n-k,alpha,a(k+1,k),1,work(k+1),1)
c Perform the symmetric rank-two update
      call ssyr2('l',n-k,-one,a(k+1,k),1,work(k+1),1,a(k+1,k+1),lda)
c Restore the co-diagonal element of R
      a(k+1,k) = beta
      end do
      end
```

Code 5.5 *Householder reduction to tridiagonal form*

without first calculating all the elements of **w** and **u**. This suggests that we can overlap the construction of the $(k+1)$th Householder vector only with that part of the rank-two update represented by **ssyr2** which does not affect column $k+1$, assuming that column $k+1$ has already been fully modified.

- The i and j loops are hidden in the calls to the BLAS and can be parallelised directly.

The debate is, therefore, whether to accept a program which simply exploits the parallelism of the inner loops, by using calls to parallel versions of **ssymv**, **ssyr2**, **snrm2**, **sscal** and **saxpy**, or to seek a more complicated code which fully utilises the parallelism of the algorithm. For a shared memory system the former approach is likely to give a reasonable increase in performance for little effort, assuming that we have available a full set of parallel BLAS. However, the latter approach will squeeze that little extra bit of performance out of the system and may be preferred, despite the additional programming effort required.

On a local memory machine the situation is, as usual, somewhat complicated because of the need to distribute the data. Suppose that A is distributed by scattered columns and consider the kth stage of the reduction process given in Code 5.5. The first step is to calculate a Householder vector $\mathbf{v}$. This involves calls to the Level 1 BLAS snrm2 and sscal and, unless communication speeds are such that it is profitable to redistribute the data so that these operations can be parallelised (a very unlikely scenario), they will be performed by the process which holds the kth column of A without requiring any communication. The next step is to form $\mathbf{w} = A\mathbf{v}$, which involves a matrix-vector product. One way to perform this is to

- broadcast the Householder vector $\mathbf{v}$ to all processes,
- form partial results for $\mathbf{w}$ in each process using the columns of A available to that process, which essentially means expressing ssymv as a sequence of saxpy operations, and then
- accumulate $\mathbf{w}$ from the partial results in a single process.

We then need to form $\mathbf{u} = \mathbf{w} - (1/2)\tau(\mathbf{w}^T\mathbf{v})\mathbf{v}$. This is in terms of sdot and saxpy operations involving $\mathbf{v}$ and $\mathbf{w}$, but since $\mathbf{v}$ has been broadcast, it does not matter in which process $\mathbf{w}$ has been accumulated. We assume, again, single-processor versions of the Level 1 BLAS being used to determine $\mathbf{u}$. Taking this approach, we have inserted a synchronisation point between the computation of $\mathbf{w}$ and $\mathbf{u}$. However, this can be removed. Using its contributions to $\mathbf{w}$, each process can form partial contributions to $\mathbf{u}$, which are then accumulated in a single process before being broadcast, along with $\mathbf{w}$. Each process then modifies its columns of A according to a symmetric rank-two update. Once the process which holds the next pivot column has so modified that column, it should construct and broadcast, asynchronously, the next Householder vector.

5.6.2   Calculating the eigenvalues of a symmetric tridiagonal matrix

On a serial machine the usual way to find all the eigenvalues and corresponding eigenvectors of a symmetric tridiagonal matrix T is to use the so-called QR iteration (with shifts) (Golub and Van Loan, 1989, p. 418) to reduce T to diagonal form. This is achieved using a sequence of Givens rotations. We return to the subject in Section 5.6.3 where we consider Jacobi's method for reducing a full matrix to diagonal form. Here we consider two alternative approaches which potentially offer more scope for parallelism.

A divide and conquer method

The first method we consider is a divide and conquer approach suggested by a number of authors, including Dongarra and Sorensen (1987b). We denote the $n \times n$ real symmetric tridiagonal matrix T by

$$
T = \begin{pmatrix}
a_1 & b_1 & & & & \\
b_1 & a_2 & b_2 & & & \\
& b_2 & \ddots & \ddots & & \\
& & \ddots & \ddots & b_{n-1} \\
& & & b_{n-1} & a_n
\end{pmatrix}.
\tag{5.30}
$$

(The notation employed here is slightly different from that used for the unsymmetric tridiagonal matrix considered in Section 5.4.1. The a_is form the diagonal, whilst the b_is are now on the off-diagonal.) We partition T as

$$
T = \begin{pmatrix}
T_1 & b_k \mathbf{e}_k \mathbf{e}_1^T \\
b_k \mathbf{e}_1 \mathbf{e}_k^T & T_2
\end{pmatrix},
\tag{5.31}
$$

where T_1 and T_2 are $k \times k$ and $(n-k) \times (n-k)$ symmetric tridiagonal matrices respectively, given by

$$
T_1 = \begin{pmatrix}
a_1 & b_1 & & & \\
b_1 & a_2 & b_2 & & \\
& b_2 & \ddots & \ddots & \\
& & \ddots & \ddots & b_{k-1} \\
& & & b_{k-1} & a_k
\end{pmatrix},
$$

$$
T_2 = \begin{pmatrix}
a_{k+1} & b_{k+1} & & & \\
b_{k+1} & a_{k+2} & b_{k+2} & & \\
& b_{k+2} & \ddots & \ddots & \\
& & \ddots & \ddots & b_{n-1} \\
& & & b_{n-1} & a_n
\end{pmatrix},
$$

$\mathbf{e}_1$ is the first column of I_{n-k}, and $\mathbf{e}_k$ is the final column of I_k. The off-diagonal components of (5.31) correspond to the coupling between T_1 and T_2. T of (5.31) can be rewritten (the eigenproblem can be *divided*) as

$$
T = \begin{pmatrix}
\hat{T}_1 & 0 \\
0 & \hat{T}_2
\end{pmatrix} + \theta b_k \begin{pmatrix}
\mathbf{e}_k \\
\theta^{-1} \mathbf{e}_1
\end{pmatrix} \left(\mathbf{e}_k^T, \ \theta^{-1} \mathbf{e}_1^T \right),
\tag{5.32}
$$

where $\hat{T}_1$ is identical to T_1 except that the (k,k) element is $\hat{a}_k = a_k - \theta b_k$, and $\hat{T}_2$ is identical to T_2 except that the $(1,1)$ element is $\hat{a}_{k+1} = a_{k+1} - \theta^{-1} b_k$. The parameter θ is chosen to avoid cancellation errors in the calculation of the modified elements $\hat{a}_k$ and $\hat{a}_{k+1}$.

Let D_i, $i = 1, 2$, be diagonal matrices whose non-zero entries are the eigenvalues of $\hat{T}_i$, $i = 1, 2$, with the columns of the orthogonal matrices Q_i, $i = 1, 2$, being the corresponding eigenvectors. From (5.24)

$$\hat{T}_i = Q_i D_i Q_i^T, \qquad i = 1, 2.$$

Substituting these decompositions into (5.32) gives

$$T = \begin{pmatrix} Q_1 D_1 Q_1^T & 0 \\ 0 & Q_2 D_2 Q_2^T \end{pmatrix} + \theta b_k \begin{pmatrix} \mathbf{e}_k \\ \theta^{-1} \mathbf{e}_1 \end{pmatrix} \begin{pmatrix} \mathbf{e}_k^T, & \theta^{-1} \mathbf{e}_1^T \end{pmatrix}.$$

This may be rewritten

$$T = Q\hat{T}Q^T, \tag{5.33}$$

where

$$Q = \begin{pmatrix} Q_1 & 0 \\ 0 & Q_2 \end{pmatrix},$$

and

$$\hat{T} = \left[\begin{pmatrix} D_1 & 0 \\ 0 & D_2 \end{pmatrix} + \theta b_k \begin{pmatrix} \mathbf{q}_1 \\ \theta^{-1} \mathbf{q}_2 \end{pmatrix} \begin{pmatrix} \mathbf{q}_1^T, & \theta^{-1} \mathbf{q}_2^T \end{pmatrix} \right],$$

with $\mathbf{q}_1 = Q_1^T \mathbf{e}_k$ and $\mathbf{q}_2 = Q_2^T \mathbf{e}_1$. Hence T is orthogonally similar to $\hat{T}$ and so we can find the eigenvalues and eigenvectors of T by finding the corresponding values of $\hat{T}$.

Expressed in slightly more general terms, we need to solve an eigenvalue problem for a matrix of the form

$$\hat{D} = D + \rho \mathbf{z} \mathbf{z}^T,$$

where D is an $n \times n$ diagonal matrix, ρ is a non-zero scalar (θb_k) and $\mathbf{z}$ is a real n-vector, $\left(\mathbf{q}_1^T, \theta^{-1} \mathbf{q}_2^T \right)^T$. It can be shown that the eigenvalues of $\hat{D}$ are the n roots of the equation

$$f(\lambda) = 1 + \rho \sum_{j=1}^{n} \frac{\xi_j^2}{d_j - \lambda} = 0, \tag{5.34}$$

where d_j is the jth diagonal element of D, and ξ_j is the jth element of $\mathbf{z}$. (5.34) is referred to as the *secular equation* and its typical form is illustrated in Figure 5.9 by the case $n = 4$, $\rho = 0.15$, $D = \text{diag}(1, 2, 3, 4)$ and $\mathbf{z} = (1, 2, 3, 4)^T$. Very efficient iterative algorithms, with guaranteed convergence, can be devised for finding the roots of (5.34). Details of these iterative algorithms and other considerations, such as how to handle zero components of $\mathbf{z}$, are dealt with by Dongarra and Sorensen (1987b).

Let λ_i be a root of $f(\lambda) = 0$, that is, an eigenvalue of $\hat{D}$ (and, hence, of T), with corresponding eigenvector $\hat{\mathbf{q}}_i$. Then

$$\mathbf{0} = \left(\hat{D} - \lambda_i I_n \right) \hat{\mathbf{q}}_i = \left(D + \rho \mathbf{z} \mathbf{z}^T - \lambda_i I_n \right) \hat{\mathbf{q}}_i.$$

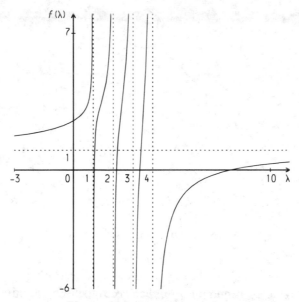

Figure 5.9 *Secular equation*

It follows that

$$\hat{\mathbf{q}}_i = -\left(D - \lambda_i I_n\right)^{-1} \rho \mathbf{z} \mathbf{z}^T \hat{\mathbf{q}}_i = \alpha_i \left(D - \lambda_i I_n\right)^{-1} \mathbf{z},$$

where $\alpha_i = -\rho \mathbf{z}^T \hat{\mathbf{q}}_i$. Since eigenvectors are unique only up to a multiplicative constant we can write

$$\hat{\mathbf{q}}_i = \gamma_i \Delta_i^{-1} \mathbf{z}, \tag{5.35}$$

where γ_i is chosen so that $\|\hat{\mathbf{q}}_i\|_2 = 1$, and the diagonal matrix $\Delta_i = D - \lambda_i I_n$. Once the eigenvectors of $\hat{D}$ (and hence of $\hat{T}$) have been determined, (5.25) indicates that the corresponding eigenvectors of T may be calculated using

$$\mathbf{x}_i = Q \hat{\mathbf{q}}_i. \tag{5.36}$$

To summarise, there are four stages to the computation of the eigenvalues and eigenvectors of T using this divide and conquer approach. We calculate

- eigenvalues and eigenvectors of the smaller tridiagonal matrices $\hat{T}_1$ and $\hat{T}_2$ (which therefore determines the matrices D_1, D_2, Q_1 and Q_2),
- the roots of the secular equation (5.34),
- the eigenvectors of $\hat{D}$ using (5.35), and
- the eigenvectors of T using (5.36).

The determination of the eigenvalues and eigenvectors of $\hat{T}_1$ and $\hat{T}_2$ constitutes two independent processes which can be performed concurrently. For a reasonably

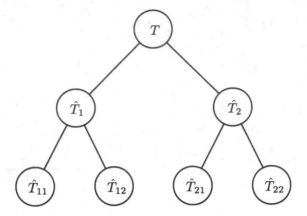

Figure 5.10 *Computational tree*

large matrix T, the grain size of each process will be accordingly large. Clearly, we can apply the divide and conquer approach recursively by dividing $\hat{T}_1$ and/or $\hat{T}_2$. Further, the calculation of the roots of the secular equation (5.34) (together with the calculation of the corresponding eigenvectors using (5.35)) can be performed concurrently (Section 6.3.2), although we note that each of these processes will have a relatively small grain size.

To illustrate how the inherent parallelism of the method might be exploited we consider the case $n = 16$. We subdivide T into T_1 and T_2, corresponding to the leading and trailing 8×8 parts of T respectively. Choosing $\theta = 1$, we have

$$T = \begin{pmatrix} \hat{T}_1 & 0 \\ 0 & \hat{T}_2 \end{pmatrix} + b_8 \begin{pmatrix} \mathbf{e}_8 \\ \mathbf{e}_1 \end{pmatrix} \begin{pmatrix} \mathbf{e}_8^T, & \mathbf{e}_1^T \end{pmatrix},$$

where $\mathbf{e}_1$ and $\mathbf{e}_8$ are the first and last columns of I_8, $\hat{T}_1$ is the same as T_1 but with $(8,8)$ entry $\hat{a}_8 = a_8 - b_8$, and $\hat{T}_2$ is the same as T_2 but with $(1,1)$ entry $\hat{a}_9 = a_9 - b_8$. $\hat{T}_1$ and $\hat{T}_2$ can be further subdivided into 4×4 matrices T_{11}, T_{12}, T_{21} and T_{22}, and corresponding matrices $\hat{T}_{11}$, $\hat{T}_{12}$, $\hat{T}_{21}$ and $\hat{T}_{22}$ with modifications to entries $(1,1)$ and/or $(4,4)$ as appropriate. We can represent this subdivision using a simple binary tree (Figure 5.10).

The calculations at the leaves of the tree (of the eigenvalues and eigenvectors of $\hat{T}_{11}$, $\hat{T}_{12}$, $\hat{T}_{21}$ and $\hat{T}_{22}$) can be performed using any standard (sequential) algorithm for calculating the eigenvalues of a symmetric tridiagonal matrix. An obvious candidate is the QR algorithm. The calculations at the other nodes of the tree use the secular equation (5.34) to find the eigenvalues of the symmetric tridiagonal matrix at that node of the tree. For example, the secular equation is used to calculate the eigenvalues and eigenvectors of $\hat{T}_2$ from those for $\hat{T}_{21}$ and $\hat{T}_{22}$. The eigenvalues and eigenvectors of T are determined, in turn, from those for $\hat{T}_1$ and $\hat{T}_2$. These data dependencies are summarised by noting that the computational tree must be traversed upwards from the leaves to the root. As we move up the computational tree we move from the coarse-grain parallelism of the computation of the eigenvalues of a tridiagonal matrix at the leaves of the tree to the finer-grain parallelism of concurrently calculating the roots of the secular equation at the other nodes of the tree. The computational tree also indicates that each subproblem is spawned from the data of its parent.

When implementing this method one decision that has to be taken is how many times to subdivide the original problem, or, equivalently, how large (or small) to make the tridiagonal matrices at the leaves of the tree. Clearly, the problem could be divided sufficiently many times that the tridiagonal matrices at the leaves of the computational tree are of such small dimension that they could be solved trivially — in practice this approach is not efficient and the optimal splitting of the problem is likely to be machine-dependent.

On a shared memory machine the implementation of this algorithm is straight-forward. On a local memory machine the implementation is, as usual, rather more complicated. One difficulty is that, for load balancing reasons, the Householder reduction to tridiagonal form leaves the matrix in a scattered column representation. These columns need to be redistributed by the divide and conquer method. A second difficulty is that the calculation of each root of the secular equation may be too fine-grain to offer much possibility of speed-up.

Multisectioning algorithms

In many circumstances we are interested in determining only a few eigenvalues and eigenvectors of a symmetric tridiagonal matrix T (or, of a general symmetric matrix A which has been reduced to tridiagonal form using similarity trans-formations). For example, we may only require those eigenvalues which lie within a given range of the real axis. In such a case an algorithm which determines all the eigenvalues may be somewhat inefficient. An alternative approach is based on the use of the *Sturm sequence property*.

Given the symmetric tridiagonal matrix (5.30) we introduce the sequence of polynomials

$$p_1(\lambda) = a_1 - \lambda,$$
$$p_k(\lambda) = (a_k - \lambda)p_{k-1}(\lambda) - b_{k-1}^2 p_{k-2}(\lambda), \qquad k = 2, 3, \ldots, n,$$

with $p_0(\lambda) = 1$. It is straightforward to prove by induction that

$$p_n(\lambda) = \det(T - \lambda I_n), \tag{5.37}$$

so that the zeros of the nth-degree polynomial $p_n(\lambda)$ are the same as the eigen-values of T. Further, the sequence of polynomials $\{p_k(\lambda), \ k = 0, 1, \ldots, n\}$ forms a Sturm sequence and satisfies the following property.

> **Sturm Sequence Property:** If, for fixed λ, $V(\lambda)$ denotes the number of agreements of sign in consecutive members of the sequence $\{p_k(\lambda), \ k = 0, 1, \ldots, n\}$, where we treat $p_l(\lambda) = 0$ as

having the opposite sign to $p_{l-1}(\lambda)$, then $V(\lambda)$ is the number
of zeros of $p_n(\lambda)$ (or eigenvalues of T) which are strictly
greater than λ. Consequently, for $a < b$, $V(a) - V(b)$ is the
number of eigenvalues of T in the half-open interval $(a, b]$.

The Sturm sequence property is often combined with bisection (or root-finding
based on (5.37)) to give a method for locating an eigenvalue of T. First we need
to find two values $\lambda_a < \lambda_b$ such that $V(\lambda_a) - V(\lambda_b) = 1$, which implies that there
is one eigenvalue, $\tilde{\lambda}$, of T in $(\lambda_a, \lambda_b]$. $V(\lambda)$ is then evaluated at the mid-point
$\lambda_c = (\lambda_a + \lambda_b)/2$; if $V(\lambda_c) = V(\lambda_b)$ then $\tilde{\lambda} \in (\lambda_a, \lambda_c]$, otherwise $\tilde{\lambda} \in (\lambda_c, \lambda_b]$. The
procedure can be repeated until the bracket known to contain $\tilde{\lambda}$ is sufficiently
small. Unfortunately, bisection has a slow (linear) convergence rate. Given an
interval $(\lambda_a, \lambda_b]$ containing a root $\tilde{\lambda}$ then either λ_a or λ_b is at most a distance
$(\lambda_a - \lambda_b)/2$ away from $\tilde{\lambda}$, and at each iteration this error bound is halved.

On a multiprocessor we can determine a number of eigenvalues in this way
concurrently. Suppose that we are interested in finding all the eigenvalues in the
interval $[a, b]$. We subdivide the interval into $p - 1$ subintervals, where p is the
number of processes, and each process evaluates $V(\lambda)$, for λ an end-point of one
of the subintervals. The number of eigenvalues in each subinterval can then be
determined. This scheme is repeated until all the eigenvalues, or clusters of very
close eigenvalues, are isolated. Once isolated, the eigenvalues can be determined
accurately using an iterative root-finding technique.

5.6.3 Jacobi's method

So far we have considered methods which determine the eigenvalues and eigenvectors of a matrix A by first using similarity transformations to reduce A to
tridiagonal form. Here, we consider a method which deals directly with the original matrix. The aim is to reduce A to diagonal form (and thus to isolate the
eigenvalues) by pre- and postmultiplication by a sequence of rotation matrices
(Section 5.2.3), and hence form a sequence of orthogonal similarity transformations. In the present context it is usual to refer to these as *Jacobi rotation
matrices*, but their construction is as for Givens matrices. On a practical note, it
is considered sufficient to reduce A to 'approximately' diagonal form, in the sense
that the off-diagonal entries of the transformed matrix are 'small'.

Let $A^{(1)} = A$, then the (p, q) element of $A^{(1)}$, $a_{pq}^{(1)}$, and by symmetry $a_{qp}^{(1)}$, can
be zeroed by forming

$$A^{(2)} = J^{(1)^T} A^{(1)} J^{(1)},$$

where the Jacobi rotation matrix $J^{(1)}$ is given by

$$\left(J^{(1)} \right)_{ij} = \begin{cases} \cos\theta, & i = j = p \text{ or } q, \\ \sin\theta, & i = p, j = q, \\ -\sin\theta, & i = q, j = p, \\ \delta_{ij}, & \text{otherwise.} \end{cases}$$

Referring back to (5.4), we see that $J^{(1)}$ is identical to the Givens rotation matrix G_{pq}. (The reader may be a little confused by the fact that we have premultiplied here by $J^{(1)^T}$, rather than $J^{(1)}$ itself. This is a detail only. With a slightly different choice of θ we could have premultiplied A by G_{ij}^T in Section 5.2.3.) $A^{(2)}$ is identical to $A^{(1)}$ except for rows and columns p and q, and the modified elements are given by

$$a_{ip}^{(2)} = a_{pi}^{(2)} = a_{ip}^{(1)}c - a_{iq}^{(1)}s, \qquad i \neq p, q,$$
$$a_{iq}^{(2)} = a_{qi}^{(2)} = a_{ip}^{(1)}s + a_{iq}^{(1)}c, \qquad i \neq p, q,$$
$$a_{pp}^{(2)} = a_{pp}^{(1)}c^2 - 2a_{pq}^{(1)}cs + a_{qq}^{(1)}s^2,$$
$$a_{qq}^{(2)} = a_{pp}^{(1)}s^2 + 2a_{pq}^{(1)}cs + a_{qq}^{(1)}c^2,$$
$$a_{pq}^{(2)} = a_{qp}^{(2)} = \left(a_{pp}^{(1)} - a_{qq}^{(1)}\right)cs + a_{pq}^{(1)}\left(c^2 - s^2\right),$$

where $c = \cos\theta$ and $s = \sin\theta$. Thus the (p,q) (and (q,p)) elements of $A^{(2)}$ are zero if

$$\left(a_{pp}^{(1)} - a_{qq}^{(1)}\right)cs + a_{pq}^{(1)}\left(c^2 - s^2\right) = 0. \tag{5.38}$$

Using the trigonometric identities $\sin 2\theta = 2\sin\theta\cos\theta$ and $\cos 2\theta = \cos^2\theta - \sin^2\theta$, (5.38) becomes

$$\left(a_{pp}^{(1)} - a_{qq}^{(1)}\right)\sin 2\theta + 2a_{pq}^{(1)}\cos 2\theta = 0,$$

that is

$$\tan 2\theta = \frac{2a_{pq}^{(1)}}{a_{qq}^{(1)} - a_{pp}^{(1)}}. \tag{5.39}$$

A preferable way of computing θ, and thus $\sin\theta$ and $\cos\theta$, follows if we let $\tau = \cot 2\theta = \left(a_{qq}^{(1)} - a_{pp}^{(1)}\right) \Big/ \left(2a_{pq}^{(1)}\right)$. Then the trigonometric identity

$$\tan^2\theta + 2\cot 2\theta \tan\theta = 1,$$

enables $\tan\theta$ to be calculated as a root of a quadratic. Stability considerations dictate that the smaller of the two roots should be taken, and so we require

$$\tan\theta = \frac{\text{sign}(\tau)}{|\tau| + \sqrt{1 + \tau^2}}.$$

Then, using $\sin^2\theta + \cos^2\theta = 1$ and $\tan\theta = \sin\theta/\cos\theta$,

$$c = \cos\theta = \frac{1}{\sqrt{1 + \tan^2\theta}},$$
$$s = \sin\theta = \tan\theta\cos\theta.$$

Clearly, this process can be repeated to zero some other off-diagonal element of $A^{(2)}$ and so on, although it should be noted that the annihilation of subsequent elements of $A^{(2)}$ is likely to make the previously zeroed element non-zero again. Jacobi's method consists of a sequence of steps of the above form, each annihilating an off-diagonal entry of the matrix. That is, starting with $A^{(1)} = A$, we form

$$A^{(k+1)} = J^{(k)^T} A^{(k)} J^{(k)}, \qquad k = 1, 2, \ldots,$$

where $J^{(k)}$ is a Jacobi rotation matrix, the aim being to reduce A to approximate diagonal form.

In the classical Jacobi method the element to be zeroed at any step is chosen to be the largest (in absolute value) off-diagonal element, and this defines the (p, q) plane in which the rotation is applied. The search for this largest element at each step is computationally expensive ($O\left(n^2\right)$, compared with $O\left(n\right)$ for the transformation itself) and alternative strategies for deciding the order of annihilation of the off-diagonal elements have been devised. In the *row-cyclic Jacobi method* the off-diagonal entries of A are zeroed in row order, that is, in the order $(1, 2), (1, 3), \ldots, (1, n), (2, 3), (2, 4), \ldots, (2, n), \ldots, (n-1, n)$, omitting any element whose size falls below some predefined threshold. (Here, and elsewhere, we give the order in which the upper triangular elements of A are zeroed; the lower triangular elements of A are zeroed by symmetry.) The process of zeroing, in turn, all $n(n - 1)/2$ off-diagonal elements of A is known as a *sweep* of Jacobi's method and the order in which the elements are zeroed (of which the above row-cyclic order is an example) is the *ordering* of the sweep. In practice the algorithm converges in a small number of sweeps, although on a serial machine it is usually slower than the QR algorithm outlined briefly at the end of this section.

As far as a parallel implementation of Jacobi's method is concerned the crucial observation is that several of the $n(n - 1)/2$ steps of a sweep could be performed concurrently. For example, for n even the rotations involved in zeroing the $(1, 2), (3, 4), \ldots, (n - 1, n)$ elements could be performed concurrently. If we define a *compound rotation matrix J* as

$$J = \begin{pmatrix} c_1 & s_1 & 0 & \cdots & \cdots & \cdots & 0 \\ -s_1 & c_1 & 0 & \ddots & & & \vdots \\ 0 & 0 & c_2 & s_2 & \ddots & & \vdots \\ \vdots & \ddots & -s_2 & c_2 & & \ddots & \vdots \\ \vdots & & \ddots & & \ddots & 0 & 0 \\ \vdots & & & \ddots & 0 & c_{n/2} & s_{n/2} \\ 0 & \cdots & \cdots & \cdots & 0 & -s_{n/2} & c_{n/2} \end{pmatrix},$$

for n even, then by choosing suitable values for the $c_i, s_i, \ i = 1, 2, \ldots, n/2$, we can simultaneously zero the $n/2$ elements, $a_{12}, a_{34}, \ldots, a_{n-1,n}$, to form

$$\tilde{A} = J^T A J.$$

$$\begin{pmatrix} * & 3 & 6 & 2 & 5 & 1 & 4 & 7 \\ & * & 2 & 5 & 1 & 4 & 7 & 6 \\ & & * & 1 & 4 & 7 & 3 & 5 \\ & & & * & 7 & 3 & 6 & 4 \\ & & & & * & 6 & 2 & 3 \\ & & & & & * & 5 & 2 \\ & & & & & & * & 1 \\ & & & & & & & * \end{pmatrix}$$

Figure 5.11 *Optimal annihilation pattern*

The premultiplication of A by J^T involves $n/2$ independent transformations, as does the postmultiplication of the result by J, but the combination of premultiplication by J^T and postmultiplication by J does not constitute $n/2$ independent transformations. For a matrix of size n the maximum number of elements which can be zeroed simultaneously (or, in other words, the maximum number of independent Jacobi rotations, in the sense that their definition, rather than their application, is independent) is $\lfloor n/2 \rfloor$, and so the minimum number of such compound steps required to complete a sweep of this parallel Jacobi method is

$$\frac{n(n-1)/2}{\lfloor n/2 \rfloor} = 2\lceil n/2 \rceil - 1 = \begin{cases} n-1, & n \text{ even,} \\ n, & n \text{ odd.} \end{cases}$$

The requirement is to define an ordering for the zeroing of the off-diagonal elements so that the sweep can be completed in $2\lceil n/2 \rceil - 1$ parallel steps. Clearly the row-cyclic ordering which we described earlier is unsuitable, since, for example, the first $n-1$ rotations all involve the index 1, and the next $n-2$ rotations all involve the index 2, and so on. However, there are orderings which do satisfy the given requirement.

As an example of an ordering which is optimal in the above sense we consider the case $n = 8$. In Figure 5.11 we indicate the number of the stage at which an element is reduced to zero. Hence, for example, at stage 3 the elements $(1,2),(3,7),(4,6)$ and $(5,8)$ may be zeroed by simultaneous post- and then premultiplications, and we observe that this ordering requires $n - 1 = 7$ compound steps to complete one sweep. There are a number of other orderings which allow the sweep to be completed in the optimal number of compound steps, and a detailed description, together with algorithms to define the orderings, is given in Modi (1988, Chapter 5) (see also Exercise 5.4 and Sameh, 1971). An alternative ordering which is almost optimal, in that it requires n (rather than the optimal $n-1$) parallel steps when n is even, is shown in Figure 5.12. Here the annihilation pattern is more clearly discernible.

Having decided on an ordering which permits simultaneous zeroing of $\lfloor n/2 \rfloor$ elements on each parallel step we now consider the implementation on a parallel machine of the corresponding parallel Jacobi method. Each of the $\lfloor n/2 \rfloor$ parallel rotations in a single compound step involves

$$
\begin{pmatrix}
* & 7 & 6 & 5 & 4 & 3 & 2 & 1 \\
  & * & 5 & 4 & 3 & 2 & 1 & 8 \\
  &   & * & 3 & 2 & 1 & 8 & 7 \\
  &   &   & * & 1 & 8 & 7 & 6 \\
  &   &   &   & * & 7 & 6 & 5 \\
  &   &   &   &   & * & 5 & 4 \\
  &   &   &   &   &   & * & 3 \\
  &   &   &   &   &   &   & *
\end{pmatrix}
$$

Figure 5.12 *Near-optimal annihilation pattern*

1. the calculation of c and s to define a Jacobi rotation matrix J,
2. the modification of rows p and q of A (which corresponds to the premultiplication of A by J^T), taking advantage of symmetry to store just the upper (or lower) triangular part of A, and
3. the modification of columns p and q of A (which corresponds to the postmultiplication of A by J).

This suggests a synchronisation point between each of the stages (1)–(3), and at the end of each parallel rotation, but if we arrange things correctly we can, for example, begin the next parallel rotation before the current one is fully complete (cf. Section 5.2.4). On a local memory system the amount of data communication required to implement a compound step should not be underestimated, particularly if only the upper or lower triangle of A is stored.

It is clear that we can employ parallelism within each individual Jacobi step. For a given p and q we can simultaneously modify rows p and q in step (2), and also columns p and q in step (3). In addition, parallelism can be applied within each row or column modification, although this will reduce the grain size of the processes and potentially reduce the speed-up that might otherwise be achieved. In either case, a considerable amount of synchronisation is required and it may well prove difficult to ensure a balanced load.

Before leaving Jacobi's method we note that the technique can equally be applied to the case of a tridiagonal matrix, in which case we refer to the QR algorithm. Each stage of the algorithm consists of determining the transformations which zero the elements below the diagonal in the order $(2, 1), (3, 2), \ldots, (n, n-1)$. Having completed these $n - 1$ premultiplications, $n - 1$ corresponding postmultiplications are performed. Implementing the operations in this order retains the tridiagonal structure of the matrix and the aim is to apply a sufficient number of sweeps so that all the non-zero co-diagonal entries are eventually reduced to zero. This algorithm, preceded by the reduction of a general symmetric matrix to tridiagonal form using Householder transformations, is the most widely used technique for determining all the eigenvalues of a matrix on a sequential machine. To improve the convergence rate of the algorithm it is usual to employ a shift before each Jacobi transformation is applied; that is, at step k we apply the transformation to $\tilde{A}^{(k)} = A^{(k)} - \mu I_n$ for some suitable μ (Golub and Van Loan, 1989, p. 373).

Parallelisation of the QR algorithm using the techniques described for the general Jacobi iteration is not possible, since we cannot guarantee to retain the tridiagonal structure. To see this, consider the rotations which zero elements (2,1) and (3,2). The first such transformation involves a rotation in the (1,2) plane, the second in the (2,3) plane. Suppose that the first transformation is delayed. Then an effect of the second rotation is to make the element (3,1) non-zero. Hence the rotations must be performed sequentially in the correct order.

5.6.4   The power method

Before closing this section on eigenvalues and eigenvectors we make brief mention of a simple technique for finding a single eigenvalue/eigenvector pair which offers parallelism at two levels.

Assume that the eigenvalues of A are ordered so that $|\lambda_1| > |\lambda_i|$, $\forall \, i > 1$, that is, λ_1 is dominant. Assume also that the corresponding eigenvectors $\mathbf{x}_i$ are linearly independent. Then any n-vector $\mathbf{v}$ may be written as $\mathbf{v} = \sum_{i=1}^n \alpha_i \mathbf{x}_i$. Now, $A\mathbf{v} = \sum_{i=1}^n \alpha_i \lambda_i \mathbf{x}_i$, and hence $A^k \mathbf{v} = \lambda_1^k \sum_{i=1}^n \alpha_i (\lambda_i/\lambda_1)^k \mathbf{x}_i$. Provided that $\alpha_1 \neq 0$, it follows that $\lim_{k\to\infty} A^k \mathbf{v} = \lim_{k\to\infty} \lambda_1^k \alpha_1 \mathbf{x}_1$.

The above result forms the basis of the *power method* which can be used to determine the dominant eigenvalue λ_1 of A and its associated eigenvector $\mathbf{x}_1$. Starting at some initial estimate $\mathbf{v}^{(1)}$ of $\mathbf{x}_1$, for $k = 1, 2, \ldots$, we

- form $\mathbf{y}^{(k+1)} = A\mathbf{v}^{(k)}$ (an **sgemv** operation),
- set $\lambda^{(k)} = \mathbf{v}^{(k)^T} \mathbf{y}^{(k+1)}$ (an **sdot** operation),
- form $\mathbf{v}^{(k+1)} = \mathbf{y}^{(k+1)} / \left\| \mathbf{y}^{(k+1)} \right\|_2$ (**snrm2** and **sscal** operations),

and, under suitable conditions, $\lambda^{(k)}$ converges to λ_1 and $\mathbf{v}^{(k)}$ converges to $\mathbf{x}_1$. In this form the power method offers parallelism within the BLAS; in a local memory environment distribution of A by scattered or block rows would be equally appropriate.

Despite the fact that the power method yields a single eigenvalue/eigenvector pair only, with a simple modification it can be employed to locate other eigenvalues and eigenvectors. Suppose that we are interested in the eigenvalue of A closest to q. Then $(A - qI_n)\mathbf{x}_i = (\lambda_i - q)\mathbf{x}_i$, and $(A - qI_n)^{-1}\mathbf{x}_i = (\lambda_i - q)^{-1}\mathbf{x}_i$. The eigenvalue of A closest to q corresponds to the dominant eigenvalue of $(A - qI_n)^{-1}$, and this can be determined by the power method. It looks like we need to invert $A - qI_n$ to form the matrix-vector product $\mathbf{y}^{(k+1)} = (A - qI_n)^{-1}\mathbf{v}^{(k)}$, but, instead, we form the LU factors of this matrix once before the iteration starts, and then find each $\mathbf{y}^{(k+1)}$ by forward and backward substitutions. Like the multisectioning technique of Section 5.6.2, this, the *inverse power method*, can be used to determine a number of eigenvalues concurrently in what we might term an embarrassingly obvious manner. Since eigenvalue estimates (values for q) could, for a symmetric tridiagonal matrix, be obtained using the Sturm sequence property, an initial reduction to tridiagonal form might be appropriate. Inverse iteration would then be performed using the tridiagonal matrix.

Exercises

5.1. Let A be an $n \times n$ matrix and $P_1, P_2, \ldots, P_r$ be $n \times n$ matrices defining the first r Householder transformations which introduce zeros below the diagonal into columns 1 to r. Show that $P = P_1 P_2 \cdots P_r$ may be written $P = I_n + WY^T$, where W and Y are $n \times r$ matrices, and the columns of Y are the scaled Householder vectors. Indicate which Level 3 BLAS operations are involved in forming PC, where C is an $n \times (n - r)$ matrix. Suggest a block algorithm based on these results for QR factorisation (Dongarra, Hammarling, *et al.*, 1987).

5.2. A factorisation of the form $A = QR$, where A and Q (orthogonal) are $m \times n$ matrices, and R is an $n \times n$ upper triangular matrix, can be determined using the modified Gram–Schmidt orthogonalisation algorithm:

For $k = 1, 2, \ldots, n$,
$$r_{kk} = \|\mathbf{a}_k\|_2 ,$$
$$\mathbf{q}_k = \mathbf{a}_k / r_{kk},$$
for $j = k + 1, k + 2, \ldots, n$,
$$r_{kj} = \mathbf{q}_k^T \mathbf{a}_j,$$
$$\mathbf{a}_j = \mathbf{a}_j - r_{kj} \mathbf{q}_k.$$

Initially $\mathbf{a}_k$ is the kth column of A. On completion, $\mathbf{q}_k$ is the kth column of Q. Express the algorithm in terms of the BLAS. How would you store Q and R? How may the algorithm be parallelised on both shared and local memory multiprocessors? (Golub and Van Loan, 1989.)

5.3. Let A be the 8×8 tridiagonal matrix

$$A = \begin{pmatrix} b_1 & c_1 \\ a_2 & b_2 & c_2 \\ & a_3 & b_3 & c_3 \\ & & \ddots & \ddots & \ddots \\ & & & a_7 & b_7 & c_7 \\ & & & & a_8 & b_8 \end{pmatrix}.$$

Verify that the elements a_2 and a_6 can be zeroed simultaneously using simple row operations akin to those employed in Gaussian elimination at the expense of fill-in at position $(6, 4)$. What is the form of the matrix if this technique is extended to zero the pairs a_3, a_7 and a_4, a_8, and then the pairs c_2, c_6 and c_1, c_5, and finally, c_4? Complete the process by indicating how the matrix may be reduced first to upper triangular form, and then to diagonal form using parallel row operations. Generalise these results and use them to derive a parallel algorithm for solving $Ax = b$, where A is an $n \times n$ tridiagonal matrix. What numerical limitations does the algorithm possess? What are the data distribution implications for a local memory multiprocessor? (Wang, 1981.)

5.4. An annihilation pattern for the parallel Jacobi method defines sets Z_k, $k = 1, 2, \ldots, 2m - 1$, of pairs (p, q), with $m = \lfloor (n + 1)/2 \rfloor$ and n the size of the symmetric matrix A. With $n = 16$ determine the annihilation patterns for the following schemes:

1. For $k = 1, 2, \ldots, m - 1$,

$$q = m - k + 1, m - k + 2, \ldots, n - k,$$

$$p = \begin{cases} 2m - 2k + 1 - q, & q \le 2m - 2k, \\ 4m - 2k - q, & 2m - 2k < q \le 2m - k - 1, \\ n, & 2m - k - 1 < q, \end{cases}$$

and for $k = m, m + 1, \ldots, 2m - 1$,

$$q = 4m - n - k, 4m - n - k + 1, \ldots, 3m - k - 1,$$

$$p = \begin{cases} n, & q < 2m - k + 1, \\ 4m - 2k - q, & 2m - k + 1 \le q \le 4m - 2k - 1, \\ 6m - 2k - 1 - q, & 4m - 2k - 1 < q. \end{cases}$$

2. For n a power of 2, $k = 1, 2, \ldots, n/2$,

$$q = 2, 4, 6, \ldots, n,$$

$$p = \begin{cases} q + n - 2k + 1, & q < 2k, \\ q - 2k + 1, & q \ge 2k. \end{cases}$$

In each case indicate the structure of the compound rotation matrix corresponding to each value of k (Sameh, 1971).

5.5. Let A be a symmetric matrix of size n, divisible by 2, with non-zeros on the principal diagonal and on the principal cross-diagonal only, that is, A has non-zero elements $a_{ii}, a_{i,n+1-i}$, $i = 1, 2, \ldots, n$. Show that the eigenvalues, λ, of A satisfy the quadratics

$$(a_{ii} - \lambda)(a_{n+1-i,n+1-i} - \lambda) - a_{i,n+1-i}^2 = 0, \qquad i = 1, 2, \ldots, n/2.$$

Hence describe a parallel algorithm based on the annihilation pattern of scheme (1) of Exercise 5.4 for locating the eigenvalues of a general matrix which first reduces it to this cross form, showing that the number of parallel steps reduces from $n-1$ to $n-2$, at the cost of computing the roots of the quadratics (Modi, 1988).

5.6. Determine the operation count of the cyclic reduction algorithm of Section 5.4.1. What form does $A^{(N-1)}$ take when $n = 2^N$? What is the maximum size of tridiagonal matrix which can be reduced to diagonal form in N stages? How many stages are required to reduce a matrix of size n, not necessarily a power of 2, to diagonal form?

5.7. How, using EPF's DO ALL construct, can the algorithm of Exercise 4.10 be implemented? What are the limitations of such an implementation? A better prospect is to allocate tasks (decomposition of a diagonal block, modification of a submatrix block, etc.) dynamically when a task is ready to be executed and a process would otherwise be idle. Develop such a task allocation strategy which, in the absence of pivoting, employs locks and events, as appropriate, to ensure that the operations take place in the correct order.

5.8. Let A be a symmetric positive definite matrix whose sparsity pattern corresponds to the matrix L of Figure 5.6, but with additional non-zero elements $a_{11,1}, a_{13,1}, a_{14,2}, a_{13,3}, a_{12,4}, a_{13,5}, a_{13,7}, a_{14,9}$. Indicate the fill-in that occurs when Cholesky factorisation is applied to this matrix. How does the elimination tree of L differ from that of Figure 5.7?

5.9. The 14×14, symmetric, positive definite matrix A has the pattern of non-zeros indicated below, where only the lower triangle is shown.

$$
A = \begin{pmatrix}
* & & & & & & & & & & & & & \\
* & * & & & & & & & & & & & & \\
 & * & * & & & & & & & & & & & \\
 & & * & * & & & & & & & & & & \\
* & & & & * & & & & & & & & & \\
 & & & & * & * & & & & & & & & \\
 & & & & & * & * & & & & & & & \\
* & & & & & & & * & & & & & & \\
 & & & & & & & * & * & & & & & \\
 & & & & & & & & * & * & & & & \\
 & * & * & & & & & & & & * & & & \\
 & & & & & & & & & & * & * & & \\
 & * & * & & & & & & & & * & & * & \\
 & & & & & & & & & & & & * & *
\end{pmatrix}.
$$

Verify that Cholesky factorisation introduces a further 24 non-zero elements and that no pivot columns can be found in parallel. Reordering the matrix according to the minimum-degree algorithm involves, at each elimination step k, searching for the diagonal element with the smallest total number of non-zeros in its row and column and, using row and column interchanges, making this the pivot element. Show that an application of this algorithm to A leads to the permutation $(10, 11, 12, 1, 8, 3, 2, 9, 5, 4, 13, 6, 14, 7)$, that is, variable (and row) 1 becomes variable 10, variable 2 becomes variable 11, etc. Verify that Cholesky factorisation applied to the result involves no fill-in and that the reordering increases the scope for parallel pivot column computation. (Ortega and Poole, 1981.)

Further reading

QR factorisation on a shared memory multiprocessor is considered by Wright (1991), who investigates dynamic task allocation for both Householder and Givens transformations. In addition to the references given earlier, parallel Givens rotations are considered by Cosnard (1991), Cosnard, Muller, *et al.* (1986), Cosnard and Robert (1986) and Sameh and Kuck (1978).

The requirement in Gauss–Seidel iteration to use updated values improves the convergence rate over Jacobi but limits the scope for parallelism. Chazan and Miranker (1969) suggest an asynchronous hybrid which updates components of the next iterate in parallel using whatever the most recent information may be.

The parallel factorisation of sparse matrices has received much attention in the literature. Jess and Kees (1982) describe the role of elimination trees in LU factorisation, as does Duff (1986). George, Heath, *et al.* (1989) examine the solution of systems in which the coefficient matrix is symmetric and positive definite. Heath, Ng, *et al.* (1990) give a detailed study of the way parallelism can be exploited in each of the stages of the Cholesky factorisation of a sparse symmetric positive definite matrix. Duff (1991) considers both direct and iterative solvers.

A comparison of three methods (QR factorisation, divide and conquer, multisection) for the determination of the eigenvalues of a symmetric matrix is given by Demmel, Du Croz, *et al.* (1988). The conclusions drawn as to which is the best depend, not surprisingly, on how much information is required (some/all eigenvalues, eigenvalues and eigenvectors).

One of the drawbacks of any parallel implementation on a local memory system of Jacobi's method for finding the eigenvalues of a symmetric matrix arises from the amount of data communication that has to take place. Further, a data distribution which is appropriate for postmultiplication by a rotation is unlikely to be appropriate for premultiplication. In a *one-sided Jacobi method* rotations are applied from one side only (say, from the right on scattered columns). See Eberlein and Park (1990) for details.

CHAPTER 6

Other areas I

6.1 What other areas and why?

At the beginning of Chapter 4 we commented that linear algebra is a core component of many numerical algorithms, whether they arise from the desire to solve a partial differential equation, to locate a minimum point of a function of several variables, etc. Hence the reader may well ask the question 'Why bother with other areas?'. If linear algebra is so important, and it forms a significant proportion of the execution time of a given algorithm, then perhaps it is sufficient to concentrate on speeding up the linear algebra part only. To some extent this is a valid comment. On the other hand, relying on this approach alone will probably lead to less than optimal results. Much depends on the respective costs of the individual components of the computation.

In this, and the next, chapter we investigate algorithms in six important areas of numerical computation, and the extent to which we are able to rely on parallelising any linear algebra component varies from nil (fast Fourier transforms which involve no significant linear algebra calculations) to a substantial amount (finite difference methods for partial differential equations which involve the solution of systems of linear equations). In the latter case we can make use of some of the material of earlier chapters, but we also introduce new methods which perform well (and parallelise well) given the particular forms of the linear equations encountered.

The division of topics between this and the next chapter is fairly arbitrary; there are loose connections within chapters and across chapters. In Chapter 7 we consider fast Fourier transforms and ordinary and partial differential equations. In this chapter we investigate methods for numerical integration, locating roots of equations and unconstrained optimisation. This by no means represents an exhaustive survey of numerical algorithms, but the aim is to illustrate various approaches to parallelism and the ease, and, sometimes, the difficulty, with which the transition from a serial to a parallel environment may be achieved.

One common theme which applies to the three topics covered in this chapter is the need to evaluate a function at many different points, the precise number and location of which may not be known a priori. The form of the function varies from topic to topic, from a real-valued function of a single real variable in the case of numerical integration, to a many-valued real function of several real variables in the case of a system of nonlinear equations. It is often the situation that the cost of function evaluations dominates, or at least is a significant component of, the overall computational cost and so one way of proceeding in a parallel environment is simply to look for ways of spreading this load. If a single function evaluation is itself very costly then decomposing the function into the sum of parts which can be evaluated separately may yield significant speed-ups. Unfortunately, this approach is very much problem-dependent and, in the cause of developing general purpose algorithms, we take the approach of attempting to perform several function evaluations concurrently. Nevertheless, any additional potential for parallelism should not be forgotten.

6.2 Numerical integration

We consider here the problem of determining an approximate value for the definite integral

$$I = \int_a^b f(x)dx,$$

where f is the *integrand* and $[a, b]$ is the *range of integration*. For simplicity we assume that both a and b are finite, although the underlying principles that we consider can equally be applied to the cases when the range of integration is semi-infinite or infinite.

The conventional way of approximating I is to use a *quadrature rule* R which is simply a linear combination of function values. We write

$$I \approx R = \sum_{i=1}^n w_i f_i, \tag{6.1}$$

where w_i, $i = 1, 2, \ldots, n$, are termed the *weights* of the quadrature rule, $f_i = f(x_i)$, and x_i, $i = 1, 2, \ldots, n$, are the *points* or *abscissae*. We define $E = I - R$ to be the *quadrature error*. Its value will depend not only on n, the number of terms in the quadrature rule, but also on the placement of the abscissae (and, by implication, the choice of the weights), and the form of the integrand. Of course, it may be that we are given the integrand in terms of a discrete set of function values, and hence have a somewhat limited choice for the abscissae. Here we take the view that f can be evaluated at any point within $[a, b]$ (although this does not necessarily imply that an analytical form for f is required) and consider a particularly simple rule, the *trapezium rule*, in which to investigate the scope for parallelism. However, the fundamental ideas of our approach are equally applicable to other rules such as Gauss–Legendre, Clenshaw–Curtis, etc.

From the user's point of view the ideal situation is that he should supply to a subprogram which computes an integral approximation no more information than is necessary. He will expect to specify the form of the integrand and the limits of the range of integration, but may be reluctant to give any details concerning the rule. Rather, he would like to supply some accuracy tolerance ϵ and leave it to the routine to determine appropriate values for n and the weights and abscissae such that some attempt is made to return an approximation which is correct to the specified tolerance. That is, the user expects a value R to be returned which satisfies the inequality

$$|I - R| \leq \epsilon.$$

It is this aim that motivates our approach to numerical quadrature and the subsequent derivation of parallel algorithms.

6.2.1 The trapezium and composite trapezium rules

One of the simplest quadrature rules for determining an integral approximation is the trapezium rule. The basic approach is to replace the integrand by a straight line joining the coordinates $(a, f(a))$ and $(b, f(b))$ and take the area under this straight line as an approximation to I. We write

$$I \approx T = \frac{b - a}{2}\{f(a) + f(b)\}.$$

In terms of our general rule (6.1), $n = 2$ (and we refer to a two-point formula), $x_1 = a$, $x_2 = b$ and $w_1 = w_2 = (b - a)/2$.

In general, T is likely to be a very crude approximation to I only, and it would be a simple matter to construct integrands for which the quadrature error E is very large. The trapezium rule as described here is, therefore, of limited practical use, but a simple extension will, potentially, yield a more accurate approximation. The range of integration $[a, b]$ is divided into $n - 1$ subintervals $[x_i, x_{i+1}]$, $i = 1, 2, \ldots, n - 1$, with $x_1 = a$ and $x_n = b$. For the moment we consider the points to be equally spaced, distance h apart, that is, $x_{i+1} - x_i = h$ with $h = (b-a)/(n-1)$. Over each subinterval the trapezium rule may be applied in turn, to give

$$I = \int_a^b f(x)dx,$$

$$= \int_{x_1}^{x_2} f(x)dx + \int_{x_2}^{x_3} f(x)dx + \cdots + \int_{x_{n-1}}^{x_n} f(x)dx,$$

$$= \sum_{i=1}^{n-1} \int_{x_i}^{x_{i+1}} f(x)dx,$$

so that $I \approx T_n$, where

$$T_n = \sum_{i=1}^{n-1} \frac{x_{i+1} - x_i}{2}\{f_i + f_{i+1}\},$$

$$= \frac{h}{2}\sum_{i=1}^{n-1}\{f_i + f_{i+1}\},$$

$$= \frac{h}{2}\left\{f_1 + 2\sum_{i=2}^{n-1} f_i + f_n\right\}.$$

We refer to T_n as the n-point *composite trapezium rule*. In terms of our general rule we now have $x_i = a + (i-1)h$, $i = 1, 2, \ldots, n$; $w_i = h$, $i = 2, 3, \ldots, n-1$, and $w_1 = w_n = h/2$. It can be shown that $E = O(h^2)$, and we say that the rule is *second order*. Further, $E = ch^2 + O(h^4)$ for some constant c (Davis and Rabinowitz, 1984).

Suppose now that we compute T_n and then T_{2n-1}. That is, having determined an n-point approximation we divide each subinterval further by introducing a new point midway between x_i and x_{i+1} (and then relabel the points and change the weights accordingly) and determine a $(2n - 1)$-point approximation. Then we have

$$I = T_n + ch^2 + O(h^4),$$

$$= T_{2n-1} + c\left(\frac{h}{2}\right)^2 + O(h^4) = T_{2n-1} + c\frac{h^2}{4} + O(h^4).$$

It follows that

$$T_{2n-1} - T_n = 3c\frac{h^2}{4} + O(h^4),$$

and hence that

$$\tilde{E} = (T_{2n-1} - T_n)/3, \tag{6.2}$$

is a second-order approximation to the error in T_{2n-1}. For h sufficiently small we can use $\tilde{E}$ as an estimate of the error in the quadrature rule. This suggests a simple, but effective, algorithm for approximating I to a specified absolute accuracy ϵ. We evaluate the sequence of composite trapezium rule approximations $T(= T_2), T_3, T_5, T_9, T_{17}, \ldots$, until the absolute value of the difference between consecutive approximations in the sequence (or, to be pedantic, one-third of this difference) is less than ϵ.

The determination of T_{2n-1} appears to involve $2n-1$ additional function evaluations. However, we observe that approximately half the function values required for T_{2n-1} are involved in the calculation of T_n. Further, a simple relationship exists between the two approximations, namely

$$T_{2n-1} = \frac{1}{2}T_n + h\sum_{i=1}^{n-1} f_{2i},$$

where h and f_{2i} refer to the spacing and function values at the points associated with T_{2n-1}. It is clear that to determine T_{2n-1} from T_n the only additional function evaluations required are at the $n-1$ newly introduced points.

A fragment of Fortran code which implements this simple-minded algorithm is given in Code 6.1. Essentially there are two loops here. The outer loop represents an iteration involving a doubling of the number of intervals in the quadrature rule. The inner loop sums the new function values. We readily recognise that there are some deficiencies in this code. In particular, the code may give incorrect results in certain circumstances (try $|\sin(x)|$ on $[-\pi, \pi]$). Such details have been ignored for two reasons: they would otherwise obscure the underlying algorithm, and we shortly introduce a somewhat better way of approximating a definite integral.

```fortran
      subroutine trap(R,a,b,f,eps)
c
c Computes a composite trapezium rule approximation R to the integral
c of f over [a,b], with eps an accuracy tolerance
c
      real half,zero,three
      parameter(half = 0.5e0,zero = 0.0e0,three = 3.0e0)
      real R,a,b,f,eps,h,t,twoh,sum,x,tnew
      integer n,nhalf,i
      external f
      h = b-a
      t = h*(f(a)+f(b))*half
c Calculate composite trapezium rule approximations until two
c succesive values agree to the specified tolerance
      n = 2
      twoh = h
      h = h*half
      nhalf = 1
   10 continue
c Sum the new function values
      sum = zero
      x = a+h
      do i = 1,nhalf
         sum = sum+f(x)
         x = x+twoh
      end do
      tnew = half*t+h*sum
c Check for convergence
      if (abs(t-tnew)/three.gt.eps) then
         nhalf = n
         n = n*2
         twoh = h
         h = h*half
         t = tnew
         goto 10
      end if
      R = tnew
      end
```

Code 6.1 *Composite trapezium rule*

It is straightforward to parallelise the composite trapezium rule algorithm. The cost of the computation is dominated by the evaluations of f within the i loop and hence some strategy is required for distributing this work. On a shared memory machine it is a simple matter to spread the i loop of Code 6.1. Essentially, the technique to employ is as for the dot product operation (Section 2.4.5). Processes determine local sums and these are accumulated in a global sum using some suitable synchronisation mechanism (say, a critical section). The same approach could be employed on a local memory system and, given that here, for a fixed n, the amount of communication is reduced, whilst the amount of computation is increased, there is likely to be some real benefit from such an approach. Once the range of integration has been broadcast each process is capable of determining, at each step of the method, the number of function evaluations it needs to perform and the points at which to perform those evaluations. All that is required is a signal from a master process to indicate that a further step is required. As with the dot product, the amount of work per process and per step depends on the length of the i loop and on the number of processes. However, since it involves evaluations of the integrand, the cost of this loop is likely to be much higher than was the case with the dot product. Once the individual sums have been computed they need to be gathered by a single process. The one reservation that we might have with this approach is that the evaluation of the integrand may be considerably more costly at some points of the range of integration than at others, and this could lead to load balancing problems if the function evaluations are apportioned in blocks.

The major drawback with any algorithm based on the composite trapezium rule is that it fails to recognise the form of the integrand and, by so doing, is likely to involve far more function evaluations than is absolutely necessary (and the number of function evaluations grows as a power of 2). For regions of $[a, b]$ in which f is difficult to integrate we need to employ a large number of integration points; elsewhere we are able to approximate the integral with sufficient accuracy using relatively few integration points. This suggests a strategy which we term *domain decomposition* and leads to embarrassingly simple parallelism as it eliminates the need for synchronisation/message-passing except at the beginning and end of the computation. All that is necessary is the imposition of some a priori discretisation of the range.

Suppose that p processes/processors are available. Then it is natural to subdivide $[a, b]$ into p equal subintervals. Each process then applies Code 6.1 to one of these subintervals. Strictly speaking, we should take care of the 'edge effects' at the subinterval interfaces. In this case the only thing that might be of concern is that two processes will evaluate f at a common point, but the synchronisation required to avoid this is probably not worthwhile. It is important to note that the tolerance given to each process should now be ϵ/p. The main difficulty with this approach again stems from load balancing; by splitting $[a, b]$ into p equal subintervals it is very likely that the numbers of subdivisions required within the different subintervals will vary considerably. A possible solution is to use many more initial subintervals than p. Each process is given a single subinterval over which to integrate and then, when this task is finished, the process selects another subinterval. Whilst effective, this approach is not as elegant as that afforded by *adaptive quadrature.* Here the placement of the quadrature points is determined by the algorithm itself in such a way that large numbers of points are employed only where necessary.

6.2.2   Adaptive quadrature

The objective of an adaptive scheme is to use an unequal mesh spacing and to determine the size of each subinterval so as to satisfy the overall accuracy requirement with the minimum number of subintervals (and consequently the minimum number of evaluations of the integrand). The underlying principles arise from a divide and conquer approach to numerical integration.

Suppose that we wish to compute an approximation to the integral of f over $[a, b]$ which is correct to some accuracy tolerance ϵ. We express the problem as $est(f, a, b, \epsilon)$ and, using the trapezium rule as a basis, an adaptive approach to its solution has three steps:

1. Calculate the two-point trapezium rule approximation

$$T_2 = \frac{b - a}{2} \{f(a) + f(b)\},$$

   and the three-point composite trapezium rule approximation

$$T_3 = \frac{b - a}{4} \left\{ f(a) + 2f\left(\frac{a + b}{2}\right) + f(b) \right\} = \frac{1}{2}T_2 + \frac{b - a}{2} f\left(\frac{a + b}{2}\right).$$

2. Estimate the absolute error in T_3 as

$$err \approx |T_3 - T_2|/3.$$

3. If $err \leq \epsilon$, then T_3 is acceptable as an approximate solution to the problem $est(f, a, b, \epsilon)$, that is, T_3 is correct to the specified accuracy tolerance. Otherwise, T_3 is not sufficiently accurate and the subdivision $[a, b]$ must be refined. We define two new problems

$$est\left(f, a, \frac{a + b}{2}, \frac{\epsilon}{2}\right), \quad \text{and} \quad est\left(f, \frac{a + b}{2}, b, \frac{\epsilon}{2}\right),$$

   on the two equal subintervals $[a, (a + b)/2]$ and $[(a + b)/2, b]$.

The algorithm described here is inherently *recursive*; either the problem is solved, or two new problems of the same type are generated. To be precise, we should refer to a *locally adaptive scheme*. In a *globally adaptive scheme* the error tolerance for each half of the subdivision is chosen to reflect an assumed distribution of the error within $[a, b]$.

Algorithms of this form are readily implemented using a language which supports recursion, such as Pascal or C. It is usual to permit some initial discretisation of $[a, b]$ as discussed at the end of Section 6.2.1. Each such subinterval is then considered in turn and sufficient levels of recursion are employed so that an integral estimate is returned which satisfies some predetermined accuracy condition. On a shared memory multiprocessor the parallelism is not restricted to the initial subdivision. If the language being used supports dynamic process creation (for example, Pascal and C on the Encore Multimax) then each recursion of the adaptive algorithm will spawn two new tasks (and the total number of tasks is not limited to the number of processors available). The parent task will be timed out until its children (which may themselves spawn tasks) have completed their execution, when it will again be available as an active task.

Despite the elegance of this approach, the cost of spawning tasks is not insignificant and, even in Pascal or C, a better way of organising things might be to avoid the use of recursion altogether. Further, the Fortran 77 programmer has little alternative since the language does not support recursion (but Fortran 90 does), and EPF does not support dynamic process creation in quite the way we require. In a serial environment the usual strategy for globally adaptive quadrature (Rice, 1975) is to employ an ordered list represented, for example, by a two-dimensional array and a one-dimensional array of pointers, through which rows of the two-dimensional array are accessed, as indicated by Figure 6.1. Each row of the two-dimensional array has four entries: the lower and upper ends of an interval; an integration rule approximation for that interval; and the corresponding error estimate. The pointer array is used to arrange the rows of the two-dimensional array in decreasing order of the size of the error estimates. When an interval subdivision occurs one of the new intervals (say, the lower half) overwrites the existing information; the other interval is added to the next free row in the two-dimensional array and the pointer array is reshuffled. Intervals for which the error estimates are within bounds may be removed from the list.

Such a data structure provides the basis for a parallel implementation in EPF of a globally adaptive scheme. Inside a **PARALLEL** block each process searches the two-dimensional array for the interval not currently being accessed by some other process which has the largest error estimate not within the specified tolerance. If no such interval exists and the entire computation is not yet complete then the process will need to continue to interrogate the data structure; that is, it must perform a busy-wait. A lock will need to be set on any row of the array while a process performs steps (1)–(3) of the adaptive algorithm on the corresponding integral, and access to the next free row and updating of the pointer array will also need to be synchronised.

The need to synchronise access to the global data structure can be avoided if the processes are allocated intervals from the initial subdivision and each process maintains its own, local, data structure. When a process becomes idle due to having determined an approximate integral over its subinterval correct to the required tolerance, and no intervals are left from the initial subdivision, that

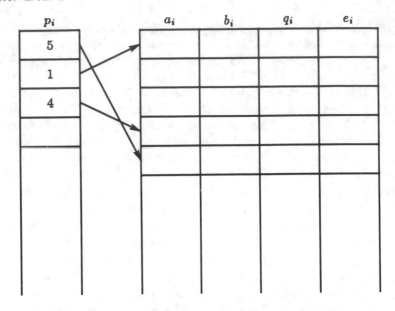

Figure 6.1 *Data structure for adaptive quadrature*

process can, in principle, 'steal' an interval from some other process which is still active; in practice this would be rather complicated to implement.

On a local memory machine we have similar problems, with the added difficulty of needing to control the amount of data traffic. A host/slave model based on the data structures outlined above could be employed but this may involve a heavy communications load, the importance of which depends on the complexity of the integrand. An alternative approach is to use an initial subdivision of the range of integration for process allocation, but again it may be necessary to impose some work redistribution strategy so as to avoid load balancing difficulties.

6.3   Nonlinear equations

We next consider the problem of finding one (or more) roots of the nonlinear equation

$$\mathbf{f}(\mathbf{x}) = \mathbf{0}, \tag{6.3}$$

where $\mathbf{f} = (f_1(\mathbf{x}), f_2(\mathbf{x}), \dots, f_n(\mathbf{x}))^T$ and $\mathbf{x}$ is an n-vector. (Equivalently this problem can be stated as finding one, or more, *zeros* of the nonlinear function $\mathbf{f}(\mathbf{x})$.) In Section 6.3.1 we consider the case $n = 1$, so that f is a function of a single real variable x. In Section 6.3.2 we consider the special case that $f(x) = P_n(x)$, where $P_n : \mathcal{C} \to \mathcal{C}$ is a polynomial of degree n, when it is known that there are exactly n (complex) roots and the usual requirement is to determine approximations to all of them. Finally, in Section 6.3.3, attention is turned to the case of a general function of several real variables, when we refer to a *system* of nonlinear equations.

6.3.1 A single nonlinear equation

We are interested here in locating a single root of the nonlinear equation

$$f(x) = 0, \tag{6.4}$$

where $f : \mathcal{R} \rightarrow \mathcal{R}$. Our principal reason for studying this special case is to provide the framework for the next two subsections. Only in exceptional cases would we consider solving this problem on a parallel machine, and then the parallelism would need to involve concurrent evaluations of f. For example, we show in Section 7.3.4 that a scheme for solving two-point boundary-value problems in ordinary differential equations (ODEs) involves root-finding techniques in which each function evaluation requires the solution of a related ODE initial-value problem. The cost of this evaluation is consequently high. If an algorithm requires several concurrent function evaluations the situation is analogous to that appertaining to the composite trapezium rule; speed-up will be achieved if the number of function evaluations is large and/or the cost of each function evaluation is high.

The simplest algorithm for locating a root of (6.4) is the *bisection algorithm* (cf. Section 5.6.2); it has the virtue of being guaranteed to converge, although the rate of convergence is slow. We start with an interval (or *bracket*) $[a, b]$ which satisfies $f(a)f(b) < 0$, so that a root of (6.4) is known to lie in $[a, b]$. (For simplicity we assume that $[a, b]$ contains one root only.) f is then evaluated at $c = (b + a)/2$, the mid-point of the interval. If $f(a)f(c) < 0$ then $[a, c]$ defines a new smaller interval in which a root of (6.4) lies, otherwise $[c, b]$ defines such an interval, and a single iteration of the algorithm is complete. To proceed we relabel either $[a, c]$ or $[c, b]$ as $[a, b]$, as appropriate, and iterate. If a and b represent approximations to a root then one of them is distance at most $(b-a)/2$ away from that root. This error bound is reduced by a constant factor $(1/2$, in this case) at each iteration and we say that the algorithm has a *linear rate of convergence*.

The bisection algorithm can be generalised to a *multisection algorithm*, similar to that discussed in Section 5.6.2. Instead of evaluating f at the mid-point c, we evaluate it at m points, $x_i = a + ih$, $i = 1, 2, \ldots, m$, equally spaced in the interval $[a, b]$, with $h = (b-a)/(m+1)$, and $x_0 = a$ and $x_{m+1} = b$. A new smaller interval $[x_r, x_{r+1}]$ in which a root of (6.4) lies can then be located. Each iteration of the algorithm reduces the length of the interval by a factor of $1/(m + 1)$, although we still have only a linear rate of convergence. It is straightforward to implement this algorithm on a p process/processor parallel machine; we simply arrange for m to be a multiple of p and spread the load of evaluating f at the m points across the processes. Each process can then investigate its own function values to detect a change of sign, and there will be the need for synchronisation or communication associated with function values at adjacent points which are dealt with by separate processes. The other communication that will be required is a broadcast (either implicitly on a shared memory system, or explicitly on a local memory system) from one process of the bracket that has been located. There is clearly no need for the other processes to continue searching. The iteration then repeats.

For this relatively simple algorithm it is possible to consider the potential parallel speed-up under the assumption that the function evaluations are the dominant computational costs and all other costs, including those associated with communication, can be ignored. Suppose that $m = p$, so that each process performs one function evaluation at each iteration. If $l_m^{(k)}$ denotes the interval length after $k - 1$ iterations using m intermediate values, then

$$l_m^{(k+1)} = \frac{l_m^{(k)}}{m + 1}.$$

For example, when $m = 31$, $l_m^{(k+1)} = l_m^{(k)}/32$. The case $m = p = 1$, which corresponds to implementing the bisection algorithm on a uniprocessor, yields $l_1^{(k+1)} = l_1^{(k)}/2$, so that five iterations are required to obtain $l_1^{(k+5)} = l_1^{(k)}/32$. Thus five iterations of the uniprocessor implementation reduces the interval length by the same amount as a single iteration of the 31 processor implementation, and the speed-up for the latter is limited to 5, whilst its efficiency is at most $100 \times (5/31) \approx 16\%$. Clearly, as p increases the efficiency of the algorithm degrades.

A remaining problem is that of locating an initial bracket for a root. The usual tactic is to start with an initial guess $x^{(1)}$ for a root and a step length h. f is evaluated at $x^{(1)}$ and $x^{(1)} + h$ and the direction of decreasing $|f|$ is determined. Further steps (of increasing magnitude) are taken in this direction of decreasing $|f|$ until a change in the sign of f is detected. Taking the multisectioning approach as our model, several concurrent function evaluations can be performed in this search for an initial bracket. The situation is similar to that of performing several function evaluations in a line search to locate a minimum point; see Section 6.4.2.

An alternative algorithm to the method of bisection is that of Newton and Raphson, often referred to as just *Newton's method*. The method, which is again iterative in nature, has a quadratic rate of convergence for a simple root. That is, at each iteration the absolute error of the approximation is proportional to the square of the absolute error at the previous iteration, provided that, at a root x^*, $f'(x^*) \neq 0$. (In the case of a repeated root, $f(x^*) = f'(x^*) = 0$, and the rate of convergence is linear only.) The method is, therefore, inherently much faster than bisection, although the cost per iteration, in terms of function evaluations, is higher. Unfortunately, convergence is guaranteed only if certain conditions are satisfied; essentially we require the initial guess $x^{(1)}$ to be 'sufficiently close' to a root. The algorithm computes the iterations

$$x^{(k+1)} = x^{(k)} - \frac{f\left(x^{(k)}\right)}{f'\left(x^{(k)}\right)}, \qquad k = 1, 2, \ldots, \tag{6.5}$$

until some appropriate accuracy condition is satisfied. Apart from the fact that $f\left(x^{(k)}\right)$ and $f'\left(x^{(k)}\right)$ could be evaluated concurrently, there is no opportunity for exploiting parallelism within this algorithm as it stands.

It is often the case that we are interested in not just one, but several roots of an equation (as, for example, in the divide and conquer method for calculating the eigenvalues and eigenvectors of a symmetric tridiagonal matrix presented in Section 5.6.2). This increases the scope for parallelism, whether we choose to use bisection, Newton's method or some other algorithm. Indeed, we potentially have embarrassingly simple parallelism. Consider Newton's method and suppose that we wish to calculate m roots using the iterations (6.5) starting with m initial approximations $x_i^{(1)}$, $i = 1, 2, \ldots, m$. Assuming p processes and m divisible by p, m/p independent Newton–Raphson iterations could be assigned to each process. Since the iterations are independent there is no data synchronisation (on a shared memory machine) or interprocess communication (on a local memory machine). There are, however, two potential difficulties with this approach:

- There may be load balancing difficulties if some of the iterations converge more quickly than others.
- For some problems it may be difficult to choose the starting values, $x_i^{(1)}$, $i = 1, 2, \ldots, m$, so that the independent iterations converge to different roots.

A solution to the former is to use some form of 'pool of tasks', as suggested for adaptive quadrature. The latter difficulty is potentially more serious and not so easily overcome.

6.3.2 Polynomial equations

Perhaps the simplest example of a nonlinear equation is that in which f is the nth-degree polynomial,

$$f(x) \equiv P_n(x) \equiv a_0 x^n + a_1 x^{n-1} + a_2 x^{n-2} + \cdots + a_{n-1}x + a_n,$$

where the coefficients, a_i, $i = 0, 1, \ldots, n$, are, in general, complex-valued. Without loss of generality we assume that $a_0 = 1$; we also assume that the polynomial has simple zeros only. The usual scenario is the requirement to find all the roots, α_i^*, $i = 1, 2, \ldots, n$, of $P_n(x) = 0$ and this is the approach adopted here. Whilst this is not usually a computationally demanding problem, the preferred algorithm for a parallel environment is different from that for a serial environment and hence is of interest.

Most serial algorithms for polynomial zero-finding take the form of an iterative method to calculate one (or two) of the zeros, followed by a *deflation* step to remove the calculated zero(s) from the polynomial. This procedure is repeated until all the zeros have been calculated. A simple algorithm of this type might use a Newton–Raphson iteration (using complex arithmetic) to calculate an approximation to a zero of the polynomial. The following two steps are repeated for $i = 1, 2, \ldots, n - 2$:

1. *Newton–Raphson iteration*
   Given a complex starting value $\alpha_i^{(1)}$, calculate

$$\alpha_i^{(k+1)} = \alpha_i^{(k)} - \frac{P_{n+1-i}\left(\alpha_i^{(k)}\right)}{P'_{n+1-i}\left(\alpha_i^{(k)}\right)}, \qquad (6.6)$$

   for $k = 1, 2, \ldots$, until $\alpha_i^{(k)}$ has converged, say to α_i^*.
2. *Deflation*
   Calculate the coefficients of the deflated polynomial

$$P_{n-i}(x) = \frac{P_{n+1-i}(x)}{x - \alpha_i^*}, \qquad (6.7)$$

which has the same zeros as $P_{n+1-i}(x)$ with the exception of α_i^*.

The final two zeros of $P_n(x)$ are calculated directly as the zeros of the quadratic $P_2(x)$ using the well-known standard formula. The special (polynomial) form of $P_{n+1-i}(x)$ is exploited in the evaluation of $P_{n+1-i}\left(\alpha_i^{(k)}\right)$ and $P'_{n+1-i}\left(\alpha_i^{(k)}\right)$ in (6.6) using Horner's rule, and in the calculation of the coefficients of the deflated polynomial $P_{n-i}(x)$ of (6.7). In the current context the difficulty with this algorithm is that it is inherently sequential and offers no possibility for significant parallelisation.

The algorithm just outlined for locating the zeros of a polynomial is fine in theory, but unless the polynomial zeros are calculated in increasing order of magnitude, the accumulation of round-off errors in the deflation process may mean that the zeros of the deflated polynomials become distant from those of $P_n(x)$. This difficulty may be overcome by performing a small number of Newton–Raphson iterations on the original polynomial using the approximate zero of a deflated polynomial as a starting value. This can be performed as a third step in the algorithm as each approximate polynomial zero is calculated, or at the end after all the approximate zeros have been found. In the latter case the situation is similar to that outlined at the end of the previous subsection. That is, from n different starting values, $\alpha_i^{(1)}$, $i = 1, 2, \ldots, n$, which are likely to be good approximations to the true zeros, perform n independent Newton–Raphson iterations

$$\alpha_i^{(k+1)} = \alpha_i^{(k)} - \frac{P_n\left(\alpha_i^{(k)}\right)}{P'_n\left(\alpha_i^{(k)}\right)}, \qquad k = 1, 2, \ldots,$$

until each iteration converges. Again, the algorithm is readily parallelisable, with the independent iterations implemented by separate processes, although each iteration will require few steps to converge.

An alternative algorithm, which is naturally parallelisable, is the *Durand–Kerner algorithm*, the kth iteration of which is given by

$$\alpha_i^{(k+1)} = \alpha_i^{(k)} - \frac{P_n\left(\alpha_i^{(k)}\right)}{Q^{[i]}\left(\alpha_i^{(k)}\right)}, \tag{6.8}$$

where

$$Q^{[i]}(x) = \prod_{\substack{j=1 \\ j \neq i}}^{n} \left(x - \alpha_j^{(k)}\right)$$

is a polynomial of degree $n-1$, and $\alpha_i^{(1)}$, $i = 1, 2, \ldots, n$, are n distinct complex starting values. Before considering a parallel implementation of the method, we consider the properties of (6.8) in more detail:

- The iteration displays second-order convergence. Assuming that, as $k \to \infty$, $\left\{\alpha_i^{(k)}\right\} \to \alpha_i^*$, then

$$\lim_{k \to \infty} Q^{[i]}\left(\alpha_i^{(k)}\right) = \lim_{k \to \infty} \prod_{\substack{j=1 \\ j \neq i}}^{n} \left(\alpha_i^{(k)} - \alpha_j^{(k)}\right) = \prod_{\substack{j=1 \\ j \neq i}}^{n} \left(\alpha_i^* - \alpha_j^*\right).$$

  But $P_n(x) = \prod_{j=1}^{n}(x - \alpha_j^*)$, so that $P_n'(x) = \sum_{k=1}^{n} \prod_{j=1, j \neq k}^{n}(x - \alpha_j^*)$, and hence $\lim_{k \to \infty} Q^{[i]}\left(\alpha_i^{(k)}\right) \to P_n'\left(\alpha_i^*\right)$. It follows that in the limit the iteration (6.8) reduces to the Newton–Raphson iteration, which is known to display second-order convergence.
- The denominator term $Q^{[i]}\left(\alpha_i^{(k)}\right)$ prevents the n separate iterations from converging to the same polynomial zero. If any two of the sequences of iterates tend to the same value then their denominator terms become very small and tend to force the sequences apart.

Hence we have an algorithm whose convergence rate in the limit approaches that of the Newton–Raphson iteration and which yields simultaneous approximations to all of the polynomial zeros provided only that the starting values are distinct.

It is a straightforward matter to parallelise the Durand–Kerner algorithm. The kth iteration consists of n independent calculations of $\alpha_i^{(k+1)}$, $i = 1, 2, \ldots, n$. There must be a synchronisation after each iteration since, for all i, $Q^{[i]}\left(\alpha_i^{(k)}\right)$ depends on all of the approximations $\alpha_j^{(k)}$, $j = 1, 2, \ldots, n$. Each cycle of the Durand–Kerner iteration thus involves

- a computation stage, where each process updates its root estimates using (6.8), followed by
- a synchronisation stage, where root estimates are made available to all other processes.

In the case of a shared memory system the synchronisation stage requires the use of a barrier. For a local memory system it requires communication; each process must broadcast its root estimates to all the other processes. In both cases the speed-up achievable will be determined by the degree of the polynomial. For a fixed number of processes p, the larger n, the more roots each process will need to locate and the larger the operation count per iteration in the calculations of (6.8) (that is, the larger the grain size).

6.3.3   Systems of nonlinear equations

We now turn to the general problem of finding a solution to the system of nonlinear equations

$$\mathbf{f}(\mathbf{x}) = \mathbf{0}, \tag{6.9}$$

where $\mathbf{f} : \mathcal{R}^n \to \mathcal{R}^n$ is a real vector-valued function of n real variables. The usual way of locating a root of (6.9) is to employ a generalisation of (6.5), which we formally write as

$$\mathbf{x}^{(k+1)} = \mathbf{x}^{(k)} - J^{(k)^{-1}}\mathbf{f}^{(k)}, \qquad k = 1, 2, \ldots,$$

where $\mathbf{x}^{(1)}$ is a given initial approximation to the solution, the superscript k on J and $\mathbf{f}$ denotes evaluation at $\mathbf{x}^{(k)}$, and $J(\mathbf{x})$ is the $n \times n$ *Jacobian matrix* of $\mathbf{f}(\mathbf{x})$ with (i, j) element $J_{ij}(\mathbf{x}) = \dfrac{\partial f_i(\mathbf{x})}{\partial x_j}$. This is usually simply referred to as *Newton's method*. Rather than invert $J^{(k)}$ explicitly we express the iteration as

$$\mathbf{x}^{(k+1)} = \mathbf{x}^{(k)} + \mathbf{p}^{(k)}, \tag{6.10}$$

where the *correction* $\mathbf{p}^{(k)}$ satisfies

$$J^{(k)}\mathbf{p}^{(k)} = -\mathbf{f}^{(k)}. \tag{6.11}$$

(6.11) is a system of simultaneous linear equations which can be solved for $\mathbf{p}^{(k)}$ using, say, Gaussian elimination followed by backward substitution.

To improve the global convergence properties, $\mathbf{p}^{(k)}$ is often used as a *search direction*. Rather than restrict ourselves to a unit step in this direction as suggested by the basic algorithm, we add to $\mathbf{x}^{(k)}$ a multiple, $\alpha^{(k)}$, of $\mathbf{p}^{(k)}$, where $\alpha^{(k)}$ is chosen so that it locates an approximate minimum of a suitable *merit function*, for example, the sum of squares function $F(\mathbf{x}) = \mathbf{f}(\mathbf{x})^T\mathbf{f}(\mathbf{x}) = \|\mathbf{f}(\mathbf{x})\|_2^2$. The kth iteration of this *damped Newton method* consists of the following steps:

1. Determine $J^{(k)}$ and $\mathbf{f}^{(k)}$ (with the latter having already been calculated in the line search of the previous iteration).
2. Solve the linear system $J^{(k)}\mathbf{p}^{(k)} = -\mathbf{f}^{(k)}$ for $\mathbf{p}^{(k)}$.
3. Perform a line search in the direction $\mathbf{p}^{(k)}$ to determine the line search parameter $\alpha^{(k)}$ and set $\mathbf{x}^{(k+1)} = \mathbf{x}^{(k)} + \alpha^{(k)}\mathbf{p}^{(k)}$.

It is straightforward to show that $\mathbf{p}^{(k)}$ is a *descent direction* for the merit function $F(\mathbf{x})$, that is, $\exists\ \alpha^{(k)} > 0$ such that $F\left(\mathbf{x}^{(k)} + \alpha^{(k)}\mathbf{p}^{(k)}\right) < F\left(\mathbf{x}^{(k)}\right)$, provided that $J^{(k)}$ is non-singular. Further, the algorithm exhibits a quadratic rate of convergence provided that $\alpha^{(k)} = 1$ for k sufficiently large.

In considering the scope for parallelisation within this damped Newton method we begin with the function evaluations required during the line search of step (3). Since $\mathbf{f}$ is vector-valued the cost of evaluating its components can be distributed. However, several components may share common subexpressions and, in a serial environment, this is exploited to reduce the overall cost of evaluating $\mathbf{f}$. In a parallel environment the same approach can be adopted but this will require synchronisation (shared memory) or communication (local memory). Appropriately grouping the components of $\mathbf{f}$ may help to reduce this synchronisation restriction. The alternative is simply to ignore common subexpressions and to re-evaluate them in individual processes. It is clearly possible also to distribute the work of evaluating the Jacobian matrix in step (1), with the same caveat applying. For example, we observe that if $\mathbf{j}_i^T$ represents the ith row of J then $\mathbf{j}_i^T = (\nabla f_i)^T$, that is, $\mathbf{j}_i$ is the *gradient vector* of f_i, the ith component of $\mathbf{f}$, and the evaluations of $\mathbf{j}_i$, $i = 1, 2, \ldots, n$, represent independent processes.

We have already remarked that step (2), the solution of a system of linear equations, is likely to involve Gaussian elimination followed by backward substitution and so we refer to the relevant material of Chapter 4. On a shared memory system there is not much more to be said, but on a local memory system there is the need to match the data distribution with that of step (1). Failure to do so will incur significant communications costs in making the transition between the two steps.

So far a synchronisation point between steps (1) and (2) has been implicitly assumed. If the rows of $J^{(k)}$ are evaluated concurrently, then the partial pivoting of Gaussian elimination imposes this synchronisation point. However, it is clear that the columns of $J^{(k)}$ can also be evaluated concurrently and, if they are, and a scattered column version of Gaussian elimination is employed, then the synchronisation point between steps (1) and (2) of the algorithm becomes rather blurred. As soon as the first column of $J^{(k)}$ is available the first pivot and the first set of multipliers can be determined. On a local memory system the multipliers can be broadcast to the other processes and the subsequent column modifications can take place as soon as the columns have been completely formed. A similar effect can be achieved on a shared memory system by maintaining a pool of tasks from which the processes select. Initially the pool will consist of tasks which involve the evaluation of the columns of $J^{(k)}$. Once the first set of multipliers has been determined, tasks involving modification of the remaining columns of $J^{(k)}$ can be added to the pool, with the constraint that a column must be fully formed before any modification can take place. Given Fortran's array storage pattern, a column-based algorithm written in EPF is likely to make the best use of data resident in a processor's cache. However, there are competing considerations. Commonality of subexpressions in the components of the Jacobian matrix is more likely across the rows than down the columns. Further, we have already remarked (Section 4.5.4) that, on a local memory multiprocessor, backward substitution is better suited to a scattered row distribution.

If an analytical form for J is not available then it will be necessary to determine a numerical approximation using, say, finite differences. We defer a discussion of this approach to Section 6.4.3 where the use of such approximations is considered in the context of optimisation algorithms.

Finally, a line search, such as that required in step (3), is usually performed by taking an initial step in the search direction, fitting a low-degree polynomial (say, a quadratic) to evaluations of the merit function F at a small number of points so encountered, and finding the minimum point of that quadratic. This is an inherently sequential algorithm which permits scope for parallelisation in the evaluation of F only. Alternatives which increase the scope for parallelism involve several concurrent evaluations of F; see Section 6.4.2.

6.4   Optimisation

In this, the final, section of the chapter we consider parallel algorithms for locating a *local minimum* of the *unconstrained optimisation problem*. That is, given $f(\mathbf{x})$, a real-valued *objective function* of n real variables, we attempt to locate a point $\mathbf{x}^*$ such that $f(\mathbf{x}^*) < f(\mathbf{x})$, $\forall\ \mathbf{x} \in \mathcal{N}(\mathbf{x}^*, \delta)$, a closed neighbourhood of $\mathbf{x}^*$. Algorithms exist which attempt to locate a *global minimum* of $f(\mathbf{x})$ but these are usually considerably different from the local minimisation algorithms and are not considered here.

Most practical numerical methods for locating $\mathbf{x}^*$ assume that $f(\mathbf{x})$ is at least twice continuously differentiable. (A notable exception is the algorithm described in Section 6.4.1.) All are iterative in nature, with a single iteration being of a common form which we refer to as the *model algorithm*:

1. Perform some function evaluations.
2. Determine a search direction.
3. Perform a search in that direction.

The 'function evaluations' we refer to in step (1) may involve evaluations of $f(\mathbf{x})$, and also its *gradient vector* and *Hessian matrix*, where appropriate. Formally, $\mathbf{g}(\mathbf{x})$, the gradient vector of $f(\mathbf{x})$, is the n-vector of first partial derivatives with components $g_i(\mathbf{x}) = \partial f(\mathbf{x})/\partial x_i$, and $G(\mathbf{x})$, the Hessian matrix of $f(\mathbf{x})$, is the $n \times n$ symmetric matrix of second partial derivatives with components $G_{ij}(\mathbf{x}) = \partial^2 f(\mathbf{x})/\partial x_i \partial x_j$. We remark that the model algorithm is a simplification of what might happen in practice. Typically, step (3) will involve function and, possibly, gradient vector evaluations which can be employed in step (1) of the next iteration.

We are primarily concerned here with methods which exploit parallelism by the simultaneous computation of function values of one form or another at a number of different points, or within the linear algebra which is encountered in step (2). It is usually a safe assumption that the function evaluations will dominate the linear algebra, and any other, costs.

For a general objective function the exploitation of parallelism in evaluating f will be problem-dependent and limited to evaluating factors of f concurrently. However, there are examples where there is natural parallelism within the function evaluation, the obvious example being furnished by a sum of squares function,

$$f(\mathbf{x}) = \sum_{i=1}^{m} \phi_i^2(\mathbf{x}) = \boldsymbol{\phi}(\mathbf{x})^T \boldsymbol{\phi}(\mathbf{x}),$$

where $\boldsymbol{\phi} : \mathcal{R}^n \rightarrow \mathcal{R}^m$. Further, evaluation of the gradient vector and Hessian matrix can be distributed in a straightforward manner; the situation is analogous to the treatment of nonlinear equations of Section 6.3.3. Given that this is ground which has, to some extent, already been covered, we concentrate on algorithms which exploit parallelism in other ways, leaving the implicit assumption that any parallelism within the evaluation of the gradient vector, say, could be used to enhance the performance of the algorithm.

6.4.1   Multi-directional search

A method which attempts to locate a minimum point using evaluations of f only is the *simplex method* of Spendley, Hext, *et al.* (1962), and Nelder and Mead (1965). The algorithm makes few demands on f; it will often work when f is discontinuous, or even if the function values are 'noisy' (f can be evaluated to a few significant figures of accuracy only). This, and the fact that derivatives are not required, means that algorithms of this class are popular in situations where f is experimentally defined only. On the down side, the convergence rate of the simplex method is slow, certainly much slower than that of the Newton-based algorithms which we consider in later subsections, and its use is generally restricted to moderate-sized problems. For larger problems the algorithm may fail to converge and hence, in this sense, it is unreliable.

A development of the simplex algorithm is the *multi-directional search algorithm* (Torczon, 1989). It is iterative in nature and based on moving a non-degenerate *simplex* around the space $\mathcal{R}^n$, changing its size but not its shape until it converges to a local minimum point $\mathbf{x}^*$. The algorithm can be described in terms of the model algorithm but this is not very natural. We begin by introducing $n + 1$ points (vertices) $\mathbf{v}_0, \mathbf{v}_1, \ldots, \mathbf{v}_n$, which are such that the set of n edges from any one vertex $\mathbf{v}_i$ to the others are linearly independent and thus span $\mathcal{R}^n$. These points are said to define a non-degenerate simplex. Examples of non-degenerate simplices are a non-empty triangle in $\mathcal{R}^2$ and a non-empty tetrahedron in $\mathcal{R}^3$.

To describe the algorithm it is necessary to indicate the three ways in which the simplex can move (and change size) at the kth iteration. The algorithm requires a suitable labelling of the vertices so that $\mathbf{v}_0^{(k)}$ is a 'best' vertex in the sense that

$$f\left(\mathbf{v}_0^{(k)}\right) \le f\left(\mathbf{v}_i^{(k)}\right), \qquad i = 1, 2, \ldots, n.$$

All remaining vertices are then moved according to one or more of the following steps:

- *Reflection step*

  A new simplex is defined by reflecting the n vertices $\mathbf{v}_i^{(k)}$, $i = 1, 2, \ldots, n$, in the best vertex $\mathbf{v}_0^{(k)}$; this defines n new vertices $\mathbf{v}_{r_i}^{(k)}$, $i = 1, 2, \ldots, n$, which satisfy

$$\mathbf{v}_{r_i}^{(k)} - \mathbf{v}_0^{(k)} = -\left(\mathbf{v}_i^{(k)} - \mathbf{v}_0^{(k)}\right),$$

  that is

$$\mathbf{v}_{r_i}^{(k)} = 2\mathbf{v}_0^{(k)} - \mathbf{v}_i^{(k)},$$

  with the subscript r_i being used to indicate a reflected vertex. Hence we have a new *reflected simplex* $\left\{\mathbf{v}_0^{(k)}, \mathbf{v}_{r_1}^{(k)}, \ldots, \mathbf{v}_{r_n}^{(k)}\right\}$. A reflection step in $\mathcal{R}^2$ is illustrated in Figure 6.2. In this and the next two figures, the original simplex is identified using dashed lines.

  The reflection step is deemed to have been successful if the new set of vertices satisfy

$$\min_{i=1,2,\ldots,n} f\left(\mathbf{v}_{r_i}^{(k)}\right) < f\left(\mathbf{v}_0^{(k)}\right), \tag{6.12}$$

  that is, the best vertex of the reflected simplex is not $\mathbf{v}_0^{(k)}$. After a successful reflection step an expansion step is tried. If the acceptance test (6.12) is not satisfied, no further action would imply that the next reflection step would simply return to the original simplex. An unsuccessful reflection step is therefore followed by a contraction step. Clearly, the function evaluations $f\left(\mathbf{v}_{r_i}^{(k)}\right)$, $i = 1, 2, \ldots, n$, are independent and hence can be performed concurrently.

- *Expansion step*

  If the reflection step is successful then further improvement (in the sense of obtaining vertices with smaller function values) may be possible by taking a larger step. The expansion step enlarges the reflected simplex by increasing

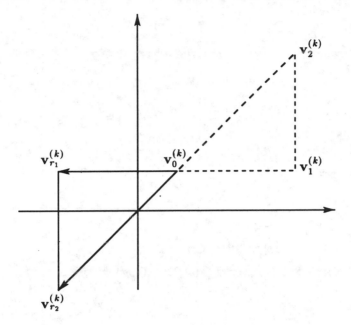

Figure 6.2 *Reflection step*

the length of each edge by a factor $\mu > 1$ to give the n new vertices $\mathbf{v}_{e_i}^{(k)}$, $i = 1, 2, \ldots, n$, which satisfy

$$\mathbf{v}_{e_i}^{(k)} - \mathbf{v}_0^{(k)} = \mu \left(\mathbf{v}_{r_i}^{(k)} - \mathbf{v}_0^{(k)} \right),$$

that is,

$$\mathbf{v}_{e_i}^{(k)} = \mu \mathbf{v}_{r_i}^{(k)} + (1 - \mu)\mathbf{v}_0^{(k)},$$

where the subscript e_i is used to indicate an expanded vertex. The new vertices, along with $\mathbf{v}_0^{(k)}$, define a new *expanded simplex* $\left\{ \mathbf{v}_0^{(k)}, \mathbf{v}_{e_1}^{(k)}, \ldots, \mathbf{v}_{e_n}^{(k)} \right\}$. An expansion step in $\mathcal{R}^2$ is illustrated in Figure 6.3.

The expansion step is deemed to have been successful if the vertices satisfy

$$\min_{i=1,2,\ldots,n} f\left(\mathbf{v}_{e_i}^{(k)} \right) < \min_{i=1,2,\ldots,n} f\left(\mathbf{v}_{r_i}^{(k)} \right),$$

so that the best vertex of the expanded simplex is better than the best vertex of the reflected simplex. The algorithm then continues with the expanded simplex,

$$\left\{ \mathbf{v}_0^{(k+1)}, \mathbf{v}_1^{(k+1)}, \ldots, \mathbf{v}_n^{(k+1)} \right\} = \left\{ \mathbf{v}_0^{(k)}, \mathbf{v}_{e_1}^{(k)}, \ldots, \mathbf{v}_{e_n}^{(k)} \right\},$$

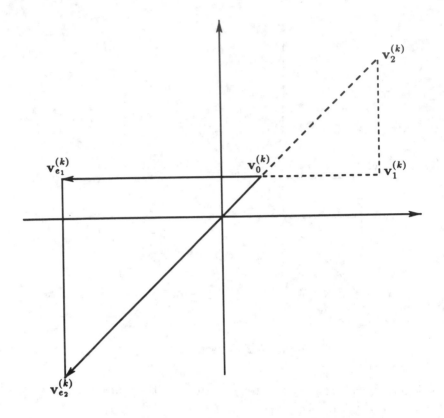

Figure 6.3 *Expansion step*

as the basis for the next iteration. If the expansion step is unsuccessful then the algorithm continues with the reflected simplex,

$$\left\{\mathbf{v}_0^{(k+1)}, \mathbf{v}_1^{(k+1)}, \ldots, \mathbf{v}_n^{(k+1)}\right\} = \left\{\mathbf{v}_0^{(k)}, \mathbf{v}_{r_1}^{(k)}, \ldots, \mathbf{v}_{r_n}^{(k)}\right\}.$$

In either case, the vertices are relabelled so that $\mathbf{v}_0^{(k+1)}$ corresponds to the best vertex of the new simplex. As with the reflection step, the function evaluations involved in an expansion step are independent and hence can be performed concurrently.

- *Contraction step*

  If the reflection step is unsuccessful then it is likely that the steps taken are too long and the original simplex needs to be reduced in size. A contraction step reduces the original simplex by decreasing the length of each edge by a factor $\theta < 1$ to give the n new vertices $\mathbf{v}_{c_i}^{(k)}$, $i = 1, 2, \ldots, n$, which satisfy

$$\mathbf{v}_{c_i}^{(k)} - \mathbf{v}_0^{(k)} = \theta\left(\mathbf{v}_i^{(k)} - \mathbf{v}_0^{(k)}\right),$$

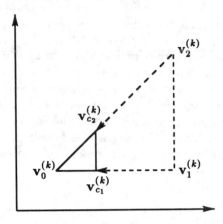

Figure 6.4 *Contraction step*

that is,

$$\mathbf{v}_{c_i}^{(k)} = (1 - \theta)\mathbf{v}_0^{(k)} + \theta\mathbf{v}_i^{(k)},$$

where the subscript c_i is used to indicate a contracted vertex. Together with $\mathbf{v}_0^{(k)}$, these new vertices define a *contracted simplex* $\left\{\mathbf{v}_0^{(k)}, \mathbf{v}_{c_1}^{(k)}, \ldots, \mathbf{v}_{c_n}^{(k)}\right\}$. A contraction step in $\mathcal{R}^2$ is illustrated in Figure 6.4.

The objective function is evaluated (concurrently) at the vertices of the contracted simplex to identify its best vertex. The algorithm then continues with the contracted simplex

$$\left\{\mathbf{v}_0^{(k+1)}, \mathbf{v}_1^{(k+1)}, \ldots, \mathbf{v}_n^{(k+1)}\right\} = \left\{\mathbf{v}_0^{(k)}, \mathbf{v}_{c_1}^{(k)}, \ldots, \mathbf{v}_{c_n}^{(k)}\right\},$$

where the vertices may have been relabelled so that $\mathbf{v}_0^{(k+1)}$ corresponds to the best vertex of the new simplex.

The fundamental difference between this simplex-based algorithm and that of Nelder and Mead (1965) is that, although the simplex is allowed to change size, it never changes shape. Under fairly mild conditions the multi-directional search algorithm can be shown to converge to a local minimum of f (Torczon, 1991). There are a number of other issues which need to be determined, such as the choice of the initial simplex and of the scaling factors μ and θ, and the stopping criterion, but these are details only and have no significant bearing on a parallel implementation of the algorithm.

In describing the algorithm we have already given the key to its parallelisation. Each step (reflection, expansion or contraction) involves the evaluation of f at n different points of $\mathcal{R}^n$ and this workload can be spread across a number of processes/processors p, say. If $p > n$ we could make use of spare processor power by evaluating f at some (or all) of the expanded and contracted vertices, as well as at the reflected vertices. We refer to this as *speculative evaluation*. The justification is that, in addition to function values at the points of the reflected simplex, values will also be required at the vertices of either the expanded or the contracted simplex. The proposal is to perform these extra evaluations speculatively before it is known precisely which are necessary. If the values are not required then we have lost nothing, assuming that the processors would otherwise be idle. Even if $p \leq n$, unless n is divisible by p we are likely to have an unbalanced load and the use of speculative evaluations may be employed to improve the situation. For $p \gg n$, Dennis and Torczon (1990) suggest performing even more speculative evaluations using the possible reflected, expanded and contracted simplices from future iterations (Exercise 6.7).

6.4.2 Newton's method

A popular way of solving the unconstrained optimisation problem is to use (the damped form of) Newton's method (Section 6.3.3), or, at least, a variation on it. For a given starting point $\mathbf{x}^{(1)}$, approximations $\mathbf{x}^{(k+1)}$ are computed according to the iteration

$$\mathbf{x}^{(k+1)} = \mathbf{x}^{(k)} + \alpha^{(k)}\mathbf{p}^{(k)}, \tag{6.13}$$

where the *Newton direction* $\mathbf{p}^{(k)}$ is obtained by solving the $n \times n$ linear system

$$G^{(k)}\mathbf{p}^{(k)} = -\mathbf{g}^{(k)}, \tag{6.14}$$

and the superscripts denote evaluation at $\mathbf{x}^{(k)}$, and where $\alpha^{(k)}$ is calculated by searching in the direction $\mathbf{p}^{(k)}$ for an approximate minimiser of f (a line search). The method has *quadratic termination* (the property of locating the minimum point of a positive definite quadratic function $f(\mathbf{x}) = \frac{1}{2}\mathbf{x}^T A\mathbf{x} + \mathbf{b}^T\mathbf{x} + c$, with A a positive definite matrix, in a single iteration); further, provided that $\alpha^{(k)} = 1$ for k sufficiently large, it has a quadratic rate of convergence so that, typically, $\mathbf{x}^*$ is determined after relatively few iterations.

In terms of the model algorithm the three steps of each iteration of the damped Newton method are as follows:

1. Evaluate the gradient vector $\mathbf{g}$ and the Hessian matrix G at the current iterate $\mathbf{x}^{(k)}$.
2. Solve the $n \times n$ linear system of equations (6.14) for $\mathbf{p}^{(k)}$.
3. Calculate the line search parameter $\alpha^{(k)}$ and determine the next iterate using (6.13).

We repeat the stated aim of deferring any potential parallelism within the evaluation of the function, gradient vector and Hessian matrix in steps (1) and (3) of each iteration. Step (2) involves the solution of a linear system whose coefficient matrix is the $n \times n$ symmetric matrix $G^{(k)}$. Note that it is sufficient to form the lower (or upper) triangular part of $G^{(k)}$ only. In the neighbourhood of a local minimum $G^{(k)}$ is positive definite, which means that the Cholesky decomposition can be employed and no pivoting is required. At iterates remote from a solution this may not be so and special action must be taken; this often involves adding a diagonal matrix to restore the positive definiteness (Gill, Murray, *et al.*, 1981). The parallelisation of the solution of general linear systems was considered in detail in Chapter 4 and the special case of a symmetric positive definite system in Exercise 4.6.

In the line search step (3) it is conventional to find a crude approximation to $\alpha^{(k)}$ only and this seldom requires more than two or three evaluations of f. To obtain greater potential for parallelism a multisectioning approach, analogous to that of Section 5.6.2, could be employed. This is likely to give a much more accurate approximation to the minimiser of f along $\mathbf{p}^{(k)}$, although it is uncertain whether this will result in an algorithm which converges significantly faster; it may even be more time-consuming. Overall the main opportunity for exploiting parallelism in the damped Newton method is in the linear algebra of step (2), the calculation of the Newton search direction.

6.4.3 Finite difference Newton methods

In the previous subsection we saw that the scope for parallelisation of the basic damped Newton method is somewhat limited. A further serious drawback, equally applicable to a serial environment, is the need for the user to provide analytical expressions for the first and second partial derivatives of f, with the more serious problem being the specification of the $n(n+1)/2$ components of the lower triangle of G. Assuming that an analytical expression for $\mathbf{g}$ is available, the components of G can be approximated using finite differences of gradient vectors. For example, an approximation $\mathbf{q}_j$ to the jth column of $G^{(k)}$ is given by the forward difference approximation

$$\mathbf{q}_j = \frac{\mathbf{g}\left(\mathbf{x}^{(k)} + h_j \mathbf{e}_j\right) - \mathbf{g}\left(\mathbf{x}^{(k)}\right)}{h_j}, \tag{6.15}$$

where $\mathbf{e}_j$ is the jth column of I_n and h_j is a suitable differencing interval. This gives a first-order, $O\left(h_j\right)$, approximation only to the columns of $G^{(k)}$, but in practice the accuracy of the Hessian matrix is not critical, given an accurate gradient vector. The use of such an approximation forms the basis of the *finite difference Newton method*, which is identical to the damped Newton method except that step (1) becomes:

1. Evaluate $\mathbf{g}$ at $\mathbf{x}^{(k)}$ and at the n displaced points $\mathbf{x}^{(k)} + h_j \mathbf{e}_j$, $j = 1, 2, \ldots, n$, and form the columns of the approximate Hessian matrix using (6.15).

Clearly, there is likely to be a price to pay, in terms of a slower convergence rate, for employing approximate derivatives only.

Although the Hessian matrix is known to be symmetric, the approximation $Q = (\mathbf{q}_1, \mathbf{q}_2, \ldots, \mathbf{q}_n)$ will, in general, be unsymmetric. There are two ways to avoid this difficulty:

- Approximate $G^{(k)}$ by $\frac{1}{2}(Q + Q^T)$, which is symmetric.
- Form finite difference approximations to the lower triangular elements of $G^{(k)}$.

In either case we emphasise that it is necessary to form only the lower triangular part of the Hessian matrix approximation. The second of these alternatives is likely to prove the least costly since fewer function evaluations (components of $\mathbf{g}$) are involved. The former requires the complete evaluation of the columns of Q; the latter requires the computation of the ith component of $\mathbf{q}_j$ for $j \leq i$ ($j \geq i$) only if the lower (upper) triangle of $G^{(k)}$ is stored, and there may be load balancing difficulties on a multiprocessor. Both alternatives involve the transposition of a matrix; the former in calculating $\frac{1}{2}(Q + Q^T)$, and both, implicitly, during the forward or backward substitution phases of the solution of the linear equations of step (2) if we work with the Cholesky factors. In a local memory environment this is likely to involve a significant amount of communication.

Step (1) of the finite difference Newton method is much better suited to parallelisation than its counterpart of Section 6.4.2, the damped Newton method. The $n + 1$ evaluations of the gradient vector can be distributed, as can the calculations of the finite difference approximations to the columns of the Hessian matrix. On a shared memory machine this is straightforward to arrange using simple loop spreading. Further, the elimination stages of step (2) can begin as soon as the first column has been completely formed; we can blur the synchronisation point between steps (1) and (2) as was the case with the Newton method of Section 6.3.3. In this context the second suggestion for ensuring symmetry of the approximate Hessian matrix is again more attractive than the first.

On a local memory machine the situation is, as usual, somewhat more complicated because of the need to avoid unnecessary data communication. The objective is to match the data distribution of the Hessian matrix approximation with that of the solution of the linear system of step (2). With a positive definite system pivoting is not an issue; nevertheless a scattered column distribution still appears to be the more appropriate. Hence $\mathbf{g}^{(k)}$ is calculated by a single master process and its value is broadcast to the remaining, slave, processes. Concurrently each slave process evaluates (possibly a portion of) $\mathbf{g}$ at those displaced points which enable it to form its columns of the lower triangular part of the approximate Hessian matrix. A scattered column version of Cholesky factorisation, modified to deal with possible indefiniteness of the approximate Hessian matrix, is then used to calculate the search direction $\mathbf{p}^{(k)}$ in step (2). The search direction is gathered by the master process which completes the iteration by performing the line search.

Further parallelisation is possible if speculative evaluations of $\mathbf{g}$ are performed during the line search. The justification for this is the observation that on most iterations the line search results in choosing $\alpha^{(k)} = 1$. This suggests that, whilst the line search of the kth iteration is performed by the master process, some of the evaluations of step (1) of the next iteration could be performed by the slave processes, using the assumed next iterate $\mathbf{x}^{(k+1)} = \mathbf{x}^{(k)} + \mathbf{p}^{(k)}$. If the line search returns a value $\alpha^{(k)} \neq 1$ then these speculative evaluations of the gradient vector are abandoned and step (1) at the next iteration is restarted using the correct value of $\mathbf{x}^{(k+1)}$. Assuming that the speculative evaluations are performed by processes which would otherwise be idle, abandoning the calculations is not a serious penalty.

A variant of the finite difference Newton method results if an analytical expression for f only is available and both $\mathbf{g}$ and G have to be approximated by finite differences. An efficient means of approximating $\mathbf{g}^{(k)}$ uses the central difference formula

$$g_i\left(\mathbf{x}^{(k)}\right) \approx \frac{f\left(\mathbf{x}^{(k)} + h_i\mathbf{e}_i\right) - f\left(\mathbf{x}^{(k)} - h_i\mathbf{e}_i\right)}{2h_i}, \qquad i = 1, 2, \ldots, n, \qquad (6.16)$$

which defines second-order, $O\left(h_i^2\right)$, approximations to the components of $\mathbf{g}^{(k)}$. The diagonal elements of the Hessian matrix may then also be approximated by a second-order central difference formula, namely

$$G_{ii}\left(\mathbf{x}^{(k)}\right) \approx \frac{f\left(\mathbf{x}^{(k)} + h_i\mathbf{e}_i\right) - 2f\left(\mathbf{x}^{(k)}\right) + f\left(\mathbf{x}^{(k)} - h_i\mathbf{e}_i\right)}{h_i^2}, \qquad i = 1, 2, \ldots, n,$$

$$(6.17)$$

and the off-diagonal elements may be approximated by the first-order forward difference formula

$$G_{ij}\left(\mathbf{x}^{(k)}\right) \approx$$

$$\frac{f\left(\mathbf{x}^{(k)} + h_i\mathbf{e}_i + h_j\mathbf{e}_j\right) - f\left(\mathbf{x}^{(k)} + h_i\mathbf{e}_i\right) - f\left(\mathbf{x}^{(k)} + h_j\mathbf{e}_j\right) + f\left(\mathbf{x}^{(k)}\right)}{h_ih_j},$$

$$i, j = 1, 2, \ldots, n, \quad i \neq j. \qquad (6.18)$$

The use of second-order central difference approximations for the off-diagonal elements of $G^{(k)}$ would involve many more function evaluations than are required by (6.18) (Exercise 6.8).

For this finite difference Newton method, step (1) of the model algorithm has the following form:

1. Evaluate f at $\mathbf{x}^{(k)}$, $\mathbf{x}^{(k)} \pm h_i\mathbf{e}_i$, $i = 1, 2, \ldots, n$, and $\mathbf{x}^{(k)} + h_i\mathbf{e}_i + h_j\mathbf{e}_j$, $i, j = 1, 2, \ldots, n$, $i \neq j$, and form approximations to $\mathbf{g}^{(k)}$ and $G^{(k)}$ using (6.16), (6.17) and (6.18).

It follows from (6.18) that the approximation to $G^{(k)}$ is symmetric and hence it is sufficient to form only the lower triangular elements. This requires the evaluation of f at $\mathbf{x} = \mathbf{x}^{(k)} + h_i \mathbf{e}_i + h_j \mathbf{e}_j$ for $i > j$ only, so that the total number of evaluations of f in step (1) is $n(n-1)/2 + 2n + 1 = n(n+3)/2 + 1$, and these can all be performed concurrently.

On a shared memory multiprocessor a simple way of parallelising this step of the algorithm is to spread the loop which computes the $n(n+3)/2 + 1$ function evaluations and then to spread the loops (the single loops of (6.16) and (6.17) and the double loop of (6.18)) which form the finite difference approximations to $\mathbf{g}^{(k)}$ and $G^{(k)}$. To ensure full processor utilisation, however, we again need to recognise the potential overlap between these loops and the linear algebra of step (2) of the algorithm.

Yet again the implementation on a local memory machine is complicated by the need to assign the computations to processes in such a way as to minimise (or at least reduce) the data communication. Assuming a scattered column distribution for the coefficient matrix in step (2), one way of achieving this is to arrange for that process which is assigned column j of $G^{(k)}$ to evaluate $f\left(\mathbf{x}^{(k)} + h_j \mathbf{e}_j\right)$, $f\left(\mathbf{x}^{(k)} - h_j \mathbf{e}_j\right)$ and $f\left(\mathbf{x}^{(k)} + h_i \mathbf{e}_i + h_j \mathbf{e}_j\right)$, for $i > j$. That process must then broadcast $f\left(\mathbf{x}^{(k)} + h_j \mathbf{e}_j\right)$ to all other processes and $f\left(\mathbf{x}^{(k)} - h_j \mathbf{e}_j\right)$ to the single process whose role is to form the approximation to $\mathbf{g}^{(k)}$. Further, the process which evaluates $f\left(\mathbf{x}^{(k)}\right)$ must broadcast its value to all other processes.

6.4.4   Quasi-Newton methods

The finite difference Newton methods of Section 6.4.3 are one way of implementing a Newton-like method when an analytical form for G is not available. An alternative approach, which has proved very popular in a serial environment, is to use a *quasi-Newton method*. Such a method is again based on the damped Newton method, but G is replaced by an approximate matrix which is modified (*updated*) from iteration to iteration rather than formed using finite differences. Our discussion of quasi-Newton methods assumes that an approximation to the inverse of G is maintained, although we note that many modern serial quasi-Newton algorithms maintain an approximation to (the LDL^T factors of) G (Gill, Murray, *et al.*, 1981).

Let the symmetric matrix $H^{(k)}$ denote an approximation to $G^{(k)^{-1}}$, and assume that $\mathbf{x}^{(1)}$ is an initial approximation to a minimum point $\mathbf{x}^*$. It is usual to choose the initial approximate inverse Hessian matrix $H^{(1)}$ as I_n, the identity matrix. The kth iteration of a quasi-Newton algorithm follows the usual basic steps with a few modifications:

1a. Evaluate $\mathbf{g}$ at the current iterate $\mathbf{x}^{(k)}$.

1b. If $k > 1$ update the approximate inverse Hessian matrix so that

$$H^{(k)} = H^{(k-1)} + \Delta H^{(k-1)}, \tag{6.19}$$

with $\Delta H^{(k-1)}$ a low-rank symmetric matrix.

2. Calculate the search direction $\mathbf{p}^{(k)}$ as

$$\mathbf{p}^{(k)} = -H^{(k)}\mathbf{g}^{(k)}. \qquad (6.20)$$

3. Calculate the line search parameter $\alpha^{(k)}$ and set $\mathbf{x}^{(k+1)} = \mathbf{x}^{(k)} + \alpha^{(k)}\mathbf{p}^{(k)}$.

There are two important differences here. The first is that, except on the first iteration, the approximate inverse Hessian matrix is updated in step (1b) rather than the Hessian matrix itself being explicitly evaluated or estimated using finite differences. The second is that the determination of the search direction using (6.20) involves a matrix-vector multiplication, rather than the solution of a system of linear equations.

BFGS and SR1 updates

Various quasi-Newton algorithms have been proposed which correspond to the use of different updating formulae of the form (6.19). Perhaps the most celebrated of these is the so-called *Broyden, Fletcher, Goldfarb, Shanno (BFGS) formula* (Broyden, 1970a; 1970b; Fletcher, 1970; Goldfarb, 1970; Shanno, 1970) which maintains a symmetric inverse Hessian approximation using a symmetric rank-two update. The formula is defined by

$$\Delta H^{(k)} = -\frac{H^{(k)}\boldsymbol{\gamma}^{(k)}\boldsymbol{\delta}^{(k)T} + \boldsymbol{\delta}^{(k)}\boldsymbol{\gamma}^{(k)T}H^{(k)}}{\boldsymbol{\gamma}^{(k)T}\boldsymbol{\delta}^{(k)}} + \left\{ 1 + \frac{\boldsymbol{\gamma}^{(k)T}H^{(k)}\boldsymbol{\gamma}^{(k)}}{\boldsymbol{\gamma}^{(k)T}\boldsymbol{\delta}^{(k)}} \right\} \frac{\boldsymbol{\delta}^{(k)}\boldsymbol{\delta}^{(k)T}}{\boldsymbol{\gamma}^{(k)T}\boldsymbol{\delta}^{(k)}},$$
$$(6.21)$$

where $\boldsymbol{\delta}^{(k)} = \mathbf{x}^{(k+1)} - \mathbf{x}^{(k)}$ and $\boldsymbol{\gamma}^{(k)} = \mathbf{g}^{(k+1)} - \mathbf{g}^{(k)}$. An alternative updating formula, which makes only a rank-one change to the approximate inverse Hessian matrix, is the *symmetric rank-one (SR1) formula* with update

$$\Delta H^{(k)} = \frac{\left(\boldsymbol{\delta}^{(k)} - H^{(k)}\boldsymbol{\gamma}^{(k)}\right)\left(\boldsymbol{\delta}^{(k)} - H^{(k)}\boldsymbol{\gamma}^{(k)}\right)^{T}}{\left(\boldsymbol{\delta}^{(k)} - H^{(k)}\boldsymbol{\gamma}^{(k)}\right)^{T}\boldsymbol{\gamma}^{(k)}}, \qquad (6.22)$$

or

$$\Delta H^{(k)} = \tau^{(k)}\mathbf{r}^{(k)}\mathbf{r}^{(k)T}, \qquad (6.23)$$

where $\mathbf{r}^{(k)} = \boldsymbol{\delta}^{(k)} - H^{(k)}\boldsymbol{\gamma}^{(k)}$ and $\tau^{(k)} = \left(\mathbf{r}^{(k)T}\boldsymbol{\gamma}^{(k)}\right)^{-1}$. The updating formulae satisfy the so-called *quasi-Newton condition*, namely

$$H^{(k+1)}\boldsymbol{\gamma}^{(k)} = \boldsymbol{\delta}^{(k)},$$

which states that $H^{(k+1)}$ should map a change in gradient vector, $\boldsymbol{\gamma}^{(k)}$, into a corresponding change in position vector, $\boldsymbol{\delta}^{(k)}$. It is justified by noting that, for $\boldsymbol{\delta}^{(k)}$ small,

$$G^{(k)-1}\boldsymbol{\gamma}^{(k)} \approx \boldsymbol{\delta}^{(k)}.$$

Parallelisation of either of these quasi-Newton algorithms essentially requires the parallelisation of the linear algebra primitives. The calculation of $\mathbf{p}^{(k)}$ in step (2) involves a Level 2 BLAS **ssymv** operation. The two updating formulae of step (1b) given here also involve Level 2 BLAS operations; **ssyr2** in the case of BFGS, **ssyr** in the case of SR1. Parallel forms of these operations have been considered in Section 3.3.2. On a local memory system we can assume that (the lower or upper triangle of) $H^{(k)}$ has already been distributed. A column distribution of $H^{(k)}$ suggests the use of the **saxpy** variant of **ssymv** and this requires an initial broadcast of $\mathbf{g}^{(k)}$ and a final gather of partial results. A row distribution still requires the initial broadcast of $\mathbf{g}^{(k)}$ and the gathering of $\mathbf{p}^{(k)}$ by the process which is to perform the line search. The low-rank matrix updates, **ssyr** and **ssyr2**, can be performed equally well with either distribution of $H^{(k)}$. For example, the SR1 update requires the broadcast of $\mathbf{r}^{(k-1)}$ and $\tau^{(k-1)}$ and then the modified rows or columns of $H^{(k)}$ can be formed without further communication. Note also that the calculation of $\mathbf{r}^{(k-1)}$ itself involves an **ssymv** operation, whilst the determination of $\tau^{(k-1)}$ requires an **sdot** operation. Given the amount of communication that is involved, only modest speed-ups can be expected, particularly if the computation is dominated by function evaluations.

As presented here, a quasi-Newton method offers no opportunity for concurrent function or gradient vector evaluations unless we expressly split the evaluation of $\mathbf{g}^{(k)}$. The situation is of course rather different if an analytical form for $\mathbf{g}$ is not available and $\mathbf{g}^{(k)}$ is approximated using a forward difference formula. Now, step (1a) becomes

1a. Evaluate $f\left(\mathbf{x}^{(k)} + h_i \mathbf{e}_i\right)$, $i = 1, 2, \ldots, n$, and form the forward difference approximations

$$g_i\left(\mathbf{x}^{(k)}\right) \approx \frac{f\left(\mathbf{x}^{(k)} + h_i \mathbf{e}_i\right) - f\left(\mathbf{x}^{(k)}\right)}{h_i}, \qquad i = 1, 2, \ldots, n.$$

These function evaluations can be performed concurrently, as can the calculation of the difference approximations to the components of $\mathbf{g}^{(k)}$. Further, the function evaluations of this step can be performed concurrently with the line search of step (3) if the speculative evaluations idea of Section 6.4.3 is adopted.

Straeter's algorithm

An alternative attempt at introducing parallelism into quasi-Newton algorithms is that suggested by Straeter (1973). The BFGS and SR1 quasi-Newton methods update an approximate inverse Hessian matrix by accumulating symmetric rank-two or rank-one updates which are based on a change in position vector $\boldsymbol{\delta}^{(k)}$ and a corresponding change in gradient vector $\boldsymbol{\gamma}^{(k)}$ from the previous iteration. The motivation for the algorithm described here is to use concurrent gradient vector evaluations to enable extra gradient information to be obtained. This additional information can be incorporated into the approximate inverse Hessian matrix to give an improved approximation, with the expectation that the resulting algorithm will converge in fewer iterations.

Let $\mathbf{x}^{(1)}$ be an initial approximation to a minimum point $\mathbf{x}^*$, the symmetric matrix $H^{(0)}$ be an initial approximation to the inverse Hessian matrix (with the usual initial choice being $H^{(0)} = I_n$), and let $\boldsymbol{\delta}_1, \boldsymbol{\delta}_2, \ldots, \boldsymbol{\delta}_n$, be a set of n linearly independent directions. Note that, here, the initial approximate inverse Hessian matrix is $H^{(0)}$ rather than $H^{(1)}$, since the first iteration of the algorithm is to use n SR1 updates to form $H^{(1)}$, an approximation to the inverse Hessian matrix at $\mathbf{x}^{(1)}$. Then the kth iteration of Straeter's parallel quasi-Newton algorithm takes the usual form with the following first step:

1a. Calculate $\mathbf{g}^{(k)}$ and $\mathbf{g}\left(\mathbf{x}^{(k)} + \boldsymbol{\delta}_j\right)$, $j = 1, 2, \ldots, n$.

1b. Calculate the gradient differences

$$\boldsymbol{\gamma}^{(k,j)} = \mathbf{g}\left(\mathbf{x}^{(k)} + \boldsymbol{\delta}_j\right) - \mathbf{g}^{(k)}, \qquad j = 1, 2, \ldots, n.$$

1c. With $V^{(k,0)} = H^{(k-1)}$, calculate, for $j = 1, 2, \ldots, n$,

$$\mathbf{r}^{(k,j)} = \boldsymbol{\delta}_j - V^{(k,j-1)} \boldsymbol{\gamma}^{(k,j)},$$

$$\tau^{(k,j)} = \left(\mathbf{r}^{(k,j)^T} \boldsymbol{\gamma}^{(k,j)}\right)^{-1},$$

$$V^{(k,j)} = V^{(k,j-1)} + \tau^{(k,j)} \mathbf{r}^{(k,j)} \mathbf{r}^{(k,j)^T}. \qquad (6.24)$$

1d. Set $H^{(k)} = V^{(k,n)}$.

It can be shown that the algorithm has quadratic termination; for any positive definite quadratic function $f(\mathbf{x}) = \frac{1}{2}\mathbf{x}^T A \mathbf{x} + \mathbf{b}^T \mathbf{x} + c$, and any starting value $\mathbf{x}^{(1)}$, $H^{(1)} = A^{-1}$ and the minimum of f is located on the first iteration. This result depends on using the SR1 updating formula in (6.24), and exploits the result that such a formula generates the true inverse Hessian matrix of a quadratic function after inexact line searches along n linearly independent directions. The result would not hold if the BFGS update were used.

The potential for parallelism in steps (1a) and (1b) is clear. Step (1a) requires $n + 1$ gradient evaluations which can be performed concurrently, and the use of finite difference approximations would increase the potential still further. Step (1b) represents n independent processes which can begin as soon as the required gradient vectors are available. The SR1 updates (6.24) must be performed sequentially, since $V^{(k,j)}$ is dependent on $V^{(k,j-1)}$. However, as before, the individual updates involve the Level 2 BLAS operations **ssymv** and **ssyr**, and the Level 1 BLAS **sdot**.

It is interesting to compare this parallel quasi-Newton algorithm with the first of the finite difference Newton algorithms of Section 6.4.3. Both schemes have quadratic termination and require $n + 1$ evaluations of the gradient vector at each iteration. The parallel quasi-Newton algorithm incorporates the gradient information into an approximate inverse Hessian matrix using n SR1 updates, $O\left(n^3\right)$ operations, and then uses a matrix-vector product, $O\left(n^2\right)$ operations, to calculate the search direction. In contrast, the finite difference Newton algorithm uses the gradient information to form a finite difference approximation to the Hessian matrix, $O\left(n^2\right)$ operations, which is then factorised to solve for the search direction, $O\left(n^3\right)$ operations. Thus the computational costs of the algorithms are similar; however, they generate different search directions and will therefore require different numbers of iterations to achieve convergence. The question as to which will converge the faster is likely to be problem-dependent.

Exercises

6.1. An approximation to

$$\int_1^2 \sum_{i=1}^q \left((x-\lambda_i)^2 + 1 \right)^{-1} dx,$$

is required for some q, where each $\lambda_i \in [1,2]$. The interval $[1,2]$ is to be subdivided into p subintervals, and p processes compute an integral approximation over each subinterval concurrently. What factors influence the way this subdivision should be formed?

6.2. Suppose that we are interested in determining $I = \int_0^1 f(x)dx$. A *Monte Carlo method* for estimating I consists of determining a set of points x_i, $i = 1, 2, \ldots, n$, randomly distributed in the range $[0, 1]$, and forming the sum of function values $R = n^{-1} \sum_{i=1}^n f(x_i)$. Then, as $n \to \infty$, $R \to I$ (Davis and Rabinowitz, 1984). (The practical use of Monte Carlo methods in quadrature is usually restricted to multi-dimensional integration.) What are the main obstacles to the parallelisation of this algorithm? Pseudo random integers can be generated by the *linear congruential algorithm* as

$$X_{n+1} = (aX_n + c) \bmod m, \qquad n \geq 1,$$

for some integers a, c, m and X_1; the X_n can then be mapped on to the interval $[0, 1]$ by dividing by $m - 1$. Verify that, for $k \geq 1$,

$$X_{n+k} = (AX_n + C) \bmod m, \qquad n \geq 1,$$

for suitable A and C. What use can be made of this result in a parallel environment? (Fox, Johnson, *et al.*, 1988.)

6.3. A weighted double summation of the form

$$S = \sum_{i=1}^m \sum_{j=1}^n w_{ij} f_{ij},$$

is required. In what ways can S be determined in parallel? Given that f_{ij} represents an evaluation of the function $f(x, y)$ at the point (x_i, y_j), indicate how an approximation to $\int_a^b \int_c^d f(x, y)\,dx\,dy$ may be determined by employing numerical quadrature techniques in both the x and y directions. How might the form of f influence your initial conclusions?

6.4. A *linear Fredholm integral equation of the second kind* takes the form

$$f(s) = g(s) + \int_a^b K(s, t)f(t)\,dt, \qquad a \le s \le b,$$

where g and K are known and f is to be determined. One way of solving the problem is to introduce a set of points s_i, $i = 1, 2, \ldots, n$, with $s_i = a + (i - 1)h$, $h = (b - a)/(n - 1)$, and solve for $f_i \approx f(s_i)$ by approximating the integral using some appropriate quadrature rule. Derive the system of equations which defines f_i, $i = 1, 2, \ldots, n$, when the trapezium rule is employed and explore the scope for parallelism in both the set-up and solution phases. How would you proceed if the integrand were of the form $K(s, t, f(t))$, that is, nonlinear in $f(t)$?

6.5. For the nth-degree polynomial $P_n(x) = x^n + a_1 x^{n-1} + \cdots + a_{n-1}x + a_n$, derive the system of n nonlinear equations

$$\alpha_1 + \alpha_2 + \cdots + \alpha_n = -a_1,$$

$$\alpha_1 \alpha_2 + \alpha_1 \alpha_3 + \cdots + \alpha_1 \alpha_n + \alpha_2 \alpha_3 + \cdots + \alpha_{n-1} \alpha_n = a_2,$$

$$\vdots$$

$$\alpha_1 \alpha_2 \cdots \alpha_n = (-1)^n a_n,$$

(the *Newton identities* or *Vieta's formulae*) which relates the unknown polynomial zeros, α_i, $i = 1, 2, \ldots, n$, to the known polynomial coefficients. Show that the Durand–Kerner algorithm is equivalent to solving this nonlinear system by Newton's method. (Freeman, 1979.)

6.6. Consider the system of nonlinear equations $\mathbf{f}(\mathbf{x}) = \mathbf{0}$, where

$$f_i(\mathbf{x}) = x_i + \sum_{j=1}^{n} x_j - (n + 1), \qquad i = 1, 2, \ldots, n - 1,$$

$$f_n(\mathbf{x}) = \left(\prod_{j=1}^{n} x_j \right) - 1.$$

Show how the storage pattern (by rows or by columns) of the Jacobian matrix affects the possibilities of processes exploiting common subexpressions in a multiprocessor implementation of Newton's method. Repeat the exercise for the nonlinear system of Exercise 6.5.

6.7.   Suppose that in the multi-directional search algorithm we choose $\mu = 2$ and $\theta = \frac{1}{2}$ and speculatively evaluate the expanded and contracted simplices of the current iteration, and all possible reflected simplices of the next iteration. Sketch this situation in $\mathcal{R}^2$.

6.8.   Consider the finite difference Newton method for minimising f in the case when an analytical expression for f only is available. Derive a second-order central difference approximation to the off-diagonal elements of $G^{(k)}$ and determine the number of function evaluations required by step (1) of each iteration of the corresponding finite difference Newton method. How would you implement the algorithm on a local memory multiprocessor?

Further reading

The possibilities for the vectorisation and parallelisation of one-dimensional quadrature codes are considered by Gladwell (1987). The implementation and the numerical performance of an adaptive quadrature code on a transputer-based local memory machine are the subjects of Burrage (1990) – in this case the very simple approach of dividing the range of integration into p (the number of processes/processors) subintervals is recommended.

A number of authors have considered the implementation and parallel performance of the Durand–Kerner algorithm on local memory machines; see Cosnard and Fraigniaud (1990) and Freeman (1989). A divide and conquer method for the calculation of polynomial zeros is given in Freeman and Brankin (1990).

The papers of Lootsma (1989), Lootsma and Ragsdell (1988) and Schnabel (1988) provide comprehensive surveys of algorithms for parallel nonlinear optimisation. An annotated bibliography on parallel numerical optimisation, covering a much wider field than considered in this chapter, is given by Zenios (1989). For different approaches to parallel quasi-Newton algorithms, see Byrd, Schnabel, *et al.* (1988a; 1988b), Freeman (1991), Schnabel (1987) and van Laarhoven (1985).

A review of concurrent stochastic methods for global optimisation is given by Byrd, Dert, *et al.* (1990) (see also Byrd, Eskow, *et al.*, 1991).

CHAPTER 7

Other Areas II

7.1 Yet more areas

The further classes of numerical problem covered in this chapter are those for determining Fourier transforms of complex-valued data (Section 7.2), and those which compute numerical approximations to the solutions of differential equations (Sections 7.3 and 7.4). There is no explicit correspondence between these two areas although, as is shown in Section 7.4.4, it is possible to solve a differential equation using Fourier transform techniques. We choose to differentiate between two subclasses of differential equation, namely those which involve only a single independent variable (ordinary differential equations), and those which involve several independent variables (partial differential equations). Not surprisingly, numerical methods for the two types of problem often have much in common. We demonstrate the increased potential for parallelism afforded by the move from a single ordinary differential equation to a system, and from a differential equation in one dimension to one in several dimensions.

7.2 Fourier transforms

7.2.1 The fast Fourier transform

Fourier transforms have a long pedigree and are widely used in problem areas such as digital signal processing, image processing and the solution of differential equations. It used to be thought that the operation count of the transform was too high for it to be of serious practical use, but the so-called *fast Fourier transform* (FFT) of Cooley and Tukey (1965) dramatically changed the situation. For an interesting discussion of the history of the FFT see Cooley (1990), where the method is traced back to Gauss.

Given a set of complex function values, f_j, $j = 0, 1, \ldots, N - 1$, the (discrete) Fourier transform (DFT) generates a new set of complex numbers, g_k, $k = 0, 1, \ldots, N - 1$, according to the equation

$$g_k = \sum_{j=0}^{N-1} e^{-2\pi i j k/N} f_j, \tag{7.1}$$

where $i = \sqrt{-1}$. We refer to this forward transformation as *Fourier analysis*. Using the identity

$$\frac{1}{N} \sum_{k=0}^{N-1} e^{2\pi i (j-l)k/N} = \delta_{jl},$$

it follows that the inverse of the analysis (7.1) takes the form

$$f_j = \frac{1}{N} \sum_{k=0}^{N-1} e^{2\pi i j k/N} g_k.$$

We refer to this inverse transformation as *Fourier synthesis*. Since the forward transformation and its inverse are so closely related it is sufficient to consider one of them only. In the following we concentrate on the forward transformation and write (7.1) as

$$g_k = \sum_{j=0}^{N-1} w_N^{jk} f_j, \tag{7.2}$$

where $w_N = e^{-2\pi i/N}$. The inclusion of the minus sign in the exponential for the forward transformation is arbitrary; we could equally have included it in the inverse transformation instead. A similar argument applies to the multiplying factor $1/N$ in the inverse transformation.

A straightforward implementation of the Fourier analysis (7.2) to generate all of the transformed coefficients, g_k, $k = 0, 1, \ldots, N - 1$, requires $O(N^2)$ operations; there are N terms to be computed and each involves a summation of length N. However, by making use of partial results, Cooley and Tukey (1965), amongst others, suggested an algorithm by which all of the transformed coefficients can be generated in $O(N \log_2 N)$ operations. For N large this represents a significant saving. Assuming N is even, (7.2) can be split into its even and odd parts to give, for $k = 0, 1, \ldots, N/2 - 1$,

$$\begin{aligned}
g_k &= \sum_{j=0}^{N/2-1} w_N^{2jk} f_{2j} + \sum_{j=0}^{N/2-1} w_N^{(2j+1)k} f_{2j+1} \\
&= \sum_{j=0}^{N/2-1} w_{N/2}^{jk} f_{2j} + w_N^k \sum_{j=0}^{N/2-1} w_{N/2}^{jk} f_{2j+1}, \\
&= g_k^{even} + w_N^k g_k^{odd}, \tag{7.3}
\end{aligned}$$

and

$$g_{N/2+k} = \sum_{j=0}^{N/2-1} w_{N/2}^{jk} f_{2j} - w_N^k \sum_{j=0}^{N/2-1} w_{N/2}^{jk} f_{2j+1},$$

$$= g_k^{even} - w_N^k g_k^{odd}, \tag{7.4}$$

where we have used the identities $w_{N/2}^{jN/2} = 1$ and $w_N^{N/2} = -1$. Thus, in order to calculate both g_k and $g_{N/2+k}$ it is necessary to calculate only the two half-length Fourier transforms g_k^{even} and g_k^{odd} at about half the computational cost. We can view this as the *divide step* in a divide and conquer algorithm. Now, if N is a power of 2 the process can be recursively applied to g_k^{even} and g_k^{odd} (the problem can be further divided) until we arrive at summations of length 1, the *conquer step*. This, the FFT, algorithm consists of combining pairs of partial results, starting at the function values. The depth of the process (the number of times the original transform (7.2) is divided) is clearly $\log_2 N$, and at each level there are N values to compute. Hence the FFT algorithm requires a total of $O(N \log_2 N)$ operations.

To describe the FFT process fully the algebra gets a little messy, but the divide step of (7.3) and (7.4) is the only real trick. For simplicity we assume $N = 2^n$ and note that the process can readily be adapted to the case where N is not a multiple of 2 (Section 7.2.2). We define a generalisation of g_k as

$$g_k(i,l) = \sum_{i=0}^{2^l-1} w_{2^l}^{jk} f_{i+jN2^{-l}},$$

and observe that, for $l \geq 1$,

$$g_{k+2^l}(i,l) = \sum_{j=0}^{2^l-1} w_{2^l}^{j(k+2^l)} f_{i+jN2^{-l}},$$

$$= \sum_{j=0}^{2^l-1} w_{2^l}^{j2^l} w_{2^l}^{jk} f_{i+jN2^{-l}},$$

$$= g_k(i,l),$$

since $w_{2^l}^{j2^l} = 1$, that is, $g_k(i,l)$ is periodic in k with period 2^l. With a little manipulation it follows that, for $k = 0, 1, \ldots, 2^l-1$, and $i = 0, 1, \ldots, N2^{-(l+1)}-1$,

$$g_k(i,l+1) = g_k(i,l) + w_{2^{l+1}}^k g_k\left(i + N2^{-(l+1)}, l\right), \tag{7.5}$$

$$g_{k+2^l}(i,l+1) = g_k(i,l) - w_{2^{l+1}}^k g_k\left(i + N2^{-(l+1)}, l\right). \tag{7.6}$$

We note that (7.5) and (7.6) are simply the generalisation of the divide step defined by (7.3) and (7.4). We also note that, for each value of l, (7.5) and

(7.6) define N values $g_*(*, l+1)$. This simple recurrence for $l = 0, 1, \ldots, n-1$, can be used to transform the original coefficients $g_0(i, 0) = f_i$, $i = 0, 1, \ldots, N-1$, into the transformed coefficients $g_k(0, n) = g_k$, $k = 0, 1, \ldots, N-1$, where the multiplying coefficients $w_{2^l+1}^k$ are evaluated before the recurrence begins. It is normal to perform the FFT in situ, with the partial results at each level (for a given value of l) overwriting those for the previous level (the previous value of l) in a one-dimensional array. If the two new partial results, $g_k(i, l+1)$ and $g_{k+2^l}(i, l+1)$, which appear on the left-hand side of (7.5) and (7.6) overwrite the old values, $g_k(i, l)$ and $g_k(i + N2^{-(l+1)}, l)$, which appear on the right then the final values are in the wrong order. The odd-even nature of the algorithm indicates that the correct order may be achieved using *bit-reversal*; that is, the value g_k is found in position k', where k', represented as a binary number, has a bit pattern which is the reverse of that for k. An alternative strategy, which gives the results in the correct order, is to use bit-reversal to rearrange the original data, f_i, $i = 0, 1, \ldots, N-1$, before it is transformed.

To illustrate the way the FFT works we consider the case $N = 4$, so that $n = 2$. We start with the initial data

$$f_0 = g_0(0, 0), \qquad f_1 = g_0(1, 0), \qquad f_2 = g_0(2, 0), \qquad f_3 = g_0(3, 0).$$

Setting $l = 0$, (7.5) and (7.6) hold for $k = 0$ and $i = 0, 1$. Using (7.5) first, and setting $i = 0$ and $i = 1$ in turn, we obtain

$$g_0(0, 1) = g_0(0, 0) + w_2^0 g_0(2, 0),$$
$$g_0(1, 1) = g_0(1, 0) + w_2^0 g_0(3, 0).$$

Similarly, (7.6) gives

$$g_1(0, 1) = g_0(0, 0) - w_2^0 g_0(2, 0),$$
$$g_1(1, 1) = g_0(1, 0) - w_2^0 g_0(3, 0),$$

and this completes the first transformation stage. We now set $l = 1$, so that $k = 0, 1$, and $i = 0$. Setting $k = 0$ and $k = 1$ in turn in (7.5) gives

$$g_0(0, 2) = g_0(0, 1) + w_4^0 g_0(1, 1),$$
$$g_1(0, 2) = g_1(0, 1) + w_4^1 g_1(1, 1),$$

whilst from (7.6) we have

$$g_2(0, 2) = g_0(0, 1) - w_4^0 g_0(1, 1),$$
$$g_3(0, 2) = g_1(0, 1) - w_4^1 g_1(1, 1).$$

This completes the second, and final, transformation stage and we now have

$$g_0 = g_0(0, 2), \qquad g_1 = g_1(0, 2), \qquad g_2 = g_2(0, 2), \qquad g_3 = g_3(0, 2).$$

$$f_0 = g_0(0,0) \qquad g_0(0,1) \qquad g_0(0,2) = g_0$$
$$f_1 = g_0(1,0) \qquad g_0(1,1) \qquad g_1(0,2) = g_1$$
$$f_2 = g_0(2,0) \qquad g_1(0,1) \qquad g_2(0,2) = g_2$$
$$f_3 = g_0(3,0) \qquad g_1(1,1) \qquad g_3(0,2) = g_3$$

Figure 7.1 *FFT for $N = 4$*

$$f_0 = g_0(0,0) \qquad g_0(0,1) \qquad g_0(0,2) = g_0$$
$$f_1 = g_0(1,0) \qquad g_0(1,1) \qquad g_2(0,2) = g_2$$
$$f_2 = g_0(2,0) \qquad g_1(0,1) \qquad g_1(0,2) = g_1$$
$$f_3 = g_0(3,0) \qquad g_1(1,1) \qquad g_3(0,2) = g_3$$

Figure 7.2 *FFT for $N = 4$ showing overwrites*

The stages outlined here are illustrated in Figure 7.1 which shows the inter-dependence of the partial results. It is interesting to compare Figure 7.1 with Figure 1.5 which illustrates a 4×4 crossbar switch implemented in terms of 2×2 crossbars. What is not immediately clear from Figure 7.1 is that the transformed data end up in the wrong places. To demonstrate this Figure 7.2 shows the transformation stages again, but now each row corresponds to a vector element. Reading from left to right the value computed at each transformation stage overwrites its predecessor. The first and last elements are correctly placed but the middle two are in the wrong order. The binary representation of 1 (using two figures) is 01. Reversing this bit pattern gives 10, the binary representation for 2. Hence, the two middle results need to be interchanged.

7.2.2 Parallel implementation

To illustrate the potential for parallelism we again consider the case $N = 4$. It is clear from Figure 7.2, where we overwrite values at each stage, that the first trans-formation stage decouples; $g_0(0,1)$ and $g_1(0,1)$ can be computed independently of $g_0(1,1)$ and $g_1(1,1)$. We can similarly decouple the second stage and compute $g_0(0,2)$ and $g_2(0,2)$ independently of $g_1(0,2)$ and $g_3(0,2)$. If we introduce the matrices

$$X = \begin{pmatrix} f_0 & f_2 \\ f_1 & f_3 \end{pmatrix} = \begin{pmatrix} g_0(0,0) & g_0(2,0) \\ g_0(1,0) & g_0(3,0) \end{pmatrix},$$

and

$$Y = \begin{pmatrix} g_0 & g_2 \\ g_1 & g_3 \end{pmatrix} = \begin{pmatrix} g_0(0,2) & g_2(0,2) \\ g_1(0,2) & g_3(0,2) \end{pmatrix},$$

then the first stage transforms X to $Y^{(1)}$, where

$$Y^{(1)} = \begin{pmatrix} g_0(0,1) & g_1(0,1) \\ g_0(1,1) & g_1(1,1) \end{pmatrix},$$

and this can be achieved by applying independent transforms to the rows of X. The second stage then transforms $Y^{(1)}$ to

$$Y^{(2)} = \begin{pmatrix} g_0(0,2) & g_1(0,2) \\ g_2(0,2) & g_3(0,2) \end{pmatrix},$$

and this can be achieved by applying independent transforms to the columns of $Y^{(1)}$. Finally, it is clear that $Y^{(2)} = Y^T$.

Generalising a little further, assume that N may be factored as $N = N_0 N_1$. We introduce the $N_0 \times N_1$ matrix X whose (l, m) entry is given as $x_{lm} = f_j$ with $j = l + mN_0$ for $l = 0, 1, \ldots, N_0 - 1$, and $m = 0, 1, \ldots, N_1 - 1$. Similarly, we introduce the $N_1 \times N_0$ matrix Y with (p, q) element $y_{pq} = g_k$, where $k = p + qN_1$ for $p = 0, 1, \ldots, N_1 - 1$, and $q = 0, 1, \ldots, N_0 - 1$. Then, from (7.2),

$$y_{pq} = \sum_{l=0}^{N_0-1} \sum_{m=0}^{N_1-1} w_N^{(l+mN_0)(p+qN_1)} x_{lm},$$

$$= \sum_{l=0}^{N_0-1} w_{N_0}^{lq} w_N^{lp} \sum_{m=0}^{N_1-1} w_{N_1}^{mp} w_1^{mq} x_{lm},$$

$$= \sum_{l=0}^{N_0-1} w_{N_0}^{lq} w_N^{lp} \sum_{m=0}^{N_1-1} w_{N_1}^{mp} x_{lm}.$$

This suggests that we compute the $N_0 \times N_1$ matrix $Y^{(1)}$ with (l, p) entry

$$y_{lp}^{(1)} = w_N^{lp} \sum_{m=0}^{N_1-1} w_{N_1}^{mp} x_{lm}, \tag{7.7}$$

and then the $N_0 \times N_1$ matrix $Y^{(2)}$ using

$$y_{qp}^{(2)} = \sum_{l=0}^{N_0-1} w_{N_0}^{lq} y_{lp}^{(1)}. \tag{7.8}$$

It follows that $y_{pq} = y_{qp}^{(2)}$, so that the results from (7.8) have to be transposed to obtain the transformed coefficients in the correct order.

Considering (7.7) in more detail, it can be seen that the computation of each row of $Y^{(1)}$ involves nothing more than a Fourier transform of the corresponding row of X; that is, we have N_0 transforms of length N_1. Similarly, the computation of each column of $Y^{(2)}$ requires a Fourier transform of the corresponding column of $Y^{(1)}$, giving N_1 transforms of length N_0. If N_1 is a small prime factor of N then we compute (7.7) directly and quickly as N_0 short transforms of length N_1, but factor N_0 further when computing the transforms (7.8).

Generalizing the procedure still further, we assume that N may be factored as the product of r factors, $N = N_0 N_1 \cdots N_{r-1}$, and define the index functions

$$J(l_0, l_1, \ldots, l_{t-1}) = l_0 + l_1 N_0 + \cdots + l_{t-1} N_0 N_1 \cdots N_{t-2},$$
$$K(p_{t-1}, p_{t-2}, \ldots, p_0) = p_{t-1} + p_{t-2} N_{t-1} + \cdots + p_0 N_{t-1} N_{t-2} \cdots N_1,$$

for $t = 0, 1, \ldots, r$. We introduce the $N_0 \times N_1 \times \cdots \times N_{r-1}$, r-dimensional array X with elements

$$x_{l_0, l_1, \ldots, l_{r-1}} = f_{J(l_0, l_1, \ldots, l_{r-1})},$$

where each $l_i = 0, 1, \ldots, N_i - 1$, and the $N_{r-1} \times N_{r-2} \times \cdots \times N_0$, r-dimensional array Y with elements

$$y_{p_{r-1}, p_{r-2}, \ldots, p_0} = g_{K(p_{r-1}, p_{r-2}, \ldots, p_0)},$$

where each $p_i = 0, 1, \ldots, N_i - 1$. Then (7.7) and (7.8) generalise to

$$y_{l_0, l_1, \ldots, l_{r-s-1}, p_{r-s}, \ldots, p_{r-1}}^{(s)}$$
$$= w_{N_0 N_1 \cdots N_{r-s}}^{J(l_0, l_1, \ldots, l_{r-s-1}) p_{r-s}} \sum_{l_{r-s}=0}^{N_{r-s}-1} w_{N_{r-s}}^{l_{r-s} p_{r-s}} y_{l_0, l_1, \ldots, l_{r-s}, p_{r-s+1}, \ldots, p_{r-1}}^{(s-1)},$$

for $s = 1, 2, \ldots, r$, with $Y^{(0)} = X$. Note that $Y^{(r)}$ is the same as Y but with the dimensions reversed.

Clearly, on a multiprocessor we can, for a fixed s, compute the individual transforms separately, and for shared memory systems this would appear to be the most sensible route to parallelism. Significant speed-ups are likely to be achievable for large enough N. At the outset a decision has to be made as to how N should be factored. We need to maintain a reasonable grain size by making sure that the individual transforms are of an appropriate length and yet, at the same time, subdivide the problem sufficiently so that the resources available may be fully exploited. On a local memory system the main problem will be the amount of data communication (effectively, a matrix transpose at each stage) that is required.

7.2.3   Transforms of real data

To determine the DFT of real data we could simply perform an FFT on complex data for which the imaginary part is zero, but this is rather wasteful. Instead, suppose that

$$f_j = f_j^{(1)} + i f_j^{(2)}, \qquad j = 0, 1, \ldots, N-1,$$

where the $f_j^{(l)}$, $l = 1, 2$, are real, and let g_k, $k = 0, 1, \ldots, N-1$, be the (complex) transforms of f_j, $j = 0, 1, \ldots, N-1$. From (7.2) and the identity $e^{i\theta} = \cos\theta + i \sin\theta$, for any θ, we have

$$g_k = \sum_{j=0}^{N-1} w_N^{jk} \left(f_j^{(1)} + i f_j^{(2)} \right),$$

$$= \sum_{j=0}^{N-1} (\cos(\theta_j) + i \sin(\theta_j)) \left(f_j^{(1)} + i f_j^{(2)} \right),$$

where $\theta_j = -2\pi jk/N$. Similarly,

$$g_{N-k} = \sum_{j=0}^{N-1} w_N^{j(N-k)} \left(f_j^{(1)} + i f_j^{(2)} \right),$$

$$= w_N^{jN} \sum_{j=0}^{N-1} w_N^{-jk} \left(f_j^{(1)} + i f_j^{(2)} \right),$$

$$= \sum_{j=0}^{N-1} (\cos(\theta_j) - i \sin(\theta_j)) \left(f_j^{(1)} + i f_j^{(2)} \right),$$

where we have used the identities $\cos(-\theta) = \cos\theta$ and $\sin(-\theta) = -\sin\theta$. It follows that, if $\operatorname{Re}(g_k)$ is the real part of g_k and $\operatorname{Im}(g_k)$ is the imaginary part, then

$$\operatorname{Re}(g_k) = \sum_{j=0}^{N-1} \left(\cos(\theta_j) f_j^{(1)} - \sin(\theta_j) f_j^{(2)} \right),$$

$$\operatorname{Re}(g_{N-k}) = \sum_{j=0}^{N-1} \left(\cos(\theta_j) f_j^{(1)} + \sin(\theta_j) f_j^{(2)} \right),$$

$$\operatorname{Im}(g_k) = \sum_{j=0}^{N-1} \left(\sin(\theta_j) f_j^{(1)} + \cos(\theta_j) f_j^{(2)} \right),$$

$$\operatorname{Im}(g_{N-k}) = \sum_{j=0}^{N-1} \left(-\sin(\theta_j) f_j^{(1)} + \cos(\theta_j) f_j^{(2)} \right).$$

Hence

$$\frac{1}{2}(\operatorname{Re}(g_k) + \operatorname{Re}(g_{N-k})) + \frac{i}{2}(\operatorname{Im}(g_k) - \operatorname{Im}(g_{N-k}))$$

$$= \sum_{j=0}^{N-1} (\cos(\theta_j) + i\sin(\theta_j)) f_j^{(1)} = g_k^{(1)}. \tag{7.9}$$

Similarly,

$$\frac{1}{2}(\operatorname{Im}(g_k) + \operatorname{Im}(g_{N-k})) - \frac{i}{2}(\operatorname{Re}(g_k) - \operatorname{Re}(g_{N-k}))$$

$$= \sum_{j=0}^{N-1} (\cos(\theta_j) + i\sin(\theta_j)) f_j^{(2)} = g_k^{(2)}. \tag{7.10}$$

Thus the transforms $g_k^{(1)}, g_k^{(2)}$, $k = 0, 1, \ldots, N-1$, of the two sets of real data $f_j^{(1)}, f_j^{(2)}$, $j = 0, 1, \ldots, N-1$, may be determined by packing them into the single complex set f_j, $j = 0, 1, \ldots, N-1$, forming the transforms g_k, $k = 0, 1, \ldots, N-1$, and then using (7.9) and (7.10) to unravel the results.

If we only have one set of real values, say, f_j, $j = 0, 1, \ldots, 2N-1$, to transform we proceed as follows. The (complex) transforms g_k, $k = 0, 1, \ldots, 2N-1$, are given by

$$g_k = \sum_{j=0}^{2N-1} w_{2N}^{jk} f_j,$$

$$= \sum_{j=0}^{N-1} w_{2N}^{2jk} f_{2j} + \sum_{j=0}^{N-1} w_{2N}^{(2j+1)k} f_{2j+1},$$

$$= \sum_{j=0}^{N-1} w_N^{jk} f_{2j} + w_{2N}^{k} \sum_{j=0}^{N-1} w_N^{jk} f_{2j+1},$$

which we may write as

$$g_k = g_k^{(1)} + w_{2N}^{k} g_k^{(2)}, \tag{7.11}$$

where the $g_k^{(1)}$ are the (complex) transforms of the (real) values f_j with even subscripts and the $g_k^{(2)}$ are the (complex) transforms of the (real) values f_j with odd subscripts. The transforms of these two sets can be calculated using the techniques outlined in the previous paragraph and the final results derived from (7.11).

7.2.4 Transforms in two dimensions

Suppose that we have a two-dimensional set of complex function values f_{jk}, $j = 0, 1, \ldots, N_x - 1$, $k = 0, 1, \ldots, N_y - 1$. Then the Fourier transforms g_{lm} of this data are defined by

$$g_{lm} = \sum_{j=0}^{N_x-1} \sum_{k=0}^{N_y-1} e^{-2\pi ijl/N_x} e^{-2\pi ikm/N_y} f_{jk},$$

for $l = 0, 1, \ldots, N_x - 1$, and $m = 0, 1, \ldots, N_y - 1$. It is a simple matter to proceed as for the one-dimensional case but with obvious coarser granularity. If F is the matrix of function values and G is the matrix of corresponding transform values then it is clear that we can

- generate an intermediate matrix $\tilde{G}$ formed by applying one-dimensional transforms to the rows of F, and
- generate G by applying one-dimensional transforms to the columns of $\tilde{G}$.

The same result can be achieved by applying the one-dimensional transforms first to the columns of F and then to the rows of the resulting intermediate matrix. In either case the transforms can be performed in situ. Since the individual transforms are independent they may be performed concurrently within each stage. In the case of a local memory multiprocessor, if we assume that F is initially distributed by block columns, then the transforms in the y-direction can be computed without communication, but the transforms in the x-direction require substantial communication of data.

7.3   Ordinary differential equations

Many physical situations can be modelled by a (system of) *ordinary differential equations* (ODEs). As a simple example we cite the celebrated *Lotka–Volterra,* or *predator–prey, equations.* Let $f(t)$ represent the population of foxes at time t and $r(t)$ the population of rabbits. If the population of foxes increases over time then the rabbits, on which the foxes feed, will decrease in number. However, if there is an insufficient number of rabbits to go round, the weakest foxes will die. Clearly the two populations are interrelated. We denote by $f' = df/dt$ the rate of change in the fox population with time, and by $r' = dr/dt$ the rate of change in the rabbit population. Then, starting with $f(a) = f_a$ and $r(a) = r_a$, some initial populations at time $t = a$, the Lotka–Volterra model states that the distributions satisfy the system of first-order ODEs

$$f' = \alpha f + \beta fr,$$
$$r' = \gamma r + \delta fr,$$

for some constants α, β, γ and δ. The term 'first order' refers to the fact that the equations involve first-order derivatives only; the term 'ordinary' (as opposed to 'partial') indicates that f and r are functions of the single independent variable, t. The above equations are said to form a *coupled system* because of the interrelationship between $f(t)$ and $r(t)$.

In general, we express a system of first-order ODEs as

$$\mathbf{y}'(t) = \mathbf{f}(t, \mathbf{y}(t)), \tag{7.12}$$

where

$$\mathbf{y}(t) = (y_1(t), y_2(t), \ldots, y_m(t))^T,$$
$$\mathbf{f}(t, \mathbf{y}(t)) = (f_1(t, \mathbf{y}(t)), f_2(t, \mathbf{y}(t)), \ldots, f_m(t, \mathbf{y}(t)))^T,$$

and m is the order of the system. Note that the term 'order' is used in this section to refer to a number of different things. This is an unfortunate, but accepted, convention. Here, the two uses of the term are equivalent, in that a single ODE involving derivatives of order m can be expressed as a coupled system of m equations involving first-order derivatives only.

A necessary condition for the system (7.12) to have a unique solution is that there be m associated conditions which specify the solution in some way for one or more values of t. In the case of the predator–prey problem these are the initial populations of foxes and rabbits. In the general case, if all conditions are specified at the same point, say

$$\mathbf{y}(a) = \boldsymbol{\alpha}, \tag{7.13}$$

then (7.12), together with the *initial conditions* (7.13), is referred to as an *initial-value problem* (IVP). If the m conditions are not all specified at the same point but are specified, for example, at two different points, say

$$
\begin{aligned}
y_i(a) &= \alpha_i, & i &= 1, 2, \ldots, s, \\
y_i(b) &= \beta_i, & i &= 1, 2, \ldots, m - s,
\end{aligned}
\tag{7.14}
$$

for some b, then (7.12), together with the *boundary conditions* (7.14), is referred to as a *boundary-value problem* (BVP) (in this case a *two-point* BVP).

In the following we concentrate mainly on IVPs and IVP techniques for BVPs. We do this principally because numerical methods for IVPs are intrinsically difficult to parallelise, and we wish to respond to the challenge. Direct solution techniques for BVPs tend to reduce to the solution of a system of linear equations and we investigate this approach with a view to parallelisation, in higher dimensions, in the context of partial differential equations in Section 7.4.

7.3.1 Initial-value problems

We proceed by considering the system of first-order ODEs (7.12), subject to the initial conditions (7.13). First we need to decide what is meant by a solution. A common requirement is to determine a value for $\mathbf{y}(b)$ for some $b > a$. Thinking of the differential equation as an evolutionary system, we wish to know the state of the system after a certain time has elapsed.

There are a variety of traditional numerical methods for solving such a problem, nearly all of which follow the same basic pattern. The domain of interest $[a, b]$ is discretised by the introduction of a number of *mesh*, or *grid*, *points* $t_0, t_1, \ldots, t_N$, with $t_0 = a$ and $t_N = b$. We refer to the distance $h_i = t_{i+1} - t_i$ as the *mesh spacing*, or *step length*, and at this point simply remark that this is chosen, preferably dynamically, in an attempt to yield a solution correct to some specified local accuracy. At each point of the grid we compute an approximation $\mathbf{Y}_i$ to $\mathbf{y}(t_i)$ in a *step-by-step* fashion; that is, starting at $\mathbf{Y}_0 \equiv \boldsymbol{\alpha}$, the given initial condition, we compute first $\mathbf{Y}_1$, then $\mathbf{Y}_2$, and so on until we determine $\mathbf{Y}_N$, an approximation to $\mathbf{y}(b)$. Ignoring possible complications at the early points of the grid due to the fact that a sufficient number of previous approximations may not be available, each $\mathbf{Y}_i$ is defined (possibly implicitly) in terms of a fixed number of previous approximations, $\mathbf{Y}_l$, $l = i - 1, i - 2, \ldots, i - k$, and we refer to a *k-step method*.

Because of their practical importance, much research work has gone into the derivation and analysis of numerical methods for IVPs, and we can only hope to give a flavour here. First we note that two main classes of method are commonly employed in practice, namely, *linear multistep* and *Runge–Kutta methods*. We look briefly at these in turn, initially in the context of the single equation $y' = f(t, y)$, subject to the initial condition $y(a) = \alpha$, where we compute approximations Y_i to $y(t_i)$. The extension to a system of first-order equations is trivial. We then explore the potential for parallelism. To distinguish between numerical methods we again use the term 'order', but in a context which refers to the method itself, rather than the problem being solved. Assuming now the grid points to be equally spaced with mesh spacing h, we define the *global truncation error* (GTE), e_n, at the point t_n to be

$$e_n = y(t_n) - Y_n,$$

that is, the difference between the true and approximate solutions at t_n. If r is the largest integer such that $e_n = O(h^r)$, we say that the method has *order r*.

Before proceeding we mention two further terms. The first is the *local truncation* or *discretisation error* (LTE), ϵ_n. Its definition, $\epsilon_n = y(t_n) - Y_n$, looks like that for the GTE, but we add the important caveat which requires that all previous values which are used in the determination of Y_n be exact, that is, they equal the true solution of the original differential equation. Normally, of course, the previous values employed are approximations only and thus the LTE represents the error per step incurred by replacing the differential equation by an appropriate discrete system. If a method is of order r then it is frequently the case that $\epsilon_n = O(h^{r+1})$. The reader may expect that a method of order r is 'better' (that is, more accurate) than one of order $q < r$, but unfortunately, this is so only in the limit as $h \to 0$. Other considerations must be taken into account, the most important of which is *stability*. This term is used to describe the effect that errors in early values of Y_i have on values computed further along the range. We say that a method is *stable* if the growth of these errors is not significant, and *unstable* otherwise. We cannot expect to do justice here to this important topic and we refer the interested reader to Hairer, Nørsett, *et al.* (1987). We simply remark that the stability of a method can have a considerably restrictive effect on the choice of step size. When choosing numerical methods for IVPs we therefore look for those which have good stability properties and have an acceptably high order.

Linear multistep methods

Assuming a constant step length h, the general form for a linear multistep method (LMM) is

$$\sum_{j=0}^{k} \alpha_j Y_{n+j} - h \sum_{j=0}^{k} \beta_j f(t_{n+j}, Y_{n+j}) = 0,$$

where the constants α_j and β_j are chosen to satisfy a number of requirements, one of which is to make the order of the method sufficiently high. In a practical application the function values $f_{n+j} = f(t_{n+j}, Y_{n+j})$, $j = 0, 1, \ldots, k-1$, are available from previous steps and, with a shift of the subscripts on the Y- and f-values, we therefore aim to solve the equation

$$Y_{n+1} = -\frac{1}{\alpha_k} \left(\sum_{j=0}^{k-1} \{\alpha_j Y_{n-k+1+j} - h\beta_j f_{n-k+1+j}\} - h\beta_k f(t_{n+1}, Y_{n+1}) \right).$$
$$(7.15)$$

For $\alpha_k \neq 0$ and $\beta_k = 0$ this yields an expression for Y_{n+1} in terms of previous values, Y_j, $j = n, n-1, \ldots, n-k+1$, only, and the method is said to be k-step *explicit*.

Using Taylor series

$$y(t_{n+1}) = y(t_n) + hy'(t_n) + O\left(h^2\right),$$

and ignoring the $O\left(h^2\right)$ term (the LTE) suggests *Euler's method*,

$$Y_{n+1} = Y_n + hf_n,$$

which is a first-order, one-step scheme. Unfortunately the method is of little practical value by itself because its order is too low and its stability properties leave much to be desired. Hence we discard it here (but return to it later) and, instead, cite the fourth-order, four-step, *Adams–Bashforth method*

$$Y_{n+1} = Y_n + \frac{h}{24}(-9f_{n-3} + 37f_{n-2} - 59f_{n-1} + 55f_n). \qquad (7.16)$$

If α_k and β_k are both non-zero then the LMM (7.15) gives a nonlinear expression for Y_{n+1} unless f happens to be linear in its second argument. We say that the method is k-step *implicit*.

Integrating the differential equation over $[t_n, t_{n+1}]$ we have

$$y(t_{n+1}) - y(t_n) = \int_{t_n}^{t_{n+1}} f(t, y(t))dt. \qquad (7.17)$$

If the integral on the right-hand side of (7.17) is approximated using the trapezium rule (Section 6.2.1) we have

$$y(t_{n+1}) - y(t_n) = \frac{h}{2}\left(f(t_n, y(t_n)) + f(t_{n+1}, y(t_{n+1}))\right) + O\left(h^3\right),$$

which suggests the second-order, one-step method

$$Y_{n+1} = Y_n + \frac{h}{2}\left(f_n + f(t_{n+1}, Y_{n+1})\right), \tag{7.18}$$

(the *trapezoidal method*) in which the LTE is $O\left(h^3\right)$. The trapezoidal method has good stability properties, but its low order is a drawback. A second example of an implicit method is the three-step *Adams–Moulton method*

$$Y_{n+1} = Y_n + \frac{h}{24}(f_{n-2} - 5f_{n-1} + 19f_n + 9f(t_{n+1}, Y_{n+1})). \tag{7.19}$$

Like (7.16), the method (7.19) is fourth-order, but its stability characteristics are inferior to those of the trapezoidal method (7.18).

Precisely how (7.19) and the explicit method (7.16), are derived is not important, but we remark that (7.17) provides the key; see Hairer, Nørsett, *et al.* (1987). On a practical note, the Adams methods are often represented in terms of backward difference formulae, or in Taylor series (Nordsieck) form; see Gear (1971) for details.

Any of the standard root-finding techniques may be employed to solve an implicit LMM for Y_{n+1}. A simple *fixed-point iteration* is adequate if accuracy requirements are the primary consideration since the step length is likely to be sufficiently small to guarantee convergence (Hairer, Nørsett, *et al.*, 1987). If we write (7.15) as

$$Y_{n+1} = \phi(Y_{n+1}), \tag{7.20}$$

then, starting with an initial approximation $Y_{n+1}^{(1)}$, further approximations can be computed using the iteration

$$Y_{n+1}^{(l+1)} = \phi\left(Y_{n+1}^{(l)}\right), \qquad l = 1, 2, \ldots. \tag{7.21}$$

Note that in this subsection we use k in its traditional role as the number of steps in an LMM and l as the iteration index. Also note that each iteration of (7.21) requires only the evaluation of $f\left(t_{n+1}, Y_{n+1}^{(l)}\right)$. To complete the specification we must decide on some mechanism for computing an initial value, $Y_{n+1}^{(1)}$, for the iteration (7.21). One way of proceeding is to employ an explicit method of the same order to *predict* this initial value, and use the implicit method to refine (*correct*) it, usually for a fixed small number (1 or 2) of iterations at each mesh point. We refer to the use of an explicit and an implicit method in this way as a *predictor–corrector pair*. In particular, Adams–Bashforth explicit and Adams–Moulton implicit methods, such as (7.16) and (7.19), are commonly used in practice. It is usual to refer to the modes PEC, PECE, PECEC, etc. in which

- P stands for an application of the predictor (the explicit method) to give an initial estimate of Y_{n+1},
- E stands for an evaluation of f in terms of the most recently available estimate of Y_{n+1}, and
- C stands for an application of the corrector (the implicit method) to give an improved estimate of Y_{n+1}.

It should be noted that the final evaluation in PECE (predict, evaluate, correct, evaluate) mode has no effect at the current grid point but, since the function value is explicitly used on the next few steps, it will influence the next, and all subsequent, approximations. Also, it is important to note that the number of corrections, and whether or not each iteration ends with an evaluation or a correction, can have a marked influence on the accuracy of the calculated values, and also on the stability characteristics of the method (Hall, 1974). For a k-step method with $k > 1$ there is the additional problem of determining the initial, or *starting values*, $Y_1, Y_2, \ldots, Y_{k-1}$, but this is a detail only; see Hairer, Nørsett, *et al.* (1987) for possible strategies.

Runge–Kutta methods

All Runge–Kutta methods are one-step methods and hence there are no problems with the provision of starting values. For a single equation and a constant step size h the general form for an s-stage Runge–Kutta method is

$$Y_{n+1} = Y_n + \sum_{i=1}^{s} w_i k_i,$$

where

$$k_i = hf\left(t_n + \alpha_i h, Y_n + \sum_{j=1}^{q_i} \beta_{ij} k_j\right).$$

The constants w_i, α_i and β_{ij} are chosen to give the method the highest possible order. If $q_i < i$, $i = 1, 2, \ldots, s$, then the method is said to be *explicit* and a sequential algorithm simply consists of, at each grid point, evaluating k_i, $i = 1, 2, \ldots, s$, in turn and then computing Y_{n+1}. When $q_i = i$, $i = 1, 2, \ldots, s$, the calculation of each k_i requires the solution of a single nonlinear equation, and the method is said to be *diagonally implicit*. If $q_i = s$, $i = 1, 2, \ldots, s$, the calculation of the vector of values, $\mathbf{k} = (k_1, k_2, \ldots, k_s)^T$, requires the solution of a system of s nonlinear equations and the method is referred to as *fully implicit*. For implicit methods some form of iteration (such as Newton's method) is required to determine k_i, $i = 1, 2, \ldots, s$, before Y_{n+1} can be calculated. Implicit Runge–Kutta methods have better stability characteristics than explicit methods, but the need to solve a system of nonlinear equations at each time step is a serious drawback to their value as practical methods.

The practical implementation of IVP solvers

Before proceeding, it is worthwhile looking at the way LMM and Runge–Kutta methods are implemented in general-purpose software for the solution of IVPs. Rather than use a constant step length throughout the range of integration, practical IVP solvers vary the step length in an attempt to guard against instability, and to produce approximate solution values which satisfy some appropriate accuracy criterion. The situation is analogous to that for adaptive quadrature (Section 6.2.2). As the integration proceeds an error estimate (an approximation to the LTE, rather than the GTE) is computed (at each step). If this estimate is considered to be too large then the step length is reduced (for example, by a factor of 2); if the error estimate is relatively small then the step length might be increased, otherwise the integration continues with the current value of the step length. Step-length control with a Runge–Kutta method is relatively straightforward. With a LMM it is a little more awkward because of the dependence of Y_{n+1} on a number of previous values. If a step-length change is required then it may be necessary to interpolate previous Y values in order that a constant step formula may continue to be employed. When considering possible parallel algorithms the need for an appropriate step-length control strategy must not be overlooked.

7.3.2   The potential for parallelism

The step-by-step philosophy which forms the basis of the algorithms introduced in the previous subsection is inherently sequential in nature. Nevertheless, there is some scope for parallelism. It is normal to categorise parallel methods according to whether they exploit

- parallelism across the system (or problem), or
- parallelism across the method (or time).

In the former case the standard LMM and Runge–Kutta methods can be employed directly. In the latter case we need to produce modifications to existing methods. At the outset we remark that the amount of parallelism available can be quite limited. In the case of parallelism across the system it is constrained by the order of the system; in the case of parallelism across the method it is typically constrained, indirectly, by the order of the method.

Parallelism across the system

We consider here the parallelisation of LMMs applied to a system of first-order ODEs; much of the discussion is also applicable to Runge–Kutta methods. In the case of an explicit method we have

$$\mathbf{Y}_{n+1} = \frac{1}{\alpha_k} \left(-\sum_{j=0}^{k-1} \alpha_j \mathbf{Y}_{n-k+1+j} + h \sum_{j=0}^{k-1} \beta_j \mathbf{f}_{n-k+1+j} \right), \qquad (7.22)$$

where $\mathbf{f}_j = \mathbf{f}(t_j, \mathbf{Y}_j)$ and $\mathbf{f}_j$ and $\mathbf{Y}_j$ are m-vectors.

Each step of the method requires

- the evaluation of $\mathbf{f}_n$, noting that $\mathbf{f}_{n-k+1+j}$, $j = 0, 1, \ldots, k-2$, are available from previous steps, and
- the computation of the new approximate solution $\mathbf{Y}_{n+1}$.

For the first of these stages we can draw from the material of Section 6.3.3. The evaluation of the component functions $f_i(t_n, \mathbf{Y}_n)$, $i = 1, 2, \ldots, m$, constitutes m independent processes which can therefore be performed concurrently. Further, the computation of each component of $\mathbf{Y}_{n+1}$ of (7.22) is an independent process, and requires knowledge of the corresponding components of $\mathbf{Y}_{n-k+1+j}$ and $\mathbf{f}_{n-k+1+j}$, $j = 0, 1, \ldots, k-1$, only. Hence, the idea is to split the system so that the components (of $\mathbf{Y}_{n-k+1+j}$ and $\mathbf{f}_{n-k+1+j}$) are grouped and associated with processes. Each process then needs to

- evaluate its components of $\mathbf{f}_n$, and
- compute its components of $\mathbf{Y}_{n+1}$.

There is a synchronisation point after each step of the method since, potentially, each function evaluation $f_i(t_n, \mathbf{Y}_n)$ depends on all the components of $\mathbf{Y}_n$. In the case of a shared memory system the synchronisation point takes the form of a barrier. For a local memory system each process must broadcast its $\mathbf{Y}_n$ values to all the other processes. The amount of data traffic thus generated can be reduced if some attempt is made to decouple the system. That is, components of the system for which the component functions f_i involve knowledge of the corresponding components of $\mathbf{Y}_n$ are grouped into processes, in an attempt to minimise the amount of interprocess communication. At the same time the distribution of the components of the system must be balanced so that processes are allotted roughly equal amounts of work, in terms of the number of components of the system, and the total cost of the corresponding function evaluations.

As an (admittedly, fairly trivial) example we consider the *orbit equations* (Hull, Enright, *et al.*, 1972)

$$\begin{aligned}
y_1' &= y_3 = f_1(t, \mathbf{y}), \\
y_2' &= y_4 = f_2(t, \mathbf{y}), \\
y_3' &= -y_1 \left(y_1^2 + y_2^2 \right)^{-3/2} = f_3(t, \mathbf{y}), \\
y_4' &= -y_2 \left(y_1^2 + y_2^2 \right)^{-3/2} = f_4(t, \mathbf{y}).
\end{aligned} \qquad (7.23)$$

Suppose that, on a local memory system, two processes are available. If the first two equations in the system are allocated to process 1 and the remaining equations allocated to process 2, then, in order for the function evaluations to be performed, four y-values must be transmitted (two from process 1 to process 2 and two from process 2 to process 1). Further, the cost of the evaluation of f_3 and f_4 is rather more than that of f_1 and f_2. We can reduce the amount of data transfer and, at the same time, balance the computational load if the first and third equations are allocated to process 1 and the second and fourth equations are allocated to process 2.

If we wish to employ an implicit method to solve a system of IVPs then the simple predictor–corrector strategy outlined in Section 7.3.1 may be inadequate for certain types of problem, particularly if function evaluations are expensive. The solution is to use Newton's method to solve the system of nonlinear equations at each step of the method. We defer a detailed study of the method to Section 7.3.3, but here observe that each iteration will involve

- a function evaluation phase, and
- a linear algebra phase,

and the former can often be expected to be the more computationally expensive.

Parallelism across the method

Parallelism across the system is straightforward, but limited to the number of equations in the system. In particular, in the case of a single first-order equation there is no parallelism of this form available. As a consequence, attention has additionally been focused on methods which, whilst still conforming to the step-by-step model, possess inherent concurrency.

Miranker and Liniger (1967) suggest a parallel predictor–corrector method which uses concurrent function evaluations. Suppose that two processes are available and a single correction per step is to be employed. Then, at each step the work is divided into two stages:

- *Prediction at time step t_{n+1}*
  Predict a value $Y_{n+1}^{(P)}$ for Y_{n+1} and evaluate $f\left(t_{n+1}, Y_{n+1}^{(P)}\right)$.
- *Correction at time step t_n*
  Compute $Y_n^{(C)}$, a correction to $Y_n^{(P)}$, the predicted value at the previous time step, and evaluate $f\left(t_n, Y_n^{(C)}\right)$.

It is clear that if these operations are to happen concurrently the prediction stage cannot make use of the corrected value at the previous step. Instead, it must employ $Y_n^{(P)}$, the predicted value at the previous step. The method is referred to as a *wavefront method* and Figure 7.3 indicates the wavefront corresponding to the use of three processes, where two corrections are employed each step. Clearly the

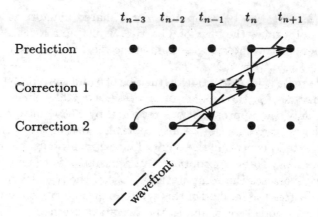

$$t_{n-3} \quad t_{n-2} \quad t_{n-1} \quad t_n \quad t_{n+1}$$

Prediction

Correction 1

Correction 2

wavefront

Figure 7.3 *Wavefront using three processes*

potential for parallelism is limited by the number of corrections employed. Given that this number is likely to be more than would be necessary in a sequential environment because of the relatively poor quality of the predicted values, and possible problems with stability, the method is not without its limitations.

A further attempt at parallelism across the method is suggested by Worland (1976) who derives an algorithm based on the block methods of Shampine and Watts (1969). For a single equation the aim is to compute a block of values in parallel. Let b be the block size. Then Worland's method uses a simple one-step method (Euler's method, Section 7.3.1) to compute initial estimates of $Y_{n+1}, Y_{n+2}, \ldots, Y_{n+b}$. In detail, for a constant step size we have

$$Y_{n+j} = Y_n + jhf_n, \qquad j = 1, 2, \ldots, b. \tag{7.24}$$

These initial estimates are then improved using some appropriate implicit LMM. For example, for $b = 2$ Worland uses

$$Y_{n+1} = Y_n + \frac{h}{12}\left(5f_n + 8f(t_{n+1}, Y_{n+1}) - f(t_{n+2}, Y_{n+2})\right),$$
$$Y_{n+2} = Y_n + \frac{h}{3}\left(f_n + 4f(t_{n+1}, Y_{n+1}) + f(t_{n+2}, Y_{n+2})\right). \tag{7.25}$$

Using a (block) fixed-point iteration the function values on the right-hand sides of (7.25) are computed in terms of the old estimates of Y_{n+1} and Y_{n+2}. Two iterations are necessary to give a fourth-order method; a third iteration potentially improves the accuracy.

For a parallel implementation of a block method, such as the above, it is natural to group the determination of Y_{n+j}, $j = 1, 2, \ldots, b$, into processes. Each process is responsible for computing those function values involving its Y_{n+j} estimates. There is then a synchronisation point since, in principle, these function values must be available to all the processes before modified estimates of

Y_{n+j}, $j = 1, 2, \ldots, b$, can be computed. For a shared memory machine this synchronisation point takes the form of a barrier, whilst on a local memory machine the requirement is the interprocess communication (broadcast) of the function values $f(t_{n+j}, Y_{n+j})$, $j = 1, 2, \ldots, b$.

In principle there is no restriction on the size of b and it would be natural to take it to be a multiple of p, the number of processes available. Since Euler's method is first-order only, the approximations returned by (7.24) have a global error of $O(jh)$, and for j large this can be unacceptable. Use of the corrector formulae will, to some extent, overcome these initial inaccuracies but only relatively small block sizes are likely to be appropriate. An alternative is to employ small values for h, but this increases the computational cost of the algorithm.

Parallelism across the method of this type can also be exploited in fully implicit Runge–Kutta methods. For example, the two-stage method

$$Y_{n+1} = Y_n + \frac{h}{2}(k_1 + k_2),$$

where

$$k_1 = f\left(t_n + \frac{3 + \sqrt{3}}{6}h, y_n + \frac{h}{4}k_1 + h\left(\frac{1}{4} + \frac{\sqrt{3}}{6}\right)k_2\right),$$

$$k_2 = f\left(t_n + \frac{3 - \sqrt{3}}{6}h, y_n + h\left(\frac{1}{4} - \frac{\sqrt{3}}{6}\right)k_1 + \frac{h}{4}k_2\right), \tag{7.26}$$

is fourth-order. The nonlinear system (7.26) can be solved for k_1 and k_2 using some appropriate iterative technique, such as Newton's method. As far as parallelism is concerned, the situation could be improved (the need for synchronisation or the amount of message-passing could be reduced) if the method was such that the computation of the k_i values could be decoupled. Iserles and Nørsett (1990) suggest the diagonally implicit Runge–Kutta method

$$Y_{n+1} = Y_n + \frac{h}{2}(-2k_1 + 3k_2 - 2k_3 + 3k_4), \tag{7.27}$$

where

$$k_1 = f\left(t_n + h/2, Y_n + hk_1/2\right),$$
$$k_2 = f\left(t_n + 2h/3, Y_n + 2hk_2/3\right),$$
$$k_3 = f\left(t_n + h/2, Y_n - 5hk_1/2 + 5hk_2/2 + hk_3/2\right),$$
$$k_4 = f\left(t_n + 2h/3, Y_n - 5hk_1/3 + 4hk_2/3 + 2hk_4/3\right).$$

Here, the computations of k_1 and k_2 are independent of each other. Likewise, the computations of k_3 and k_4 are independent of each other, but each depends on k_1 and k_2. Thus, on a parallel machine, each step of the method could be implemented as five stages:

- Determine k_1 and k_2 in separate processes using some appropriate iterative technique.

- Synchronise (shared memory), or send k_1 to the process which has computed k_2 and vice versa (local memory).
- Determine k_3 and k_4 in separate processes, again using some appropriate iterative technique.
- Synchronise, or collect k_i, $i = 1, 2, 3, 4$, in one process.
- In one process determine Y_{n+1} from (7.27) and, in the case of a local memory machine, send Y_{n+1} to the other process ready for the next step.

Iserles and Nørsett suggest employing Newton's method for the solution of the nonlinear equations for k_i, $i = 1, 2, 3, 4$. Again, the potential benefits from parallelism are limited. We can expect to exploit more processes by increasing s, the number of stages of the Runge–Kutta method, but this does not necessarily increase the order of the method by a corresponding amount. Further, imposing constraints on the Runge–Kutta coefficients so as to facilitate the use of parallelism may have the effect of increasing the constant coefficient appearing in the expression for the local truncation error and, as a consequence, reduce the potential accuracy of the method.

7.3.3   Systems and linear algebra

Several references have been made in the above to the use of Newton's method, which has already been considered in Section 6.3.3. The point to be made here is that the simple iterative techniques so far employed will be inadequate for certain types of problem; the convergence rates of these iterations will be too slow and/or the conditions for convergence of the iterations will impose too strict a condition on the step length h.

Typically, we wish to solve the system of nonlinear equations

$$\mathbf{Y}_{n+1} = \mathbf{\Phi}(\mathbf{Y}_{n+1}),$$

and Newton's method takes the form

$$\mathbf{Y}_{n+1}^{(l+1)} = \mathbf{Y}_{n+1}^{(l)} - J^{(l)^{-1}} \mathbf{\Phi}\left(\mathbf{Y}_{n+1}^{(l)}\right),$$

where $J^{(l)}$ is the Jacobian matrix of $\mathbf{\Phi}$ evaluated at $\mathbf{Y}_{n+1}^{(l)}$ (cf. Section 6.3.3) with (i, j) element

$$\left(J^{(l)}\right)_{ij} = \frac{\partial \phi_i}{\partial z_j}\left(\mathbf{Y}_{n+1}^{(l)}\right),$$

and where the components, ϕ_i, of $\mathbf{\Phi}$ are given by

$$\phi_i(\mathbf{z}) = -\frac{1}{\alpha_k}\left(\sum_{j=0}^{k-1}\left\{\alpha_j\left(\mathbf{Y}_{n-k+1+j}\right)_i - h\beta_j\left(\mathbf{f}_{n-k+1+j}\right)_i\right\} - h\beta_k f_i(t_{n+1}, \mathbf{z})\right).$$

As in Section 6.3.3, rather than invert $J^{(l)}$ explicitly we solve the system of equations

$$J^{(l)} \, \delta \mathbf{Y}_{n+1}^{(l)} = -\mathbf{\Phi}\left(\mathbf{Y}_{n+1}^{(l)}\right), \qquad (7.28)$$

for the correction $\delta \mathbf{Y}_{n+1}^{(l)}$ and set $\mathbf{Y}_{n+1}^{(l+1)} = \mathbf{Y}_{n+1}^{(l)} + \delta \mathbf{Y}_{n+1}^{(l)}$. At each iteration, therefore, we need to

- form the Jacobian matrix $J^{(l)}$ (or a finite difference approximation to it) and the right-hand side vector $\mathbf{\Phi}\left(\mathbf{Y}_{n+1}^{(l)}\right)$, and
- solve the system of linear equations (7.28) (using, say, Gaussian elimination).

Which of these stages turns out to be the more expensive will depend critically on the form of $\mathbf{f}$. We refer to Section 6.3.3 for a discussion of the issues involved in the parallelisation of Newton's method.

7.3.4   Boundary-value problems

As indicated earlier, IVPs usually represent evolutionary systems with the independent variable t on a time axis. BVPs, on the other hand, usually represent steady-state systems with the independent variable (often denoted x) on a space axis. For consistency we continue to use t as the independent variable. Again we are interested in a solution to the system of m differential equations

$$\mathbf{y}'(t) = \mathbf{f}(t, \mathbf{y}(t)),$$

but now specify (boundary) conditions at more than one point. For simplicity we assume s of the boundary conditions to be imposed at $t = a$ and the remaining $m - s$ at $t = b$, these conditions being of the form

$$\begin{aligned} y_i(a) &= \alpha_i, & i &= 1, 2, \ldots, s, \\ y_i(b) &= \beta_i, & i &= 1, 2, \ldots, m - s. \end{aligned} \qquad (7.29)$$

The approach to the solution of a BVP is somewhat different to that for an IVP. With an IVP we are usually interested in a solution at a single point, say $t = b$, whereas with a BVP we are given information about the solution at $t = a$ and $t = b$ and are interested in the behaviour of the solution at points in between. A consequence is that numerical methods which compute a continuous approximation to the solution are often popular. However here, as with IVPs, attention is restricted to those methods which compute a discrete approximation, that is, a solution whose value is known at a discrete set of points only. We consider two types of method in particular. The first, termed a *finite difference method*, is based on a direct approach to the BVP and we give a brief discussion only. The second, termed a *shooting method*, requires the BVP to be recast as an IVP and then makes use of the material of earlier subsections.

Finite difference methods

We again introduce a set of mesh points $t_i = t_0 + ih$, $i = 1, 2, \ldots, N$, with $t_0 = a$, $t_N = b$ and $h = (b - a)/N$. Adapting the trapezoidal method (7.18) to a system yields

$$\mathbf{Y}_{n+1} - \mathbf{Y}_n = \frac{h}{2} \left(\mathbf{f}(t_n, \mathbf{Y}_n) + \mathbf{f}(t_{n+1}, \mathbf{Y}_{n+1}) \right), \qquad n = 0, 1, \ldots, N - 1. \quad (7.30)$$

At the ends of the range ($n = 0$ and $n = N - 1$) we include only those components of (7.30) which correspond to the known boundary conditions. Then, (7.30), together with the boundary conditions (7.29), defines approximations $\mathbf{Y}_n$ to $\mathbf{y}(t_n)$, $n = 1, 2, \ldots, N - 1$.

The system (7.30) and the boundary conditions (7.29) constitute a system of $(N - 1)m$ nonlinear equations for $\mathbf{Y}_n$, $n = 1, 2, \ldots, N - 1$. We write this system as

$$\mathbf{F}(\mathbf{Y}) = \mathbf{0},$$

where $\mathbf{Y}$ is a block vector with ith block $\mathbf{Y}_i$, and $\mathbf{F}$ is a corresponding block vector-valued function. Again we employ Newton's method to determine a solution. Each iteration involves $N - 1$ independent vector-valued function evaluations to determine $\mathbf{F}\left(\mathbf{Y}^{(l)}\right)$, where $\mathbf{Y}^{(l)}$ is the previous iterate, plus additional function evaluations, of $O\left(N^2 m^2\right)$, to define the Jacobian matrix of $\mathbf{F}$ or a finite difference approximation to it. Clearly, there is considerable scope to exploit parallelism here.

Shooting methods

Perhaps surprisingly, one of the more successful numerical techniques for solving BVPs is based on the idea of reformulating the problem as a (sequence of) IVPs. To do this all the conditions must be specified at one point. Suppose we choose to impose m initial conditions at $t = a$, where s boundary conditions are already known. We guess the remaining $m - s$ conditions at this point and, for the moment, ignore the known $m - s$ boundary conditions at $t = b$. We now have an IVP which can be solved using a linear multistep, Runge–Kutta, or any other appropriate method to obtain a numerical solution at $t = b$. These numerical values are then compared with the known boundary conditions at $t = b$. If the guessed initial conditions are correct then there are no discrepancies with the known boundary conditions at $t = b$ and the solution to the IVP will be the solution to the BVP. If not, we need to modify the guessed initial conditions at $t = a$. We refer to the method as a *shooting method* for obvious reasons.

Formally, we introduce the vector $\boldsymbol{\lambda}$ with the components λ_i, $i = 1, 2, \ldots, m - s$, being estimates of $y_{s+i}(a)$. The solution to the corresponding IVP

$$\mathbf{y}'(t) = \mathbf{f}(t, \mathbf{y}(t)),$$

with known initial conditions

$$y_i(a) = \alpha_i, \qquad i = 1, 2, \ldots, s,$$

and guessed initial conditions

$$y_{s+i}(a) = \lambda_i, \qquad i = 1, 2, \ldots, m - s,$$

is denoted by $\mathbf{y}\,(t, \boldsymbol{\lambda})$, where we have made the dependence on $\boldsymbol{\lambda}$ explicit. This IVP is solved numerically to determine $\mathbf{Y}_N\,(\boldsymbol{\lambda})$, an approximate solution at $t = b = t_N$. We are then interested in the difference

$$\mathbf{F}(\boldsymbol{\lambda}) = \boldsymbol{\beta} - \tilde{\mathbf{Y}}_N\,(\boldsymbol{\lambda}),$$

where $\tilde{\mathbf{Y}}_N\,(\boldsymbol{\lambda})$ comprises the first $m - s$ components of $\mathbf{Y}_N\,(\boldsymbol{\lambda})$ and $\boldsymbol{\beta}$ is the vector of known boundary conditions at $t = b$. Again, this is a system of nonlinear equations which can be solved by Newton's method. If we let $\boldsymbol{\lambda}^{(l)}$ be an iterate, then at each iteration we need to integrate an IVP to determine the residual vector $\mathbf{F}\left(\boldsymbol{\lambda}^{(l)}\right) = \boldsymbol{\beta} - \tilde{\mathbf{Y}}_N\left(\boldsymbol{\lambda}^{(l)}\right)$, and this can be parallelised using the techniques of Section 7.3.2. Further, the components of the Jacobian matrix of $\mathbf{F}(\boldsymbol{\lambda})$ are themselves the solutions to IVPs, and these can be computed concurrently with that for the residual vector.

When introducing numerical methods for IVPs we made a brief mention of stability. In recasting a BVP as an IVP stability may become a problem which it is sometimes possible to overcome simply by shooting in the opposite direction. That is, we guess initial values at $t = b$ and integrate towards $t = a$. This may increase or decrease the order of the nonlinear system to be solved, depending on the size of s. An alternative strategy is to choose some point c between a and b, shoot from both ends, and match at $t = c$. Now the order of the nonlinear system is m, but we observe that the two IVPs, one integrating forwards from a, the other integrating backwards from b, are not coupled and can therefore be solved entirely independently; the only problem on a multiprocessor is that of load balancing. This suggests that there may be something to be gained from a further subdivision, and *multiple shooting* provides the key.

The original motivation for multiple, or parallel, shooting was to overcome any stability problems that may be present in the basic shooting method. We begin with the discretisation points t_n, $n = 0, 1, \ldots, q$, with $t_0 = a$ and $t_q = b$. Over each subinterval $[t_n, t_{n+1}]$, $n = 0, 1, \ldots, q - 1$, we solve, independently, an IVP, integrating forwards from t_n to t_{n+1}, making use of known conditions at $t = a$ and guessing all other initial conditions that are required. The discrepancies between adjacent approximations at the internal grid points t_n, $n = 1, 2, \ldots, q - 1$, and with the known boundary conditions at $t = b$, are measured and the guessed initial conditions are updated. This completes one iteration of the multiple shooting method. The situation is illustrated in Figure 7.4.

In a shooting method parallelism can be exploited in

- the solution of the independent IVPs, and
- the linear algebra for updating the guesses at the initial conditions,

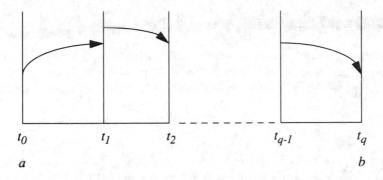

t_0 t_1 t_2 t_{q-1} t_q

a b

Figure 7.4 *Multiple shooting*

and it is likely that the first of these will dominate. Again, problems with load balancing are likely to be experienced on a multiprocessor system. Unfortunately, as with adaptive integration (Section 6.2.2), simply using an equal spacing for the initial discretisation will probably not result in a balanced load, since the adaptive integrator which is used to solve the IVPs will require different numbers of steps to integrate over the different subintervals. Further, we cannot subsequently redefine the mesh to balance the load without restarting from scratch.

7.4 Partial differential equations

The solution of an ODE is a function of one independent variable only. With a *partial differential equation* (PDE) we attempt to determine the characteristics of a function of two or more independent variables which may be in space or time. A simple example is furnished by *Laplace's equation*

$$\left(\frac{\partial^2}{\partial x^2} + \frac{\partial^2}{\partial y^2}\right) u(x, y) = 0,$$

or

$$\nabla^2 u(x, y) = 0, \tag{7.31}$$

where $\nabla = (\partial/\partial x, \partial/\partial y)^T$. Here, we wish to determine the form of the unknown function $u(x, y)$ in some region $\Omega \subset \mathcal{R}^2$, say, the unit square $[0, 1] \times [0, 1]$, in which the differential equation holds. As with an ODE, the specification of the differential equation alone is not sufficient to define a unique solution and additional conditions are required. Depending on the type of the differential equation these conditions can be of initial or boundary type. For example, for Laplace's equation we must have boundary conditions. These can be a combination of *Dirichlet type*

$$u(x, y) = g(x, y), \qquad (x, y) \in \Gamma_1,$$

and *Neumann type*

$$\nabla u(x, y).\mathbf{n} = 0, \qquad (x, y) \in \Gamma_2,$$

where $\Gamma_1 \cup \Gamma_2 = \Gamma$, Γ is the boundary of Ω, **n** is the unit outward normal to Γ_2, and $\mathbf{v}.\mathbf{w} = \mathbf{v}^T\mathbf{w}$. We refer to this as an example of an *elliptic equation*.

PDEs are often used to model some physical situation. For example, if the unit square represents a thin flat plate, then the differential equation

$$\frac{\partial u}{\partial t} = c\nabla^2 u, \tag{7.32}$$

where $u \equiv u(x, y, t)$ is the temperature at the point (x, y) at time t and c is a constant, models the variation in temperature across the plate with time. A Dirichlet boundary condition corresponds to a temperature source along the edge of the plate; a Neumann boundary condition corresponds to a part of the edge at which there is no heat loss. We refer to (7.32) as an example of a *parabolic equation* which is solved for $(x, y) \in [0, 1] \times [0, 1]$ and $t > t_0$ for some initial time t_0. Thus, in addition to the boundary conditions at the edges of the plate (which may now be dependent on t), we need to know the temperature distribution at time $t = t_0$, which corresponds to an initial condition. If the heat source on the edges of the plate is constant in time then, eventually, the temperature distribution will reach a *steady state*, that is, it is invariant with time. (7.32) then reduces to Laplace's equation (7.31).

The equations (7.31) and (7.32) are two simple, but typical, examples of the types of PDE we may come across. More complicated examples could involve space coordinates in three-dimensional space, or the equation could be defined over a more complicated region, or could involve cross derivatives (such as $\partial^2 u/\partial x \partial y$). In a book of this nature we can consider relatively simple examples only but, fortunately, numerical methods for all types of problem can be loosely categorised as

- finite difference,
- finite element, or
- series,

and each method has, in one form or another,

- a matrix set-up phase, and
- a solution phase,

although the line between these phases may be a little blurred. We consider the methods in turn, concentrating mainly on the scope for parallelism within the solution phase. However, it is worth noting that the set-up phase itself can, for certain problems, be significant and even dominate the rest of the calculation. Much depends on the form of the coefficients in the differential equation and the cost of their evaluation.

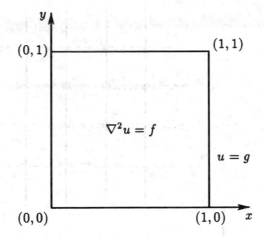

Figure 7.5 *Poisson's equation and Dirichlet boundary conditions on the unit square*

7.4.1 Finite differences and elliptic equations

The five-point star
We consider the finite difference solution of *Poisson's equation*

$$\nabla^2 u(x, y) = f(x, y), \tag{7.33}$$

(which is Laplace's equation with an inhomogeneous right-hand side f) on the unit square, $\Omega = [0, 1] \times [0, 1]$, subject to the Dirichlet boundary condition

$$u(x, y) = g(x, y),$$

for all $(x, y) \in \Gamma$, the boundary of Ω (Figure 7.5). If the region is less regular then the situation becomes more complicated, although the underlying principles remain unchanged.

To solve this problem numerically we discretise in both the x and y directions by introducing the grid points x_i, $i = 0, 1, \ldots, n + 1$, and y_i, $i = 0, 1, \ldots, n + 1$, each distance $h = 1/(n + 1)$ apart, with $x_0 = y_0 = 0$, $x_{n+1} = y_{n+1} = 1$ and $x_i = y_i = ih$. Then, Taylor's series gives

$$u(x_i + h, y_j) = u(x_i, y_j) + hu_x(x_i, y_j) + \frac{h^2}{2} u_{xx}(x_i, y_j) + \frac{h^3}{6} u_{xxx}(x_i, y_j) + O\left(h^4\right),$$
$$\tag{7.34}$$

where, for example, $u_x(x_i, y_j)$ is the first partial derivative of u with respect to x evaluated at the point (x_i, y_j). If $u(x_i - h, y_j)$ is also expanded about (x_i, y_j), then

$$u(x_i - h, y_j) = u(x_i, y_j) - hu_x(x_i, y_j) + \frac{h^2}{2} u_{xx}(x_i, y_j) - \frac{h^3}{6} u_{xxx}(x_i, y_j) + O\left(h^4\right).$$
$$\tag{7.35}$$

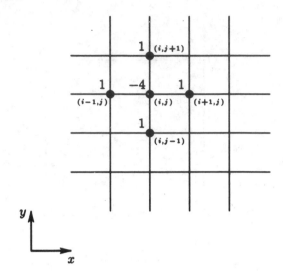

Figure 7.6 *Five-point finite difference star*

Combining (7.34) and (7.35) we have

$$u(x_i + h, y_j) + u(x_i - h, y_j) = 2u(x_i, y_j) + h^2 u_{xx}(x_i, y_j) + O\left(h^4\right).$$

Similarly,

$$u(x_i, y_j + h) + u(x_i, y_j - h) = 2u(x_i, y_j) + h^2 u_{yy}(x_i, y_j) + O\left(h^4\right).$$

Hence

$$\nabla^2 u_{ij} = \frac{1}{h^2}\left(u_{i+1,j} + u_{i-1,j} + u_{i,j+1} + u_{i,j-1} - 4u_{ij}\right) + O\left(h^2\right), \qquad (7.36)$$

where $u_{ij} = u(x_i, y_j)$ and $\nabla^2 u_{ij}$ is $\nabla^2 u(x, y)$ evaluated at (x_i, y_j). Using U_{ij} to denote an approximation to u_{ij}, this suggests that the differential equation (7.33) be replaced by

$$U_{i+1,j} + U_{i-1,j} + U_{i,j+1} + U_{i,j-1} - 4U_{ij} = h^2 f_{ij}, \qquad i, j = 1, 2, \ldots, n, \quad (7.37)$$

where $f_{ij} = f(x_i, y_j)$. We refer to this difference replacement as a *five-point finite difference star*, since it relates the solution value at an interior grid point to the solution values of its neighbours in the North, East, South and West directions (Figure 7.6). Equations in the system (7.37) corresponding to $i, j = 1$ or n involve the boundary conditions

$$U_{ij} = g_{ij}, \qquad i, j = 0 \text{ or } n + 1,$$

where $g_{ij} = g(x_i, y_j)$. The method is termed second-order because of the $O\left(h^2\right)$ discretisation error incurred in the transition from (7.33) to (7.37) via (7.36).

The relation (7.37) represents a system of linear equations for the n^2 unknowns U_{ij}, $i, j = 1, 2, \ldots, n$. Using some appropriate ordering, the system may be written as $A\mathbf{x} = \mathbf{b}$, where the coefficient matrix A is sparse but structured. It would be possible to determine $\mathbf{x}$ using some direct technique, such as Gaussian elimination, but the sparsity of A suggests that the use of an iterative method should be considered. Rearranging (7.37) we have

$$U_{ij} = \tfrac{1}{4}\left(U_{i+1,j} + U_{i-1,j} + U_{i,j+1} + U_{i,j-1} - h^2 f_{ij}\right), \qquad (7.38)$$

for which the Jacobi iteration (Section 5.3.1) is

$$U_{ij}^{(k+1)} = \tfrac{1}{4}\left(U_{i+1,j}^{(k)} + U_{i-1,j}^{(k)} + U_{i,j+1}^{(k)} + U_{i,j-1}^{(k)} - h^2 f_{ij}\right),$$
$$i = 1, 2, \ldots, n, \quad j = 1, 2 \ldots, n, \qquad (7.39)$$

in which the superscript denotes an iteration number.

As noted in Section 5.3.2, from the point of view of parallelism the Jacobi iteration is very attractive since, within an iteration, the computation of each approximation $U_{ij}^{(k+1)}$ is independent of that of every other. Since we are working with a two-dimensional array it is natural to subdivide this data structure into block columns and distribute work accordingly, so that, for example, process $l = 0, 1, \ldots, p - 1$, is given the task of computing values of U_{ij} for all i and for $j = n_l, n_l + 1, \ldots, n_l + q_l$, where $n_l + q_l + 1 = n_{l+1}$, $n_0 = 1$, $n_{p-1} + q_{p-1} = n$, and p is the number of processes. Making sure that the first and last columns in each block are updated first will help to reduce the effect of the synchronisation between iterations, although it is likely that we may wish to retain a synchronisation in order to test for convergence. In a local memory environment, at the end of each iteration, process l needs to receive values of $U_{ij}^{(k)}$ for all i, and for $j = n_l - 1$ from process $l - 1$, and for all i and for $j = n_{l+1}$ from process $l + 1$. Further, process l must send values of $U_{ij}^{(k)}$ for all i, and for $j = n_l$ to process $l - 1$ and for all i and for $j = n_l + q_l$ to process $l + 1$.

We can regard a block column (or block row) strategy as an example of domain decomposition. Essentially we have subdivided Ω into p subdomains Ω_l, with interfaces $\Gamma_{l-1,l}$ between Ω_{l-1} and Ω_l. For example, in the case of a block column distribution of U, $\Omega_l = \{(x, y), 0 \le x \le 1, \tilde{y}_l \le y \le \tilde{y}_{l+1}\}$ and $\Gamma_{l-1,l} = \{(x, \tilde{y}_l), 0 \le x \le 1\}$, for some $\tilde{y}_l \in (y_{n_{l-1}+q_{l-1}}, y_{n_l})$, the precise point being unimportant here. The interprocess communication required by a local memory implementation of the algorithm arises from a need to impose consistency across the subdomain interfaces.

Successive overrelaxation

Unfortunately the convergence rate of the iteration (7.39) is too slow for it to be of practical use, even in a parallel environment. We observed in Section 5.3.2 that the situation can be improved by the use of the Gauss–Seidel iteration. Here we attempt to improve the convergence rate still further by employing *successive overrelaxation* (SOR). For any $\omega \neq 0$, (7.38) may be rewritten as

$$U_{ij} = U_{ij} + \frac{\omega}{4}\left(U_{i+1,j} + U_{i-1,j} + U_{i,j+1} + U_{i,j-1} - 4U_{ij} - h^2 f_{ij}\right). \qquad (7.40)$$

If we update the U_{ij} estimates according to increasing values of the subscripts i and j, (7.40) suggests the iterative scheme

$$U_{ij}^{(k+1)} = U_{ij}^{(k)} + \frac{\omega}{4}\left(U_{i+1,j}^{(k)} + U_{i-1,j}^{(k+1)} + U_{i,j+1}^{(k)} + U_{i,j-1}^{(k+1)} - 4U_{ij}^{(k)} - h^2 f_{ij}\right),$$
$$i = 1, 2, \ldots, n, \quad j = 1, 2, \ldots, n,$$

$$(7.41)$$

in which the most recently computed values $U_{*,*}^{(k+1)}$ are used as soon as they are available. The choice $\omega = 1$ corresponds to the Gauss–Seidel iteration. Varga (1962, p. 77), shows that the optimal value of ω lies in $(0, 2)$.

At first sight the computation of the U_{ij} by (7.41) appears to be an inherently sequential process. However, suppose that we regard the finite difference grid as a chess board (with the points representing squares) and label the points accordingly. That is, point (i, j) is denoted white (or, more commonly these days, red) or black according to whether $i + j$ is odd or even (Figure 7.7). Then the evaluation of each $U_{ij}^{(k+1)}$ corresponding to a red grid point involves black values only, and vice versa. Thus, all the red values, $U_{ij}^{(k+1)}$ for $i + j$ odd, can be computed concurrently, followed by all the black values, $U_{ij}^{(k+1)}$ for $i + j$ even (again, concurrently). All that we have done is simply update the approximations in a different order to that suggested by (7.41). For $i + j$ odd we have

$$U_{ij}^{(k+1)} = U_{ij}^{(k)} + \frac{\omega}{4}\left(U_{i+1,j}^{(k)} + U_{i-1,j}^{(k)} + U_{i,j+1}^{(k)} + U_{i,j-1}^{(k)} - 4U_{ij}^{(k)} - h^2 f_{ij}\right),$$

whilst for $i + j$ even,

$$U_{ij}^{(k+1)} = U_{ij}^{(k)} + \frac{\omega}{4}\left(U_{i+1,j}^{(k+1)} + U_{i-1,j}^{(k+1)} + U_{i,j+1}^{(k+1)} + U_{i,j-1}^{(k+1)} - 4U_{ij}^{(k)} - h^2 f_{ij}\right).$$

Updating all the black values first, followed by all the red values, gives a different order again. For each half-iteration a different value for the relaxation parameter ω could be employed so as to achieve the optimal convergence rate.

As with the Jacobi iteration, if, on a local memory multiprocessor, the work is apportioned according to a block column strategy, then it is necessary to synchronise at the end of each (half-)iteration so that data can be passed across the subdomain boundaries.

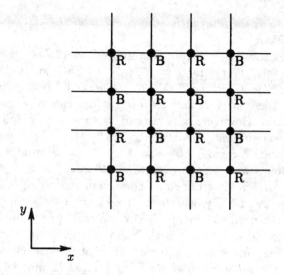

Figure 7.7 *Red/black ordering*

Alternating directions

Returning to (7.40), an alternative to (7.41) is first to employ a half-iteration in the x direction

$$U_{ij}^{(k+1/2)} = U_{ij}^{(k)} + \frac{\omega}{4}\left(U_{i+1,j}^{(k+1/2)} + U_{i-1,j}^{(k+1/2)} + U_{i,j+1}^{(k)}\right.$$

$$\left. + U_{i,j-1}^{(k)} - 2U_{ij}^{(k+1/2)} - 2U_{ij}^{(k)} - h^2 f_{ij}\right), \qquad (7.42)$$

$$i,j = 1, 2, \ldots, n,$$

followed by a half-iteration in the y direction

$$U_{ij}^{(k+1)} = U_{ij}^{(k+1/2)} + \frac{\omega}{4}\left(U_{i+1,j}^{(k+1/2)} + U_{i-1,j}^{(k+1/2)} + U_{i,j+1}^{(k+1)}\right.$$

$$\left. + U_{i,j-1}^{(k+1)} - 2U_{ij}^{(k+1)} - 2U_{ij}^{(k+1/2)} - h^2 f_{ij}\right), \qquad (7.43)$$

$$i,j = 1, 2, \ldots, n.$$

Rearranging (7.42) and (7.43) we have

$$-U_{i-1,j}^{(k+1/2)} + 2\left(1 + 2\omega^{-1}\right)U_{ij}^{(k+1/2)} - U_{i+1,j}^{(k+1/2)}$$

$$= U_{i,j-1}^{(k)} + 2\left(2\omega^{-1} - 1\right)U_{ij}^{(k)} + U_{i,j+1}^{(k)} - h^2 f_{ij}, \qquad (7.44)$$

and

$$-U_{i,j-1}^{(k+1)} + 2\left(1 + 2\omega^{-1}\right)U_{ij}^{(k+1)} - U_{i,j+1}^{(k+1)}$$

$$= U_{i-1,j}^{(k+1/2)} + 2\left(2\omega^{-1} - 1\right)U_{ij}^{(k+1/2)} + U_{i+1,j}^{(k+1/2)} - h^2 f_{ij}, \qquad (7.45)$$

which is referred to as the *alternating direction implicit* (ADI) *method* (Peaceman and Rachford, 1955).

It is natural to solve the system represented by (7.44) by columns; that is, for a fixed j we solve (7.44) for $i = 1, 2, \ldots, n$. Each such subsystem involves a tridiagonal coefficient matrix T which is diagonally dominant for positive ω. If ω is constant then so is T and its LU factors can be computed once only in $O(n)$ operations. However, it is normal to vary ω each full iteration. Since the solution of each subsystem is independent of that of every other, (7.44) can be solved in parallel for values of $j = 1, 2, \ldots, n$. Similarly, (7.45) is solved by rows; subsystems are formed according to values of $i = 1, 2, \ldots, n$. Each subsystem again involves a tridiagonal coefficient matrix of order n and can be solved independently of every other subsystem. On a shared memory system the main concern is the need to synchronise at the end of each of the half-iterations (7.44) and (7.45). Clearly, (7.44) is the more amenable to Fortran's storage by columns; (7.45) may involve a significant amount of data movement between a local cache and the memory modules. On a local memory system we have a similar situation. A distribution of the unknowns U_{ij} by columns is suitable for the first half-iteration, (7.44), but a distribution by rows is more suitable for the second, (7.45). This suggests that a transposition of the matrix of solution values at the end of each half-iteration is required, and this is very expensive in terms of communication.

7.4.2   Finite differences and parabolic equations

Consider the one-dimensional heat equation

$$\frac{\partial u}{\partial t} = c \frac{\partial^2 u}{\partial x^2},$$

where $u \equiv u(x, t)$ and c is a constant, on $(x, t) \in [0, 1] \times [0, \infty)$, subject to the initial temperature distribution $u(0, x) = g(x)$ and the boundary conditions $u(0, t) = \alpha(t)$, $u(1, t) = \beta(t)$. In a finite difference method we discretise in both space and time. For reasons which will soon become clear, the discretisation intervals will almost certainly be different (perhaps by one or two orders of magnitude). Let Δt be the time interval and, as before, let $h = 1/(n + 1)$ be the space discretisation. Then, using Taylor's series again,

$$u(x, t) = u(x, t - \Delta t) + \Delta t u_t(x, t - \Delta t) + O\left(\Delta t^2\right),$$

so that, for some time step t_{m-1}, and space ordinate x_i,

$$\frac{\partial u(x_i, t_{m-1})}{\partial t} = \frac{1}{\Delta t}\left(u(x_i, t_m) - u(x_i, t_{m-1})\right) + O\left(\Delta t\right). \tag{7.46}$$

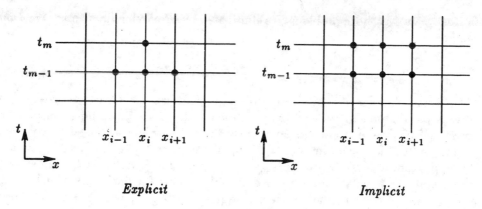

Explicit   **Implicit**

Figure 7.8 *Explicit and implicit finite difference methods*

To replace the space derivative we again use Taylor series approximations. For example, the second-order finite difference scheme of Section 7.4.1 gives

$$\frac{\partial^2 u(x_i, t_{m-1})}{\partial x^2} = \frac{1}{h^2} \left(u(x_{i+1}, t_{m-1}) - 2u(x_i, t_{m-1}) + u(x_{i-1}, t_{m-1}) \right) + O\left(h^2\right),$$
(7.47)

and, combining (7.46) with (7.47) and ignoring the discretisation errors, we have

$$U_{im} - U_{i,m-1} + \mu \left(U_{i+1,m-1} - 2U_{i,m-1} + U_{i-1,m-1} \right),$$
(7.48)

for computing U_{im}, an approximation to $u(ih, m\Delta t)$, where $\mu = c\Delta t/h^2$. (See the left-hand diagram of Figure 7.8 for the interdependence of solution values.) It is natural to solve (7.48) for $m = 1, 2, \ldots$, in turn, with $i = 1, 2, \ldots, n$, at each time step (each value of m). Since the method is explicit, for a fixed m the approximate solution values U_{im}, $i = 1, 2, \ldots, n$, are independent of each other, and may be computed concurrently. A natural work sharing strategy is to employ some form of domain decomposition in the space dimension. Typically, we would organise the x-grid points into blocks. The beginning of the computation at each time step then imposes a synchronisation point which, in a local memory environment, requires process intercommunication of the approximate solution values at the edges of the space discretisation blocks.

It is not too difficult to see intuitively that the simple explicit method (7.48) is second-order in space but only first-order in time. Essentially we have employed Euler's method (Section 7.3.1) to integrate forward over $[t_{m-1}, t_m]$. Further, despite its inherent attractions in the context of parallel numerical algorithms, the method has a serious drawback, namely that it is stable only for $\mu \in [0, 0.5]$ (Smith, 1978, p. 83), and this can place a severe restriction on the size of the time interval Δt. This latter stability problem can be overcome by using a fully implicit method. The problem of the low-order accuracy of Euler's method could, of course, be overcome by employing a higher-order finite difference approximation in time. However, both problems are eliminated by employing the *Crank–Nicolson method*

$$U_{im} = U_{i,m-1} + \frac{\mu}{2} \left(U_{i+1,m} - 2U_{im} + U_{i-1,m} + U_{i+1,m-1} - 2U_{i,m-1} + U_{i-1,m-1} \right),$$
(7.49)

(Crank and Nicolson, 1947) which averages second-order space differences at two time steps (Figure 7.8). The method represented by (7.49) is second-order in both space and time, and is unconditionally stable. It is used to advance the solution in time, so that for each time step, that is, for a fixed m, (7.49) is used to calculate the approximate solution values U_{im}, $i = 1, 2, \ldots, n$. Thus, at each time step it is necessary to solve a symmetric tridiagonal system of n linear equations which are of a similar form to those which arise in the ADI method. Parallel methods for such systems were discussed in Section 5.4.1. An alternative approach is to solve (7.49) by iteration. If the Jacobi iteration,

$$U_{im}^{(k+1)} = U_{i,m-1}^{(k)} + \frac{\mu}{2} \left(U_{i+1,m}^{(k)} - 2U_{im}^{(k)} + U_{i-1,m}^{(k)} \right.$$

$$\left. + U_{i+1,m-1}^{(k)} - 2U_{i,m-1}^{(k)} + U_{i-1,m-1}^{(k)} \right),$$

is used then the components corresponding to a single iteration, $U_{im}^{(k+1)}$, $i = 1, 2, \ldots, n$, can be computed concurrently. Alternative approaches are to use Gauss–Seidel or SOR iterations in the space dimension. Whether the explicit or the implicit method is the more efficient depends on the balance between the cost of solving (7.49), say by iteration, and the restriction on the time interval required to maintain stability of the explicit method.

7.4.3   Finite elements

The *finite element method* (FEM) is a member of a class of methods collectively referred to as *projection*, or *expansion*, *methods*. In a finite difference method we look for a solution on a set of mesh points; in a projection method we express an approximate solution as a linear combination of known basis functions and aim to determine values for the coefficients in this expansion. It so happens that with the FEM these coefficients are often approximate solution values at discrete points.

The FEM can be derived by a reformulation of the differential equation using a *variational principle*. Again, we use a simple problem, Poisson's equation, (7.33), on $\Omega \subset \mathcal{R}^2$, subject to Dirichlet boundary conditions $u(x, y) = 0$ on Γ, to illustrate the technique. If the discussion is restricted to members of the space of twice differentiable functions which satisfy the given boundary conditions, it can be shown that the solution to the differential equation minimises the *functional*

$$I(u) = (\mathcal{L}u, u) - 2(u, f),$$
(7.50)

where $\mathcal{L}$ is the *differential operator* ∇^2 and, for any functions $u(x, y)$ and $v(x, y)$,

$$(u, v) = \int\int_\Omega uv\,dxdy,$$

(Mikhlin, 1964). It can further be shown that a minimum point of (7.50) must satisfy

$$(\mathcal{L}u, v) = (f, v), \tag{7.51}$$

for all twice differentiable functions v which satisfy the boundary conditions on Γ. (7.51) is referred to as the *weak form* of the differential equation, and it follows that its solution, the so-called *weak solution*, satisfies the original differential equation. Using Green's Theorem (integration by parts) the weak form may be written as

$$a(u, v) = (f, v), \tag{7.52}$$

where

$$a(u, v) = \int\int_\Omega \left(\frac{\partial u}{\partial x}\frac{\partial v}{\partial x} + \frac{\partial u}{\partial y}\frac{\partial v}{\partial y} \right) dxdy,$$

and we refer to (7.52) as the *Galerkin form*.

To proceed we approximate the solution $u(x, y)$ by $U(x, y) = \sum_{i=1}^N \alpha_i \phi_i(x, y)$, where ϕ_i, $i = 1, 2, \ldots, N$, is a set of *basis*, or *trial*, *functions* which satisfy the same boundary conditions as u, and α_i, $i = 1, 2, \ldots, N$, is a set of *expansion coefficients*. Further, we restrict the choice of v in (7.52) to a finite-dimensional space spanned by a set of *test functions* $\psi_j(x, y)$, $j = 1, 2, \ldots, N$. Then the Galerkin form becomes

$$a\left(\sum_{i=1}^N \alpha_i \phi_i, \psi_j \right) = (f, \psi_j), \qquad j = 1, 2, \ldots, N,$$

which is a system of N linear equations in α_i, $i = 1, 2, \ldots, N$. Different choices of ψ_j lead to different classes of method. The particular choice $\psi_j = \phi_j$ leads to the *Ritz method*, and this is equivalent to substituting the approximate solution U in the functional (7.50), and minimising with respect to the expansion coefficients. See Wait and Mitchell (1985) for a more detailed discussion.

In the finite element method Ω is discretised using a number of polygonal elements. For example, the unit square could be discretised using rectangles or triangles, with the element sizes reflecting the expected behaviour of the solution (Figure 7.9). The trial functions ϕ_i are chosen to have local support only (that is, they are non-zero only over a small number of neighbouring elements) and are defined so that the expansion coefficients correspond to approximate solution values at the nodes. (Vertices of the elements are usually chosen as nodes for this purpose; depending on the order of the approximation, points along element sides and in the interior of elements may also be selected as nodes.) The important thing here is that the solution to the Ritz problem reduces to a system of linear equations of the form $A\mathbf{x} = \mathbf{b}$ where

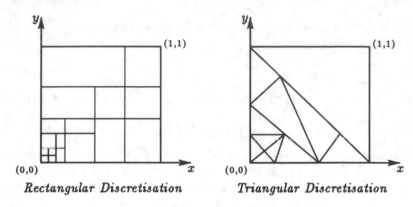

Figure 7.9 *Polygonal discretisations of the unit square*

- the components of the coefficient matrix A (which is sparse, due to the local support of the basis functions, symmetric and positive definite) are surface integrals over the elements involving the differential operator,
- the components of the right-hand side vector are surface integrals involving f, and
- the solution vector defines the expansion coefficients.

The sparsity pattern of A, the *global stiffness matrix*, will depend on the order in which the element nodes are defined and an appropriate node reordering phase is often a precursor to the main set-up and solve phases. Depending on the strategy employed, the non-zero elements can be grouped around the principal diagonal, or spread out. Precisely which is appropriate will depend on the scheme used to solve the resulting system.

Ignoring the possibility of node reordering, the main computational components of the FEM are the setting up of the coefficient matrix A and right-hand side vector **b**, and the solution of the resulting system of linear equations. The set-up phase offers straightforward parallelism since the volume integrals constitute independent computations provided that contributions to A are assembled in local (element) submatrices. Expansion coefficients which correspond to points common to more than one element involve contributions to the stiffness matrix from each such element and it is therefore essential that the assembly of the local submatrices into the global stiffness matrix is synchronised. If a direct method is to be employed for solving the linear system, the solution phase can begin before the matrix set-up phase is complete, and we refer to a *frontal method*. The situation is similar to that considered in Section 6.3.3 in the context of systems of nonlinear equations, with the additional consideration that the sparsity of A may mean that, as in Section 5.4.2, several pivots can become available at once. We refer to this as a *multifrontal method* in which parallelism is exploited to overlap matrix assembly with elimination both within and across the stages of the inclusion of local submatrices; see Duff (1986). Alternatively, A is fully assembled as a banded matrix and the system of equations is then solved by iteration, with the most popular technique being the *conjugate gradient method*.

Conjugate gradient method

We consider the objective function

$$f(\mathbf{x}) = \frac{1}{2}(A\mathbf{x}, \mathbf{x}) - (\mathbf{b}, \mathbf{x}), \qquad (7.53)$$

where A is a symmetric, positive definite, banded matrix and $(\cdot, \cdot)$ is the usual dot product of two vectors. Then, at the minimum point $\mathbf{x}^*$ of f the gradient vector $A\mathbf{x} - \mathbf{b}$ is zero, that is,

$$A\mathbf{x}^* = \mathbf{b}. \qquad (7.54)$$

Hence, minimising (7.53) is equivalent to solving the system of linear equations (7.54). We have a standard unconstrained minimisation problem which can be solved using the conjugate gradient (CG) method, taking special note of the quadratic form of f. The derivation of the method is lengthy and we refer the interested reader to Gill, Murray, *et al.* (1981) for full details. The reader is asked to accept a description of the method without a mathematical justification.

Starting with the initial values $\mathbf{r}^{(1)} = \mathbf{p}^{(1)} = \mathbf{b} - A\mathbf{x}^{(1)}$, where $\mathbf{x}^{(1)}$ is an initial approximation to $\mathbf{x}^*$, $\mathbf{r}^{(k)}$ denotes the negative gradient vector at $\mathbf{x}^{(k)}$, and $\mathbf{p}^{(k)}$ is a search direction, the kth iteration of the CG method is given by

1. $\mathbf{q}^{(k)} = A\mathbf{p}^{(k)}$,
2. $\alpha^{(k)} = \left(\mathbf{r}^{(k)}, \mathbf{r}^{(k)}\right) / \left(\mathbf{p}^{(k)}, \mathbf{q}^{(k)}\right)$,
3. $\mathbf{r}^{(k+1)} = \mathbf{r}^{(k)} - \alpha^{(k)}\mathbf{q}^{(k)}$,
4. $\mathbf{x}^{(k+1)} = \mathbf{x}^{(k)} + \alpha^{(k)}\mathbf{p}^{(k)}$,
5. $\beta^{(k)} = \left(\mathbf{r}^{(k+1)}, \mathbf{r}^{(k+1)}\right) / \left(\mathbf{r}^{(k)}, \mathbf{r}^{(k)}\right)$,
6. $\mathbf{p}^{(k+1)} = \mathbf{r}^{(k+1)} + \beta^{(k)}\mathbf{p}^{(k)}$.

Although expressed here as an iterative method, the CG method converges to $\mathbf{x}^*$ in at most n iterations in the absence of round-off errors.

The CG method is rich in Levels 1 and 2 BLAS operations. In detail, step (1) is a matrix-vector multiplication (**ssbmv**, since A is banded and symmetric), steps (2) and (5) require the determination of dot products (**sdot**), whilst steps (3), (4) and (6) all involve adding a multiple of one vector to another (**saxpy**). Note that each iteration involves the calculation of two dot products only; $\left(\mathbf{p}^{(k)}, \mathbf{q}^{(k)}\right)$ and $\left(\mathbf{r}^{(k+1)}, \mathbf{r}^{(k+1)}\right)$. The latter value is used as $\left(\mathbf{r}^{(k)}, \mathbf{r}^{(k)}\right)$ of the next iteration.

Parallelisation of the CG method appears to be straightforward since there is potential parallelism to be exploited at each step. Much of it is fine-grain, with the operation count of the matrix-vector multiplication of step (1) offering the largest grain size. The form in which we have expressed the algorithm suggests a number of synchronisation points. For example, the calculation of the vector $\mathbf{q}^{(k)}$ of step (1) needs to be completed before the inner product $\left(\mathbf{p}^{(k)}, \mathbf{q}^{(k)}\right)$ of step (2)

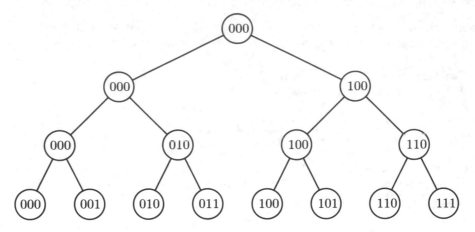

Figure 7.10 *Tree structure for inner product accumulation*

can be fully formed, and this inner product must, in turn, be available for $\mathbf{r}^{(k+1)}$ of step (3) to be determined. There is, however, some concurrency amongst the steps. For example, step (4) can be performed concurrently with steps (3), (5) and (6).

On a local memory machine we could distribute A by block rows with all the vectors blocked conformally. The matrix-vector product of step (1) requires a certain amount of interprocess communication, whilst the inner product required to determine the denominator in step (2) involves the accumulation of partial dot products. The model implementation of **sdot** described in Section 3.3.1 requires all the processes to send their partial results to a single host process where the final sum is formed, and this accumulation of the final result requires $O\,(p)$ sequential steps, where p is the number of processes. A further $O\,(p)$ sequential steps might be required to distribute the result to all the processes. Assuming the usual one-to-one correspondence of processes to processors, this can be improved by imposing a tree structure on the connectivity of the processors. For example, in Figure 7.10 the binary labels for the nodes of the tree indicate the processor number for a hypercube of dimension 4. The process at each node receives the partial results from its successors (one of which is itself), accumulates these partial results and passes the new partial result to its predecessor. Now the number of time steps in the accumulation of the final result is $O\,(\log_2 p)$. The same structure can be employed to distribute the result, again in $O\,(\log_2 p)$ time steps. An alternative on a hypercube is to employ a neighbour exchange approach similar to that suggested in Exercise 1.6.

The **saxpy** operations of steps (3) and (4) represent, individually, independent operations according to our data distribution. Step (5) requires a further inner product accumulation and broadcast, followed by, in step (6), another **saxpy** operation.

We recall that, on a local memory machine, each communication involves a start-up cost and a transmission cost (Section 3.3.1). Aykanat, Özgüner, *et al.* (1988) suggest a simple reorganisation of the CG method which reduces the former. It can be shown that the CG method generates orthogonal gradients (Gill, Murray, *et al.*, 1981) so that here the vectors $\mathbf{r}^{(k)}$ are mutually orthogonal, that is,

$$\left(\mathbf{r}^{(k)}, \mathbf{r}^{(j)}\right) = 0, \qquad k \neq j.$$

Hence

$$\left(\mathbf{r}^{(k+1)}, \mathbf{r}^{(k+1)}\right) = \left(\mathbf{r}^{(k+1)}, \mathbf{r}^{(k)} - \alpha^{(k)}\mathbf{q}^{(k)}\right) = -\alpha^{(k)} \left(\mathbf{r}^{(k+1)}, \mathbf{q}^{(k)}\right).$$

It follows from the definitions of $\alpha^{(k)}$ and $\beta^{(k)}$ in steps (2) and (5) that

$$\beta^{(k)} = \frac{\left(\mathbf{r}^{(k+1)}, \mathbf{r}^{(k+1)}\right)}{\left(\mathbf{r}^{(k)}, \mathbf{r}^{(k)}\right)} = -\alpha^{(k)} \frac{\left(\mathbf{r}^{(k+1)}, \mathbf{q}^{(k)}\right)}{\left(\mathbf{r}^{(k)}, \mathbf{r}^{(k)}\right)} = -\frac{\left(\mathbf{r}^{(k+1)}, \mathbf{q}^{(k)}\right)}{\left(\mathbf{p}^{(k)}, \mathbf{q}^{(k)}\right)}. \qquad (7.55)$$

Now

$$\left(\mathbf{r}^{(k+1)}, \mathbf{q}^{(k)}\right) = \left(\mathbf{r}^{(k)}, \mathbf{q}^{(k)}\right) - \alpha^{(k)} \left(\mathbf{q}^{(k)}, \mathbf{q}^{(k)}\right), \qquad (7.56)$$

and

$$\begin{aligned}
\left(\mathbf{r}^{(k)}, \mathbf{q}^{(k)}\right) &= \left(\mathbf{p}^{(k)}, \mathbf{q}^{(k)}\right) - \beta^{(k-1)} \left(\mathbf{p}^{(k-1)}, \mathbf{q}^{(k)}\right), \\
&= \left(\mathbf{p}^{(k)}, \mathbf{q}^{(k)}\right) \quad \beta^{(k-1)} \left(\mathbf{p}^{(k-1)}, A\mathbf{p}^{(k)}\right), \\
&= \left(\mathbf{p}^{(k)}, \mathbf{q}^{(k)}\right), \qquad (7.57)
\end{aligned}$$

where we have used the fact that $\mathbf{p}^{(l)}$, $l = 1, 2, \ldots, k$, are A-conjugate, that is, $\left(\mathbf{p}^{(i)}, A\mathbf{p}^{(j)}\right) = 0$, $i, j = 1, 2, \ldots, k$, $i \neq j$ (Gill, Murray, *et al.*, 1981). Combining (7.55), (7.56) and (7.57) we have

$$\beta^{(k)} = \alpha^{(k)} \frac{\left(\mathbf{q}^{(k)}, \mathbf{q}^{(k)}\right)}{\left(\mathbf{p}^{(k)}, \mathbf{q}^{(k)}\right)} - 1. \qquad (7.58)$$

Once the matrix-vector product $\mathbf{q}^{(k)} = A\mathbf{p}^{(k)}$ of step (1) has been formed the inner products $\left(\mathbf{p}^{(k)}, \mathbf{q}^{(k)}\right)$ and $\left(\mathbf{q}^{(k)}, \mathbf{q}^{(k)}\right)$, the latter replacing $\left(\mathbf{r}^{(k+1)}, \mathbf{r}^{(k+1)}\right)$, can be formed concurrently. Whilst the amount of data transmitted remains the same, the transfer can be performed with only one start-up cost instead of the two that were required before. $\beta^{(k)}$ is formed from (7.58) rather than as shown in step (5). Indeed, the role of this step has been reversed. Before we needed to calculate $\left(\mathbf{r}^{(k+1)}, \mathbf{r}^{(k+1)}\right)$ in order to determine $\beta^{(k)}$; now we form $\beta^{(k)}$ from (7.58) and then calculate $\left(\mathbf{r}^{(k+1)}, \mathbf{r}^{(k+1)}\right)$ from

$$\left(\mathbf{r}^{(k+1)}, \mathbf{r}^{(k+1)}\right) = \beta^{(k)} \left(\mathbf{r}^{(k)}, \mathbf{r}^{(k)}\right).$$

This modified CG method is mathematically equivalent to the original method although, in the presence of round-off errors, the two methods may not be numerically equivalent.

7.4.4   Series methods

For problems defined on simple domains we can make use of Fourier series (and, in particular, the FFT). For example, consider again Poisson's equation (7.33) on the unit square, but this time subject to *periodic* boundary conditions of the form $u(x,0) = u(x,1)$ and $u(0,y) = u(1,y)$. We express the five-point finite difference replacement (7.37) as

$$U_{j+1,k} + U_{j-1,k} + U_{j,k+1} + U_{j,k-1} - 4U_{jk} = h^2 f_{jk}. \tag{7.59}$$

Now, using the results of Section 7.2, and the periodicity of the boundary conditions, we write the Fourier analysis of U_{jk} as

$$V_{lm} = \sum_{j=0}^{n-1} \sum_{k=0}^{n-1} w_n^{-(jl+km)} U_{jk}, \qquad l, m = 0, 1, \ldots, n-1,$$

with the corresponding synthesis

$$U_{jk} = \frac{1}{n^2} \sum_{l=0}^{n-1} \sum_{m=0}^{n-1} w_n^{jl+km} V_{lm}, \qquad j, k = 0, 1, \ldots, n-1. \tag{7.60}$$

Similarly, the Fourier analysis of f_{jk} gives

$$g_{lm} = \sum_{j=0}^{n-1} \sum_{k=0}^{n-1} w_n^{-(jl+km)} f_{jk}, \qquad l, m = 0, 1, \ldots, n-1,$$

with the corresponding synthesis

$$f_{jk} = \frac{1}{n^2} \sum_{l=0}^{n-1} \sum_{m=0}^{n-1} w_n^{jl+km} g_{lm}, \qquad j, k = 0, 1, \ldots, n-1. \tag{7.61}$$

Substituting (7.60) and (7.61) into (7.59) and using the identity $e^{i\theta} = \cos\theta + i\sin\theta$, we obtain the equations

$$2\left(\cos(2\pi l/n) + \cos(2\pi m/n) - 2\right) V_{lm} = g_{lm}. \tag{7.62}$$

Hence, the solution of the differential equation involves the following steps:

- Evaluate $f(x, y)$ on the two-dimensional grid defined by the mesh points x_j, $j = 0, 1, \ldots, n - 1$, and y_k, $k = 0, 1, \ldots, n - 1$, to give $f_{jk} = f(x_j, y_k)$. Note that the labelling must be such that $x_0 = y_0 = 0$ and $x_n = y_n = 1$. With a suitable mapping of the grid to processes (for example, block columns) these function evaluations may be performed concurrently.
- Fourier analyse f_{jk}, $j, k = 0, 1, \ldots, n - 1$, in parallel, using the techniques outlined in Section 7.2.4, to give the coefficients g_{lm}, $l, m = 0, 1, \ldots, n - 1$.
- Use (7.62) to calculate V_{lm}, $l, m = 0, 1, \ldots, n - 1$. Again, given an appropriate grid to process mapping, the required divisions may be performed concurrently.
- Fourier synthesise the resulting values V_{lm} to give the approximate solution values U_{jk}, $j, k = 0, 1, \ldots, n - 1$. These transforms are again performed in parallel using the techniques outlined in Section 7.2.4.

On a local memory multiprocessor there is likely to be significant data communication at the end of each analysis and synthesis step.

Exercises

7.1. Construct diagrams, equivalent to Figures 7.1 and 7.2, corresponding to the Fourier analysis of $N = 8$ complex values. In a local memory environment two processes/processors are available and f_j is allocated to process number j mod 2. What interprocess communications are involved? How would you distribute the calculations so that g_k is resident on processor k mod 2? (Hockney and Jesshope, 1988.)

7.2. A numerical solution to the second-order ODE

$$y''(t) = f\left(t, y(t), y'(t)\right), \qquad a < t < b,$$

subject to the boundary conditions $y(a) = \alpha$ and $y(b) = \beta$, is required. Given a uniform discretisation $\{t_i\}$, with $t_i = t_{i-1} + h$, use Taylor's series to show that

$$y''(t_i) = \frac{y(t_{i+1}) - 2y(t_i) + y(t_{i-1})}{h^2} + O\left(h^2\right),$$

$$y'(t_i) = \frac{y(t_{i+1}) - y(t_{i-1})}{2h} + O\left(h^2\right).$$

Indicate how these results can be used to derive a system of nonlinear equations defining approximations Y_i to $y(t_i)$. How could a red/black ordering be employed in the solution of this system by iteration?

7.3. In a finite difference approximation to an elliptic partial differential equation U_{ij} is expressed as a linear combination of its eight nearest neighbours. Show that if SOR is employed the points may be coloured so that the next set of iterates may be computed in four stages, with each stage consisting of independent operations.

7.4. Derive an SOR method for solving the Crank–Nicolson finite difference replacement to the one-dimensional heat conduction equation. Indicate the iteration based on a red/black ordering in the space dimension.

7.5. A five-point finite difference star is to be used to solve Poisson's equation on the unit square, with four interior grid points in each space dimension. A *speedometer ordering* of the nodes is obtained by first labelling all the nodes corresponding to $y = y_1$ in increasing order of the x-subscript, then labelling all the nodes corresponding to $y = y_2$ in the same way, and so on. Derive the form of the coefficient matrix which results from this ordering. Verify that the matrix is symmetric, diagonally dominant and banded. Derive the form of the coefficient matrix resulting from a red/black ordering, taking, on a first sweep, (x_i, y_1), $i = 1, 3$; (x_i, y_2), $i = 2, 4$; (x_i, y_3), $i = 1, 3$; (x_i, y_4), $i = 2, 4$; and on a second sweep, (x_i, y_1), $i = 2, 4$; (x_i, y_2), $i = 1, 3$; and so on. Which matrix is the more appropriate on a multiprocessor for (a) Cholesky factorisation, and (b) the CG method? (Ortega and Poole, 1981.)

Further reading

The FFT algorithm is of considerable interest, both as a practical tool and as an interesting problem with which to exploit parallel architectures. One of the earliest attempts to describe a parallel implementation is given by Pease (1968). Swartztrauber (1987) considers implementations on both scalar and vector shared memory systems, and on hypercubes. Van Loan (1992) considers many aspects of FFTs, including their implementation on shared and local memory systems.

One of the first overviews of parallel numerical methods for first-order IVPs in ODEs is that of Gear (1987). Besides parallelism across the problem/method, other approaches for initial-value problems have been suggested which employ spare processor power to perform tasks which, although significant, are peripheral to the main thrust of a method. The central theme of many of these methods is error estimation, although they differ as to the use to which such an error estimate is put. For example, the error estimates can be used to yield improved approximations using Richardson extrapolation (van der Houwen and Sommeijer, 1990). Keller and Nelson (1989a; 1989b) consider hypercube implementations of parallel shooting.

For an overview of parallel numerical methods for PDEs see Ortega and Voigt (1985). Cook (1992) considers a further reformulation of the CG method which, on a transputer array, permits communication of the dot products to be overlapped with computation.

In practical implementations of the CG method a modification is usually made which improves its convergence characteristics. This involves the use of a *preconditioning matrix M*, an approximation to A^{-1}, and the method is referred to as a *preconditioned conjugate gradient* (PCG) *method.* In a PCG method we determine $\hat{\mathbf{r}}^{(k+1)}$ by solving $M\hat{\mathbf{r}}^{(k+1)} = \mathbf{r}^{(k+1)}$ and calculate $\beta^{(k)} = \left(\mathbf{r}^{(k+1)}, \hat{\mathbf{r}}^{(k+1)}\right) / \left(\mathbf{r}^{(k)}, \hat{\mathbf{r}}^{(k)}\right)$, and $\mathbf{p}^{(k+1)} = \hat{\mathbf{r}}^{(k+1)} + \beta^{(k)}\mathbf{p}^{(k)}$. The definition of $\alpha^{(k)}$ remains unchanged. See Ortega and Voigt (1985) for a discussion of the basic method and the references therein for parallel implementations. See also Saad (1989) for an overview of parallel implementations of the PCG method.

Sweet, Briggs, *et al.* (1991) consider a hypercube implementation of an FFT solution of Poisson's equation in three dimensions.

Multigrid methods are a relatively recent, and efficient, means of solving the equations which arise in the discretisation of PDEs. The parallelisation of multigrid methods has been considered by a number of authors, including Decker (1991) and Frederickson and McBryan (1991); see also the papers in McCormick (1988) and Mandel, McCormick, *et al.* (1989).

The parallel solution of systems of nonlinear equations arising in the areas of optimisation and ordinary and partial differential equations is considered by Dixon (1991).

CHAPTER 8

Recent developments

8.1 An ever changing world

Inevitably with a book of this nature, as soon as it appears certain aspects of it will already be dated. We remarked in Chapter 1 that each five-year period between 1950 and 1980 was witness to a tenfold increase in the achievable performance of computer systems. That this trend continues, apparently unabated, is partly due to yet further improvements in the hardware of single-processor systems, but it is also a result of the increasing reliance on parallelism in one form or another.

In this final chapter we attempt to look into the near future and the not so near future and attempt to predict, in a general way, how we believe parallel systems may evolve. Throughout this book we have carefully, even painfully, drawn a distinction between shared and local memory multiprocessors, and have employed two particular systems, the Encore Multimax and the Intel Hypercube, as examples which are, in one way or another, 'typical' of the architectures which are commercially available at the beginning of the 1990s. That such systems will continue to be available in the near future appears to be unquestionable and in Section 8.2 we describe the most recent developments at Encore and Intel. In Section 8.3 we review the principles which underpin these systems and question whether the shared memory/local memory distinction is entirely necessary from a software point of view. Without it, a unified approach to the development of parallel numerical algorithms is, perhaps, possible, and in Section 8.4 we highlight the features of several past, current and near-future systems which attempt to blur the shared memory/local memory division, providing a shared memory software model on a local memory architecture. Finally, in Section 8.5 we consider developments in the longer term; in particular, we look at theoretical models which may have a major influence on the design of future parallel systems, and on the way such systems will be programmed.

For the most part we concentrate on developments in hardware. It goes without saying that if future systems are to have commercial viability the software support must follow.

8.2   Developments at Encore and Intel

Following the Multimax range of machines, Encore produced the Encore 91 and 93 series of machines based on Motorola 88100 RISC processors. (RISC stands for *reduced instruction set computer*. The idea is that if the instruction set is limited to a core of fast instructions, operating system and compiler writers can hope to produce high-performance object codes.) Each processor card on an Encore 93 series machine has two or four 88100 processors, each capable of a single-precision peak performance of 20 Mflops, two Motorola 88200 cache/memory management units and 1 Mbyte of secondary cache, all connected together by a local bus. Processor cards are connected to shared memory cards, with 32 or 64 Mbytes memory, by a nanobus. Up to eight processor cards and up to ten memory cards can be included in an Encore 93, so that a maximally configured system has 32 processors and 640 Mbytes of global memory, giving a peak performance of about 600 mips (compared to the maximum figure of 171 mips for a Multimax 520). The nanobus of the Encore 93 is essentially unchanged from that used in the Multimax, yet the processors are much more powerful. This imbalance will eventually be rectified by a faster bus; we understand that Encore is aiming at an order-of-magnitude increase in the performance of the nanobus in the medium term.

The top-of-the-range system which is currently being shipped by Intel remains the iPSC/860; recent developments have concentrated on the software. Intel is heavily involved in the Touchstone Program, which is a research and development project involving industry, a number of universities and the US government, and is co-funded by DARPA (Defense Advanced Research Projects Agency) and the Intel Corporation. It seems reasonable to speculate that future products from Intel will build on the experiences of the Touchstone Program which has involved the development of four prototype hardware systems: IOTA, GAMMA, DELTA and SIGMA. IOTA (September 1989) and GAMMA (December 1989) consist of up to 128 nodes connected by a hypercube interconnection network; GAMMA employs the i860 processor. DELTA (June 1990) consists of up to 512 nodes which are grid, rather than hypercube, connected. Each node consists of an i860 processor and a mesh routing chip which supports *wormhole routing* (Dally and Seitz, 1987; Seitz and Flaig, 1987). A worm is a message packet whose destination is defined by a header; it is processed by the mesh routing chips in a pipeline fashion. The use of wormhole routing limits the amount of storage necessary per node for messages. Further, Athas and Seitz (1988) claim that this makes the message latency largely insensitive to the distance in the network that a message has to travel. The final prototype in the Touchstone Program, SIGMA, should scale to at least 2048 processors. In addition to these hardware developments the Touchstone Program has addressed software issues, such as operating systems and memory management techniques.

8.3   The shared/local divide

Throughout this book we have drawn a clear distinction between shared and local memory multiprocessors. When deriving numerical algorithms for such machines we were, to some extent, able to ignore the differences, but when implementation details were addressed, the various features of the two architectures greatly influenced the way we were able to proceed. The reader may wonder whether we have overdone things. From a software portability point of view it would clearly be desirable to write parallel applications programs which are independent of the underlying architecture. The question then arises as to whether either of the two models is acceptable as the 'standard'.

Although we can think of a local memory system as a shared memory system in which there happens to be a one-to-one correspondence between processors and memory modules, we are very unlikely to be able to maintain the attribute that all memory accesses are of unit cost. Consequently, any code developed to run on a local memory system must be written in such a way that the data distribution of the code allows as much use as possible of references to a processor's own local memory, as opposed to the local memory of some other processor. In a shared memory system, however, the data distribution is, in principle, almost irrelevant and of little concern to the programmer who may write codes without giving any thought to such issues. Further, as we have seen, the programmer often has no control over the distribution of his data, or even the allocation of processes to processors. In practice the situation is a little more complicated since processors will possess a certain amount of cache memory, and the program writer will need to write code which makes full use of data stored in the cache. Indeed, as caches become larger it could be argued that shared memory systems are, in fact, local memory systems which provide interprocessor communication by using page swaps with global memory.

In Chapter 1 we attempted a brief comparison of shared and local memory systems. We recall that in terms of software support (compilers, debuggers, etc.) shared memory systems offer a much more user-friendly environment. On the other hand, because of the difficulties associated with building shared memory systems with more than thirty or forty processors, local memory systems offer the only real hope of a scalable architecture in which the number of processors may be in the hundreds or thousands. The question is, therefore, whether it is possible to get the best of both worlds in a single machine. There have already been several attempts in this direction and we now look at some currently available commercial systems.

8.4 Virtual shared memory multiprocessors

The three systems described in this section are, from a hardware point of view, local memory systems but, from a software point of view, shared memory systems. They are scalable and hence systems can be constructed of an almost unlimited number of processors. Although all of the memory is physically distributed, there is some form of global addressing of the memory (we refer to a *single address space*), so that, as far as the applications programmer is concerned, the machine can be treated as though it were a shared memory machine. These machines are therefore said to support *virtual shared memory* (VSM). A system designed on this basis offers the relative ease of use of a shared memory machine combined with the scalability of a local memory machine.

The first member of the Butterfly family of parallel systems which appeared from BBN Advanced Computers Inc. was the Butterfly I in the early 1980s. The latest product (end 1991) is the BBN TC2000. In some ways it is similar to the Encore 93 series; in particular, it uses the same processors. Where the Butterfly differs from the Encore is in the interconnection (a switch in place of a bus) and the distribution of the main memory.

Each processing element of the BBN TC2000 consists of a Motorola 88100 processor, together with two 88200 cache/memory management units and 16 Mbytes of main memory, bus-connected. As with the Encore, one of the 88200s provides a 16-kbyte instruction cache, the other a 16-kbyte data cache. Drawing a direct comparison, therefore, on the BBN there are two 88200s per 88100, whereas there are two 88200s per board (of two or four 88100s) on the Encore, and the BBN has 16 Mbytes of main memory compared with the 1 Mbyte of secondary cache per board of the Encore. The processing elements are interconnected using a *butterfly switch*, a multistage switch built from 8 × 8 crossbar switches (Feng, 1981). A two-stage TC2000 butterfly switch can have up to 64 input/output ports and a three-stage switch up to 512 ports. Even though the memory is physically distributed it is globally addressable and the software supports a VSM. For more details of the hardware and software of the TC2000, see BBN (1989).

High-level language support for the TC2000 consists of compilers for Fortran, Ada and C, and a set of tools to aid the debugging and performance tuning of parallel programs. TC2000 Fortran conforms to the ANSI Fortran 77 standard and contains extensions for parallel programming which follow the spirit of PCF (Section 2.4.6). In particular, the compiler supports the constructs `PARALLEL DO`, with qualifiers additional to those given in Section 2.4.6, `PARALLEL REGION`, and `PARALLEL SECTIONS`. There are further extensions which permit variables to be declared as shared among all processes or private to one process, and which allow arrays to be scattered across processor memories to exploit the scalable bandwidth of the butterfly switch.

The next generation of BBN machine is scheduled to be the Coral. Some research and development on the Coral project has been undertaken, but it is our understanding that the cost of further development cannot be supported from the current revenue stream and that the project is on hold until external funding can be found. TC2000 systems continue to be delivered.

Also based on the VSM concept is the KSR1, developed and manufactured by Kendall Square Research Corporation. The first four systems were delivered towards the end of 1991. The KSR1 is built from a number of processor cells, each of which has a high-performance floating-point unit capable of 40 Mflops peak performance, a 0.5 Mbyte cache and 32 Mbytes of memory. The cells are connected in a hierarchy of clusters of processor cells. The lowest level, termed Search Engine:0, consists of a ring of up to 32 processor cells; at the next level, Search Engine:1, up to 34 of the lower level, Search Engine:0, clusters, are interconnected. In principle this hierarchy can be continued, with a Search Engine:2 consisting of a ring of a number of Search Engine:1s, and so on; thus the architecture is scalable. The Search Engines pass data around from one station to the next in a conveyor-belt fashion in packets of 128 bytes.

To indicate how the KSR1 supports a VSM we suppose that a process which is executing on a processor labelled A requires access to a data item which is stored on a different processor, labelled B, attached to the same Search Engine:0 as processor A. When the access is requested, processor A recognises that the data must be imported from a different processor by checking the data which is stored in its own memory. A packet containing the data request is assembled by processor A and sent onto the Search Engine:0. It passes through the different processor cells on Search Engine:0 until it finds processor B which contains the requested data. A response packet containing a 128-byte block which includes the requested data is assembled by processor B and returned, via Search Engine:0, to processor A. If processor B is on a different Search Engine:0 from processor A then the Ring Routing Cells (RRCs) play a crucial role. Each Search Engine:0 has an RRC which knows the contents of the memories of all its processor cells. When the data request from processor A reaches the RRC of its own Search Engine:0 a decision is made as to whether or not the requested data is on this Search Engine:0. In the case that the data is on some other Search Engine:0 the data request moves onto Search Engine:1 and passes through the different RRCs until it finds the Search Engine:0 which holds the requested data. The request then moves around this Search Engine:0 to find processor B, collects the 128-byte block containing the requested data, and returns to processor A via processor B's Search Engine:0, Search Engine:1, and finally processor A's Search Engine:0.

We have described how a request for access to data is satisfied by searching for the physical memory location of that data. The other major technical problem which is solved by the KSR1 is to ensure that all the processor cells have access to the latest values of data. If data is acquired for reading only, a local copy is made in the cache of the processor that needs the data; the original copy remains for access by other processors. On a write access the data is brought to the processor which wishes to write and all other copies of the data are invalidated, thus maintaining memory coherence. The precise technical details of this so-called ALLCACHE memory architecture are outside the scope of this book and we refer to KSR (1991a) for further information.

In addition to the Fortran 77 standard, KSR Fortran (KSR, 1991b) also supports part of the array syntax of Fortran 90 (Section 2.4.2) and some of the parallel programming extensions of PCF Fortran (Section 2.4.6). For example, it is possible to reference multiple elements of an array (either the complete array or a section of it) in a single statement, or to apply an intrinsic function to multiple elements of an array. The extensions include **PARALLEL REGIONS** and **PARALLEL SECTIONS**. KSR Fortran also supports a relatively sophisticated DO loop parallelisation facility called *tiling*; it allows the parallelisation of multiple loops and permits constraints on the ordering of the loop iterations to be imposed. The language includes constructs for controlling the explicit placement of tasks on processors; a typical application consists of a sequence of parallel tasks and to exploit the ALLCACHE memory architecture efficiently it is important to ensure that, where possible, tasks which access the same data are mapped on to the same processor. Nevertheless, the Fortran compiler generates instructions which anticipate remote memory accesses in order to reduce latency.

Besides providing parallel constructs, the KSR Fortran compiler additionally supports automatic parallelisation. This, together with the parallel programming extensions, enables KSR Fortran to support a combination of automatic and manual parallelisation in a manner previously only available on true shared memory systems. The KSR1 also has a C compiler and a debugger which supports both serial and parallel applications written in C, Fortran and Assembler.

A third system worthy of mention is the Myrias SPS-2, a 1024-node machine built on a three-level hierarchical bus system. Each node processor is a Motorola 68020 with 4 Mbytes of memory and giving 150 kflops. At the lowest level is a *board* of four bus-connected nodes. Boards are grouped into *cages* of sixteen bus-connected boards, and, at the highest level, a *system* consists of 16 bus-connected cages. Again, the software supports a VSM, with the Fortran compiler supporting a parallel loop construct (**PARDO**) and parallel blocks (**PARBEGIN**). The first SPS-2 appeared in 1987. Sadly, the company went into receivership in 1990, although it appears to be showing signs of life again.

The VSM machines described here appear to offer one way forward for parallel computing. They are scalable and, on the surface, they appear to be no more difficult to program than a shared memory machine; there is certainly no need to worry about explicit message-passing. However, the machines do not support the shared memory principle (rarely achieved in practice, even on 'true' shared memory machines) that all memory accesses are of unit cost; the latency in accessing data depends on the remoteness of the memory which holds the data. Thus to obtain the maximum performance from these machines the user needs to take account of the relationship between processes and the data which they use, and, as far as possible, to maintain data in memory locations close to the processes which require the data. The provision of large caches and their careful management, supported by the hardware and software, helps to alleviate this difficulty to some extent.

8.5   Theory into practice?

The systems described in the previous section attempt to bridge the shared/local memory divide but rely on what we might term traditional approaches to parallelism. The speed of interprocessor communications remains the limiting factor and, despite the very real progress that has been made, the goal of a uniform data access time has yet to be achieved. Of course, one way of satisfying this requirement would be to increase the latency of local data access, but this is hardly a realistic scenario.

For some time theoreticians have studied the principle of the parallel random access machine (PRAM) (Valiant, 1989) and this may provide the basis of parallel architectures in the medium- to long-term future. A PRAM consists of

- an unbounded number of identical random access machines (RAMs), and
- an unbounded number of global memory cells,

with each RAM able to access any given memory cell in one time step. Access to data can take the form of

- exclusive read, exclusive write (EREW), ensuring that there are no read or write conflicts,
- concurrent read, exclusive write (CREW), ensuring that there are no write conflicts, and
- concurrent read, concurrent write (CRCW),

with CRCW being further subdivided into

- identical CRCW, in which the data of each write is the same,
- arbitrary CRCW, in which the data of each write may be different and an arbitrary one succeeds, or
- priority CRCW, in which the lowest-numbered processor succeeds with the write when more than one are competing.

Faced with an abstraction of this form, the applications programmer can develop an algorithm without regard to the cost of data access. The problem is then to simulate the PRAM model on a physical architecture.

Associated with the PRAM model is the bulk synchronous parallel (BSP) model (Valiant, 1990), which has three parts,

- a number of *components* (for example, processing elements),

- a *router* providing point-to-point communication between components, and
- facilities for synchronising some, or all, of the components at regular, fixed intervals.

A program is broken down into a number of *supersteps*, in each of which all components are allocated a task of one form or another. Supersteps are broken down further into time intervals. At the end of each such interval a check is made as to whether the current superstep has completed, when the computation moves on to the next superstep, or not. In the latter case the computation within the current superstep continues for the next time interval.

The PRAM model has been used to determine many elegant results concerned with complexity and has proved to be an interesting abstraction. Whether efficient physical realisations of this, or the BSP, model are possible remains to be seen, although there has recently been some promising activity in this area.

One way or another, the way forward inevitably appears to be the simulation of global memory on a local memory system, providing flexibility with ease of use. This convergence has to be welcomed by the applications programmer in the same way as he appreciated the arrival of high-level languages and their compilers. It points the way forward to true trans-architectural portability.

Further reading

For a more detailed review of the merits of shared and local memory multiprocessors see the report by Jordan (1991). Mayr (1990) gives an overview of theoretical aspects of parallel computing and is a useful first reference. The collection of Fincham and Ford (1992) contains several papers which address the above issues. It is also a useful source for an overview of parallel numerical algorithms.

Appendix
The single precision real BLAS

A.1 Level 1 BLAS

- SUBROUTINE SSWAP(N,X,INCX,Y,INCY)
  sswap interchanges two vectors $(\mathbf{x} \leftrightarrow \mathbf{y})$ of length n.
- SUBROUTINE SSCAL(N,ALPHA,X,INCX)
  sscal scales a vector by α $(\mathbf{x} \leftarrow \alpha\mathbf{x})$.
- SUBROUTINE SCOPY(N,X,INCX,Y,INCY)
  scopy copies one vector into another $(\mathbf{y} \leftarrow \mathbf{x})$.
- SUBROUTINE SAXPY(N,ALPHA,X,INCX,Y,INCY)
  saxpy adds a multiple of one vector to another $(\mathbf{y} \leftarrow \alpha\mathbf{x} + \mathbf{y})$.
- FUNCTION SDOT(N,X,INCX,Y,INCY)
  sdot forms the dot product of two vectors $(\text{SDOT} \leftarrow \mathbf{x}^T\mathbf{y})$.
- FUNCTION SNRM2(N,X,INCX)
  snrm2 forms the two-norm of a vector $(\text{SNRM2} \leftarrow \|\mathbf{x}\|_2 \equiv \sqrt{\sum_{i=1}^{n} x_i^2}\,)$.
- FUNCTION SASUM(N,X,INCX)
  sasum forms the one-norm of a vector $(\text{SASUM} \leftarrow \|\mathbf{x}\|_1 \equiv \sum_{i=1}^{n} |x_i|\,)$.
- FUNCTION ISAMAX(N,X,INCX)
  isamax finds the lowest index of the element of maximum modulus in a vector (the first k such that $|x_k| \geq |x_i|$, $\forall\, i$).
- SUBROUTINE SROTG(A,B,C,S)
  srotg computes the cosine (c) and the sine (s) of the plane rotation involving a and b such that
  $$\begin{pmatrix} c & s \\ -s & c \end{pmatrix} \begin{pmatrix} a \\ b \end{pmatrix} = \begin{pmatrix} r \\ 0 \end{pmatrix}.$$
- SUBROUTINE SROT(N,X,INCX,Y,INCY,C,S)
  srot applies the plane rotation defined by c and s to consecutive pairs of

the components of the vectors $\mathbf{x}$ and $\mathbf{y}$, that is,

$$\begin{aligned} x_i &\leftarrow \quad cx_i + sy_i, \\ y_i &\leftarrow -sx_i + cy_i, \end{aligned} \qquad i = 1, 2, \ldots, n,$$

with the assignments taking place simultaneously.

A.2   Level 2 BLAS

- SUBROUTINE SGEMV(TRANS,M,N,ALPHA,A,LDA,X,INCX,BETA,Y,INCY)
  sgemv forms the matrix-vector product $\mathbf{y} \leftarrow \alpha \mathrm{op}(A)\mathbf{x} + \beta\mathbf{y}$, where $\mathrm{op}(A) = A$ or A^T.
- SUBROUTINE SGBMV(TRANS,M,N,KL,KU,ALPHA,A,LDA,X,INCX,BETA,Y,
                   INCY)
  Version of **sgemv** for a banded matrix A.
- SUBROUTINE SSYMV(UPLO,N,ALPHA,A,LDA,X,INCX,BETA,Y,INCY)
  Version of **sgemv** for a symmetric matrix A.
- SUBROUTINE SSBMV(UPLO,N,K,ALPHA,A,LDA,X,INCX,BETA,Y,INCY)
  Version of **sgemv** for a symmetric banded matrix A.
- SUBROUTINE SSPMV(UPLO,N,ALPHA,AP,X,INCX,BETA,Y,INCY)
  Version of **sgemv** for a symmetric matrix A stored in packed form.
- SUBROUTINE STRMV(UPLO,TRANS,DIAG,N,A,LDA,X,INCX)
  strmv forms the matrix-vector product $\mathbf{x} \leftarrow \mathrm{op}(A)\mathbf{x}$, where $\mathrm{op}(A) = A$ or A^T, with A triangular.
- SUBROUTINE STBMV(UPLO,TRANS,DIAG,N,K,A,LDA,X,INCX)
  Version of **strmv** for a triangular banded matrix A.
- SUBROUTINE STPMV(UPLO,TRANS,DIAG,N,AP,X,INCX)
  Version of **strmv** for a triangular matrix A stored in packed form.
- SUBROUTINE STRSV(UPLO,TRANS,DIAG,N,A,LDA,X,INCX)
  strsv forms the matrix-vector product $\mathbf{x} \leftarrow \mathrm{op}(A^{-1})\mathbf{x}$ with A triangular (that is, solves the system of linear equations $\mathrm{op}(A)\mathbf{y} = \mathbf{x}$, with the solution $\mathbf{y}$ overwriting $\mathbf{x}$).
- SUBROUTINE STBSV(UPLO,TRANS,DIAG,N,K,A,LDA,X,INCX)
  Version of **strsv** for a triangular banded matrix A.
- SUBROUTINE STPSV(UPLO,TRANS,DIAG,N,A,X,INCX)
  Version of **strsv** for a triangular matrix A stored in packed form.
- SUBROUTINE SGER(M,N,ALPHA,X,INCX,Y,INCY,A,LDA)
  sger forms the general rank-one update $A \leftarrow \alpha\mathbf{x}\mathbf{y}^T + A$.
- SUBROUTINE SSYR(UPLO,N,ALPHA,X,INCX,A,LDA)
  ssyr forms the symmetric rank-one update $A \leftarrow \alpha\mathbf{x}\mathbf{x}^T + A$.
- SUBROUTINE SSPR(UPLO,N,ALPHA,X,INCX,AP)
  Variant of **ssyr** for a symmetric matrix A stored in packed form.

- SUBROUTINE SSYR2(UPLO,N,ALPHA,X,INCX,Y,INCY,A,LDA)
  ssyr2 forms the symmetric rank-two update $A \leftarrow \alpha \mathbf{x}\mathbf{y}^T + \alpha \mathbf{y}\mathbf{x}^T + A$.
- SUBROUTINE SSPR2(UPLO,N,ALPHA,X,INCX,Y,INCY,AP)
  Variant of ssyr2 for a symmetric matrix A stored in packed form.

A.3   Level 3 BLAS

- SUBROUTINE SGEMM(TRANSA,TRANSB,M,N,K,ALPHA,A,LDA,B,LDB,BETA,C,
                   LDC)
  sgemm forms the matrix-matrix product $C \leftarrow \alpha \mathrm{op}(A)\mathrm{op}(B) + \beta C$.
- SUBROUTINE SSYMM(SIDE,UPLO,M,N,ALPHA,A,LDA,B,LDB,BETA,C,LDC)
  ssymm forms the matrix-matrix product $C \leftarrow \alpha AB + \beta C$ or $C \leftarrow \alpha BA + \beta C$.
- SUBROUTINE SSYRK(UPLO,TRANS,N,K,ALPHA,A,LDA,BETA,C,LDC)
  ssyrk forms the symmetric rank-k update $C \leftarrow \alpha AA^T + \beta C$ or $C \leftarrow \alpha A^T A + \beta C$.
- SUBROUTINE SSYR2K(UPLO,TRANS,N,K,ALPHA,A,LDA,B,LDB,BETA,C,LDC)
  ssyr2k forms the symmetric rank-$2k$ update $C \leftarrow \alpha AB^T + \alpha BA^T + \beta C$ or $C \leftarrow \alpha A^T B + \alpha B^T A + \beta C$.
- SUBROUTINE STRMM(SIDE,UPLO,TRANSA,DIAG,M,N,ALPHA,A,LDA,B,LDB)
  strmm forms the matrix-matrix product $B \leftarrow \alpha \mathrm{op}(A)B$ or $B \leftarrow \alpha B\mathrm{op}(A)$, where A is a triangular matrix.
- SUBROUTINE STRSM(SIDE,UPLO,TRANSA,DIAG,M,N,ALPHA,A,LDA,B,LDB)
  strsm forms the matrix-matrix product $B \leftarrow \alpha \mathrm{op}(A^{-1})B$ (solves a system of equations with the same coefficient matrix but several right-hand sides) or $B \leftarrow \alpha B\mathrm{op}(A^{-1})$, where A is a triangular matrix.

Bibliography

Ahuja, S., Carriero, N. and Gelernter, D. (1986). Linda and friends. *IEEE Comput.* **19**, 8 (August), pp. 26–34.

Allan, R. J. (1990). Numerical algorithm libraries for multicomputers. In Pritchard and Scott (1990), pp. 335–340.

Almasi, G. S. and Gottlieb, A. (1988). *Highly Parallel Computing*. Benjamin/ Cummings, Redwood City, CA.

Amdahl, G. M. (1967). The validity of the single processor approach to achieving large scale computing capabilities. *AFIPS Conf. Proc. Spring Jt. Conf.* **30**, pp. 483–485.

AMT (1990). *FORTRAN–PLUS enhanced language reference manual*. AMT Tech. Pub. man 102.01, AMT Ltd., Reading.

Anderson, E., Bai, Z., Bischof, C., Demmel, J., Dongarra, J., Du Croz, J., Greenbaum, A., Hammarling, S., McKenney, A. and Sorensen, D. (1992). *LAPACK Users' Guide*. SIAM, Philadelphia, PA.

ANSI (1978). *American National Standard Programming Language FORTRAN, X3.9-1978*. ANSI, New York.

Aspray, W. and Burks, A. (eds) (1987). *Papers of John von Neumann on Computing and Computer Theory*. MIT Press, Cambridge, MA.

Athas, W. C. and Seitz, C. L. (1988). Multicomputers: message-passing concurrent computers. *IEEE Comput.* **21**, 8 (August), pp. 9–24.

Aykanat, I., Özgüner, F., Ercal, F. and Sadayappan, P. (1988). Iterative algorithms for solution of large sparse systems of linear equations on hypercubes. *IEEE Trans. Comput.* **37**, pp. 1554–1567.

Banahan, M., Brady, D. and Doran, M. (1991). *The C Book* (2nd edn). Addison-Wesley, Wokingham.

Barnes, J. G. P. (1989). *Programming in Ada* (3rd edn). Addison-Wesley, Wokingham.

BBN (1989). *TC2000 Technical Product Summary*. BBN Advanced Computers Inc., Cambridge, MA.

Beauvaris, R. and de Groen, P. (1992). *Proceedings of the IMACS International Symposium on Iterative Methods in Linear Algebra*. Elsevier, New York.

Ben-Ari, M. (1990). *Principles of Concurrent and Distributed Programming.* Prentice Hall, New York.

Bisseling, R. H. and van de Vorst, J. G. G. (1989). Parallel LU decomposition on a transputer network. In van Zee and van de Vorst (1989), pp. 61–77.

Bomans, L., Roose, D. and Hempel, R. (1990). The Argonne/GMD macros in FORTRAN for portable parallel programming and their implementation on the Intel iPSC/2. *Parallel Computing* **15**, pp. 119–132.

Bomans, L. and Roose, D. (1989). Benchmarking the iPSC/2 hypercube multiprocessor. *Concurrency: Practice and Experience* **1**, pp. 3–18.

Brawer, S. (1989). *Introduction to Parallel Programming.* Academic Press, Boston, MA.

Brown, N. G., Delves, L. M., Howard, G., Downing, S. and Phillips, C. (1990). Numerical library development for transputer arrays. In Freeman and Phillips (1990), pp. 103–112.

Broyden, C. G. (1970a). The convergence of a class of double-rank minimization algorithms. 1. General considerations. *J. Inst. Math. Applic.* **6**, pp. 76–90.

Broyden, C. G. (1970b). The convergence of a class of double-rank minimization algorithms. 2. The new algorithm. *J. Inst. Math. Applic.* **6**, pp. 222–231.

Burns, A. (1988). *Programming in occam 2.* Addison-Wesley, Wokingham.

Burrage, K. (1990). An adaptive numerical integration code for a chain of transputers. *Parallel Comput.* **16**, pp. 305–312.

Bustard, D., Elder, J. and Welsh, J. (1988). *Concurrent Program Structures.* Prentice Hall, New York.

Byrd, R. H., Dert, C. L., Rinnooy Kan, A. H. G. and Schnabel, R. B. (1990). Concurrent stochastic methods for global optimization. *Math. Prog.* **46**, pp. 1–29.

Byrd, R. H., Eskow, E. H., Schnabel, R. B. and Smith, S. L. (1991). *Parallel global optimization: numerical methods, dynamic scheduling methods, and application to molecular configuration.* Tech. Rept. CU-CS-553-91, Dept. of Comput. Sci., Univ. of Colorado at Boulder, CO.

Byrd, R. H., Schnabel, R. B. and Shultz, G. A. (1988a). Parallel quasi-Newton methods for unconstrained optimization. *Math. Prog.* **42**, pp. 273–306.

Byrd, R. H., Schnabel, R. B. and Shultz, G. A. (1988b). Using parallel function evaluations to improve Hessian approximations for unconstrained optimization. *Annals Oper. Res.* **14**, pp. 167–193.

Cannon, L. E. (1969). *A cellular computer to implement the Kalman filter algorithm.* Ph. D. thesis, Montana State Univ., Bozeman, MT.

Carriero, N. and Gelernter, D. (1990). *How to Write Parallel Programs. A First Course.* MIT Press, Cambridge, MA.

Chandy, K. M. and Misra, J. (1988). *Parallel Program Design. A Foundation.* Addison-Wesley, Reading, MA.

Chazan, D. and Miranker, W. (1969). Chaotic relaxation. *J. Lin. Alg. Applic.* **2**, pp. 199–222.

Chu, E. and George, A. (1987). Gaussian elimination with partial pivoting and load balancing on a multiprocessor. *Parallel Comput.* **5**, pp. 65–74.

Clementi, E. (1989). Global scientific and engineering simulation on scalar, vector and parallel LCAP-type supercomputers. In Elliott and Hoare (1989), pp. 89–114.

Cok, R. S. (1991). *Parallel Programs for the Transputer.* Prentice Hall, Englewood Cliffs, NJ.

Commission of the European Communities (eds) (1987). *ESPRIT '87. Achievements and Impact. Part 1.* North-Holland, Amsterdam.

Cook, R. (1992). A reformulation of preconditioned conjugate gradients suitable for a local memory multi-processor. In Beauvaris and de Groen (1992), pp. 313–322.

Cooley, J. W. (1990). How the FFT gained acceptance. In Nash (1990), pp. 133–140.

Cooley, J. W. and Tukey, J. W. (1965). An algorithm for the machine calculation of complex Fourier series. *Math. Comput.* **19**, pp. 297–301.

Cosnard, M. (1991). Scheduling parallel factorization algorithms on a shared memory computer. In Spedicato (1991), pp. 217–252.

Cosnard, M. and Fraigniaud, P. (1990). Asynchronous Durand–Kerner and Aberth polynomial root finding methods on a distributed memory multicomputer. In Evans, Joubert, *et al.* (1990), pp. 79–84.

Cosnard, M., Muller, J.-M. and Robert, Y. (1986). Parallel QR decomposition of a rectangular matrix. *Numer. Math.* **48**, pp. 239–249.

Cosnard, M. and Robert, Y. (1986). Complexity of parallel QR decomposition. *J. Assoc. Comput. Mach.* **33**, pp. 712–723.

Cosnard, M., Robert, Y., Quinton, P. and Tchunete, M. (eds) (1986). *International Workshop on Parallel Algorithms and Architectures.* North-Holland, Amsterdam.

Cowell, W. R. (ed.) (1984). *Sources and Development of Mathematical Software.* Prentice Hall, Englewood Cliffs, NJ.

Crank, J. and Nicolson, P. (1947). A practical method for numerical evaluation of solutions of partial differential equations of the heat-conduction type. *Proc. Camb. Phil. Soc.* **43**, pp. 60–67.

Curnow, H. J. and Wichman, B. A. (1976). A synthetic benchmark. *Comput. J.* **19**, pp. 43–49.

Dally, W. (1990). Network and processor architecture for message-driven computers. In Suaya and Birtwistle (1990), pp. 140–222.

Dally, W. J. and Seitz, C. L. (1987). Deadlock-free message routing in multiprocessor interconnection networks. *IEEE Trans. Comput.* **C-36**, pp. 547–553.

Davis, P. J. and Rabinowitz, P. (1984). *Methods of Numerical Integration* (2nd edn). Academic Press, Orlando, FL.

DeCegama, A. L. (1989). *The Technology of Parallel Processing. Parallel Processing Architectures and VLSI Hardware. Volume 1.* Prentice Hall, Englewood Cliffs, NJ.

Decker, N. H. (1991). Note on the parallel efficiency of the Frederickson-McBryan multigrid algorithm. *SIAM J. Sci. Statist. Comput.* **12**, pp. 208–220.

DeGroot, D. (1985). *Proceedings of the 1985 International Conference on Parallel Processing*. IEEE, Washington, DC.

Demmel, J., Dongarra, J. J., Du Croz, J., Greenbaum, A., Hammarling, S. and Sorensen, D. (1987). *LAPACK working note #1. Prospectus for the development of a linear algebra library for high-performance computers*. Tech. Mem. 97, Math. and Comput. Sci. Division, Argonne Nat. Lab., Argonne, IL.

Demmel, J., Du Croz, J., Hammarling, S. and Sorensen, D. (1988). *LAPACK working note #4. Guidelines for the design of symmetric eigenroutines, SVD, and iterative refinement and condition estimation for linear systems*. Tech. Mem. 111, Math. and Comput. Sci. Division, Argonne Nat. Lab., Argonne, IL.

Demmel, J. W. and Higham, N. J. (1991). *Stability of block algorithms with fast Level 3 BLAS*. Numer. Anal. Rept. No. 188, Univ. of Manchester. Also in *ACM Trans. Math. Software*, to appear.

Dennis, J. E. Jr and Torczon, V. J. (1990). *Direct search methods on parallel machines*. Tech. Rept. 90-19, Rice Univ., Houston, TX.

Diaz, J. C. (ed.) (1989). *Mathematics for Large Scale Computing*. Lecture Notes in Pure and Applied Mathematics **120**, Marcel Dekker, New York.

Dijkstra, E. W. (1971). Hierarchical ordering of sequential processes. *Acta Informatica* **1**, pp. 115–138.

Dixon, L. C. W. (1991). The solution of partially separable linear equations on parallel processing systems. In Spedicato (1991), pp. 299–338.

Dodson, D. S. and Grimes, R. G. (1982). Remark on algorithm 539. *ACM Trans. Math. Software* **8**, pp. 403–405.

Dodson, D. S., Grimes, R. G. and Lewis, J. G. (1991a). Sparse extensions to the Fortran basic linear algebra subprograms. *ACM Trans. Math. Software* **17**, pp. 253–263.

Dodson, D. S., Grimes, R. G. and Lewis, J. G. (1991b). Model implementation and test package for the sparse basic linear algebra subprograms. *ACM Trans. Math. Software* **17**, pp. 264–272.

Dongarra, J. J. (1988). The LINPACK benchmark: an explanation. In Houstis, Papatheodorou, *et al.* (1988), pp. 456–474.

Dongarra, J. J. (1991a). *Performance of various computers using standard linear equations software*. Rept. No. CS-89-85, Comput. Sci. Dept., Univ. of Tennessee, Knoxville, TN.

Dongarra, J. J. (1991b). *Workshop on the BLACS*. LAPACK Working Note #34, Rept. No. CS-91-134, Comput. Sci. Dept., Univ. of Tennessee, Knoxville, TN.

Dongarra, J., Brewer, O., Kohl, J. A. and Fineberg, S. (1990). A tool to aid in the design, implementation and understanding of matrix algorithms for parallel processors. *J. Parallel Distributed Comput.* **9**, pp. 185–202.

Dongarra, J. J., Bunch, J. R., Moler, C. B. and Stewart, G. W. (1979). *LINPACK Users' Guide*. SIAM, Philadelphia, PA.

Dongarra, J. J., Du Croz, J., Hammarling, S. and Duff, I. (1990a). A set of Level 3 basic linear algebra subprograms. *ACM Trans. Math. Software* **16**, pp. 1–17.

Dongarra, J. J., Du Croz, J., Hammarling, S. and Duff, I. (1990b). Algorithm 679: A set of Level 3 basic linear algebra subprograms. *ACM Trans. Math. Software* **16**, pp. 18–28.

Dongarra, J. J., Du Croz, J., Hammarling, S. and Hanson, R. (1988a). An extended set of Fortran basic linear algebra subprograms. *ACM Trans. Math. Software* **14**, pp. 1–17.

Dongarra, J. J., Du Croz, J., Hammarling, S. and Hanson, R. (1988b). Algorithm 656: An extended set of Fortran basic linear algebra subprograms. *ACM Trans. Math. Software* **14**, pp. 18–32.

Dongarra, J., Duff, I., Gaffney, P. and McKee, S. (1989). *Vector and Parallel Computing*. Ellis Horwood, Chichester.

Dongarra, J. J. and Grosse, E. (1987). Distribution of mathematical software by electronic mail. *Commun. Assoc. Comput. Mach.* **30**, pp. 403–407.

Dongarra, J. J., Hammarling, S. J., and Sorensen, D. C. (1987). *LAPACK working note #2. Block reduction of matrices to condensed forms for eigenvalue computations*. Tech. Mem. 99, Math. and Comput. Sci. Division, Argonne Nat. Lab., Argonne, IL.

Dongarra, J. J. and Sorensen, D. C. (1987a). A portable environment for developing parallel Fortran programs. *Parallel Comput.* **5**, pp. 175–186.

Dongarra, J. J. and Sorensen, D. C. (1987b). A fully parallel algorithm for the symmetric eigenproblem. *SIAM J. Sci. Statist. Comput.* **8**, pp. 139–154.

Dongarra, J. J., Sorensen, D. C., Connolly, K. and Patterson, J. (1988). Programming methodology and performance issues for advanced computer architectures. *Parallel Comput.* **8**, pp. 41–58.

Dongarra, J. J. and van Geijn, R. A. (1991). *Two dimensional basic linear algebra communication subprograms*. LAPACK Working Note #37, Rept. No. CS-91-138, Comput. Sci. Dept., Univ. of Tennessee, Knoxville, TN.

Duff, I. S. (1986). Parallel implementation of multifrontal schemes. *Parallel Comput.* **3**, pp. 193–204.

Duff, I. S. (1991). Parallel algorithms for general sparse systems. In Spedicato (1991), pp. 277–297.

Duff, I. S., Erisman, A. M. and Reid, J. K. (1986). *Direct Methods for Sparse Matrices*. Oxford University Press, New York.

Eberlein, P. J. and Park, H. (1990). Efficient implementation of Jacobi algorithms and Jacobi sets on distributed memory architectures. *J. Parallel Distributed Comput.* **8**, pp. 358–366.

Elliott, R. J. and Hoare, C. A. R. (eds) (1989). *Scientific Applications of Multiprocessors*. Prentice Hall, New York.

Ellis, T. M. R. (1990). *Fortran 77 Programming. With an Introduction to the Fortran 90 Standard*. Addison-Wesley, Wokingham.

Encore (1988a). *Encore Parallel Threads Manual.* Ref. No. 724-06210, Encore
Computer Corporation, Fort Lauderdale, FL.

Encore (1988b). *Encore Parallel Fortran.* Ref. No. 724-06785, Encore Computer
Corporation, Fort Lauderdale, FL.

Evans, D. J., Joubert, G. R. and Peters, F. J. (1990). *Parallel Computing 89.*
Elsevier, Amsterdam.

Fairweather, G. and Keast, P. M. (eds) (1987). *Numerical Integration. Recent
Developments, Software and Applications.* NATO ASI Series **C203**, D. Rei-
del, Dordrecht.

Feng, T. Y. (1981). A summary of interconnection networks. *IEEE Comput.* **14**,
12 (December), pp. 12–27.

Feo, J. T. (1988). An analysis of the computational and parallel complexity of
the Livermore Loops. *Parallel Comput.* **7**, pp. 163–185.

Fincham, A. E. and Ford, B. (eds) (1992). *Parallel Computation.* Oxford University Press, Oxford.

Fletcher, R. (1970). A new approach to variable metric algorithms. *Comput. J.*
13, pp. 317–322.

Flynn, M. J. (1972). Some computer organizations and their effectiveness. *IEEE
Trans. Comput.* **C-21**, pp. 948–960.

Fox, G. C. and Furmanski, W. (1989). The physical structure of concurrent
problems and concurrent computers. In Elliott and Hoare (1989), pp. 55–
88.

Fox, G. C., Johnson, M. A., Lyzenga, G. A., Otto, S. W., Salmon, J. K. and
Walker, D. W. (1988). *Solving Problems on Concurrent Processors. Vol-
ume 1. General Techniques and Regular Problems.* Prentice Hall, Englewood
Cliffs, NJ.

Frederickson, P. O. and McBryan, O. A. (1991). Normalized convergence rates
for the PSMG method. *SIAM J. Sci. Statist. Comput.* **12**, pp. 221–229.

Freeman, L. and Phillips, C. (eds) (1990). *Applications of Transputers 1.* IOS,
Amsterdam.

Freeman, T. L. (1979). A method for computing all the zeros of a polynomial
with real coefficients. *BIT* **19**, pp. 321–333.

Freeman, T. L. (1989). Calculating polynomial zeros on a local memory parallel
computer. *Parallel Comput.* **12**, pp. 351–358.

Freeman, T. L. (1991). A parallel unconstrained quasi-Newton algorithm and its
performance on a local memory parallel computer. *Appl. Numer. Math.* **7**,
pp. 369–379.

Freeman, T. L. and Brankin, R. W. (1990). A divide and conquer method for
polynomial zeros. *J. Comput. Appl. Math.* **30**, pp. 71–79.

Gallivan, K. A., Heath, M. T., Ng, E., Ortega, J. M., Peyton, B. W., Plemmons,
R. J., Romine, C. H., Sameh, A. H. and Voigt, R. G. (1990). *Parallel
Algorithms for Matrix Computations.* SIAM, Philadelphia, PA.

Gallivan, K. A., Plemmons, R. J. and Sameh, A. H. (1990). Parallel algorithms for dense linear algebra computations. In Gallivan, Heath, *et al.* (1990), pp. 1–82. Also in *SIAM Rev.* **32**, 1990, pp. 54–135.

Gear, C. W. (1971). *Numerical Initial Value Problems in Ordinary Differential Equations.* Prentice Hall, Englewood Cliffs, NJ.

Gear, C. W. (1987). *Parallel methods for ordinary differential equations.* Tech. Rept. UIUCDCS-R-87-1369, Dept. of Comput. Sci., Univ. of Illinois at Urbana-Champaign, IL.

Geist, G. A. and Heath, M. T. (1986). Matrix factorization on a hypercube multiprocessor. In Heath (1986), pp. 161–180.

Gentleman, W. M. (1973). Least squares computation by Givens transformations without square roots. *J. Inst. Math. Applic.* **12**, pp. 329–336.

George, A. and Liu, J. W.-H. (1981). *Computer Solution of Large Sparse Positive Definite Systems.* Prentice Hall, Englewood Cliffs, NJ.

George, J. A., Heath, M. and Liu, J. (1986). Parallel Cholesky factorization on a shared-memory multiprocessor. *Lin. Alg. Applic.* **77**, pp. 165–187.

George, J. A., Heath, M., Liu, J. and Ng, E. (1989). Solution of sparse positive definite systems on a hypercube. *J. Comput. Appl. Math.* **27**, pp. 129–156.

Gill, P. E., Murray, W. M. and Wright, M. H. (1981). *Practical Optimization.* Academic Press, London.

Gladwell, I. (1987). Vectorisation of one dimensional quadrature codes. In Fairweather and Keast (1987), pp. 230–238.

Goldfarb, D. (1970). A family of variable-metric methods derived by variational means. *Math. Comput.* **24**, pp. 23–26.

Golub, G. H. and Van Loan, C. F. (1989). *Matrix Computations* (2nd edn). Johns Hopkins University Press, Baltimore, MD.

Hairer, E., Nørsett, S. P. and Wanner, G. (1987). *Solving Ordinary Differential Equations 1: Nonstiff Problems.* Springer-Verlag, Berlin.

Hall, G. (1974). Stability analysis of predictor-corrector algorithms of Adams type. *SIAM J. Numer. Anal.* **11**, pp. 494–505.

Harp, J. G. (1987). Phase 2 of the reconfigurable transputer project (P1085). In Commission of the European Communities (1987), pp. 583–591.

Heath, M. T. (ed.) (1986). *Hypercube Multiprocessors 1986.* SIAM, Philadelphia, PA.

Heath, M. T., Ng, E. and Peyton, B. W. (1990). Parallel algorithms for sparse linear systems. In Gallivan, Heath, *et al.* (1990), pp. 83–124. Also in *SIAM Rev.* **33**, 1991, pp. 420–460.

Hey, A. J. G. (1989). Reconfigurable transputer networks: practical concurrent computation. In Elliott and Hoare (1989), pp. 39–54.

Higham, N. J. (1990). Exploiting fast matrix multiplication within the Level 3 BLAS. *ACM Trans. Math. Software* **16**, pp. 352–368.

Hoare, C. A. R. (1978). Communicating sequential processes. *Commun. Assoc. Comput. Mach.* **21**, pp. 666–677.

Hoare, C. A. R. (1986). *Communicating Sequential Processes*. Prentice Hall, Englewood Cliffs, NJ.

Hockney, R. W. and Jesshope, C. R. (1988). *Parallel Computers 2*. Adam Hilger, Bristol.

Houstis, E. N., Papatheodorou, C. D. and Polychronopoulos, C. D. (1988). *Supercomputing. Proceedings, First International Conference, Athens*. Lecture Notes in Computer Science **297**, Springer-Verlag, Berlin.

Hull, T. E., Enright, W. II., Fellen, B. M. and Sedgwick, A. E. (1972). Comparing numerical methods for ordinary differential equations. *SIAM J. Numer. Anal.* **9**, pp. 603–637.

Hwang, K. and Briggs, F. A. (1984). *Computer Architecture and Parallel Programming*. McGraw-Hill, New York.

Inmos (1988). *Transputer Reference Manual*. Prentice Hall, New York.

Inmos (1990). *Transputer Development System* (2nd edn). Prentice Hall, New York.

Inmos (1991). *The T9000 Transputer Products Overview Manual*. Inmos, Bristol.

Intel (1991). *iPSC/2 and iPSC/860 user's guide*. Order No. 311532-007, Intel Scientific Computers, Beaverton, OR.

Iserles, A. and Nørsett, S. P. (1990). On the theory of parallel Runge–Kutta methods. *IMA J. Numer. Anal.* **10**, pp. 463–488.

Jess, J. A. G. and Kees, H. G. M. (1982). A data structure for parallel L/U decomposition. *IEEE Trans. Comput.* **C-31**, pp. 231–239.

Jordan, H. F. (1991). *Shared versus distributed memory multiprocessors*. Rept. No. 91-7, ICASE, NASA Langley Res. Cent., Hampton, VA.

Kahaner, D., Nash, S. and Moler, C. (1989). *Numerical Methods and Software*. Prentice Hall, Englewood Cliffs, NJ.

Keller, H. B. and Nelson, P. (1989a). Hypercube implementations of parallel shooting. *Appl. Math. Comput.* **31**, pp. 574–603.

Keller, H. B. and Nelson, P. (1989b). A comparison of hypercube implementations of parallel shooting. In Diaz (1989), pp. 49–79.

Kernighan, B. W. and Ritchie, D. M. (1988). *The C Programming Language* (2nd edn). Prentice Hall, Englewood Cliffs, NJ.

Krishnamurthy, E. V. (1989). *Parallel Processing. Principles and Practice*. Addison-Wesley, Sydney.

Krumme, D. W., Couch, A. L. and Cybenko, G. (1989). Debugging support for parallel programs. In Dongarra, Duff, *et al.* (1989), pp. 205–214.

KSR (1991a). *Kendall Square Product Description*. Kendall Square Res. Corp., Waltham, MA.

KSR (1991b). *KSR Fortran Programming*. Kendall Square Res. Corp., Waltham, MA.

Kung, H. T. (1989). Computational models for parallel computers. In Elliott and Hoare (1989), pp. 1–15.

Lakshmivarahan, S. and Dhall, S. K. (1990). *Analysis and Design of Parallel Algorithms: Arithmetic and Matrix Problems*. McGraw-Hill, New York.

Lawrie, D. H. (1975). Access and alignment of data in an array processor. *IEEE Trans. Comput.* **C-24**, pp. 1145–1155.

Lawson, C., Hanson, R., Kincaid, D. and Krogh, F. (1979a). Basic linear algebra subprograms for Fortran usage. *ACM Trans. Math. Software* **5**, pp. 308–323.

Lawson, C., Hanson, R., Kincaid, D. and Krogh, F. (1979b). Algorithm 539: Basic linear algebra subprograms for Fortran usage. *ACM Trans. Math. Software* **5**, pp. 324–325.

Lazou, C. (1986). *Supercomputers and their Uses.* Clarendon Press, Oxford.

Liu, J. W. H. (1986). A compact row storage scheme for Cholesky factors using elimination trees. *ACM Trans. Math. Software* **12**, pp. 127–148.

Lootsma, F. A. (1989). *Parallel non-linear optimization.* Rept. 89-45, Faculty of Tech. Math. and Informatics, Delft Univ. of Tech., Delft.

Lootsma, F. A. and Ragsdell, K. M. (1988). State-of-the-art in parallel nonlinear optimization. *Parallel Comput.* **6**, pp. 133–155.

McCormick, S. F. (ed.) (1988). *Multigrid Methods: Theory, Applications and Supercomputing.* Lecture Notes in Pure and Applied Mathematics **110**, Marcel Decker, New York.

McLean, M. and Rowland, T. (1985). *The INMOS Saga.* Frances Pinter, London.

Mandel, J., McCormick, S. F., Dendy, J. E., Farhat, C., Lonsdale, G., Parter, S. V., Ruge, J. W. and Stüben, K. (eds) (1989). *Proceedings of the Fourth Copper Mountain Conference on Multigrid Methods.* SIAM, Philadelphia, PA.

May, D. (1989). The influence of VLSI technology on computer architecture. In Elliott and Hoare (1989), pp. 21–37.

Mayr, E. W. (1990). Theoretical aspects of parallel computation. In Suaya and Birtwistle (1990), pp. 85–139.

Meiko (1989). *CS Tools Reference Manual.* Meiko, Bristol.

Metcalf, M. and Reid, J. (1991). *Fortran 90 Explained.* Oxford University Press, Oxford.

Mikhlin, S. G. (1964). *Variational Methods in Mathematical Physics.* Macmillan, London.

Miranker, W. L. and Liniger, W. (1967). Parallel methods for the numerical integration of ordinary differential equations. *Math. Comput.* **21**, pp. 303–320.

Modi, J. J. (1988). *Parallel Algorithms and Matrix Computation.* Clarendon Press, Oxford.

Modi, J. J. and Clarke, M. R. B. (1984). An alternative Givens ordering. *Numer. Math* **43**, pp. 83–90.

Moler, C. (1986). Matrix computation on distributed memory multiprocessors. In Heath (1986), pp. 181–195.

Moré, J. J., Sorensen, D. C., Garbow, B. S. and Hillstrom, K. E. (1984). The MINPACK project. In Cowell (1984), pp. 88–111.

Nash, S. G. (ed.) (1990). *A History of Scientific Computing.* Addison-Wesley, Reading, MA.

Nelder, J. A. and Mead, R. (1965). A simplex method for function minimization. *Comput. J.* **7**, pp. 308–313.

Ortega, J. M. (1988). *Introduction to Parallel and Vector Solution of Linear Systems.* Plenum Press, New York.

Ortega, J. M. and Poole, W. G. Jr (1981). *An Introduction to Numerical Methods for Differential Equations.* Pitman, Marshfield, MA.

Ortega, J. M. and Romine, C. H. (1988). The *ijk* forms of factorization methods II. Parallel systems. *Parallel Comput.* **7**, pp. 149–162.

Ortega, J. M. and Voigt, R. G. (1985). *Solution of Partial Differential Equations on Vector and Parallel Computers.* SIAM, Philadelphia, PA.

Ortega, J. M., Voigt, R. G. and Romine, C. H. (1990). A bibliography on parallel and vector numerical algorithms. In Gallivan, Heath, *et al.* (1990), pp. 125–197.

Peaceman, D. W. and Rachford, H. H. (1955). The numerical solution of parabolic and elliptic differential equations. *J. Soc. Indust. Appl. Math.* **3**, pp. 28–41.

Pease, M. C. (1968). An adaption of the fast Fourier transform for parallel processing. *J. Assoc. Comput. Mach.* **15**, pp. 252–264.

Pease, M. C. (1977). The indirect binary *n*-cube microprocessor array. *IEEE Trans. Comput.* **C-26**, pp. 458–473.

Perrott, R. H. (1987). *Parallel Programming.* Addison-Wesley, Wokingham.

Phillips, C. (1991). The performance of the BLAS and LAPACK on a shared memory scalar multiprocessor. *Parallel Comput.* **17**, pp. 751–761.

Piessens, E., de Doncker-Kapenga, E., Überhuber, C. W. and Kahaner, D. K. (1983). *QUADPACK – A Subroutine Package for Automatic Integration.* Springer-Verlag, New York.

Plachy, E. C. and Kogge, P. M. (eds) (1989). *Proceedings of the 1989 International Conference on Parallel Processing. Vol II. Software.* Pennsylvania State University Press, University Park, PA.

Polychronopoulos, C. D., Girkar, M., Haghighat, M. R., Lee, C. L., Leung, B. and Schouten, D. (1989). Parafrase-2: an environment for parallelizing, partitioning, synchronizing, and scheduling programs on multiprocessors. In Plachy and Kogge (1989), pp. 39–48.

Pountain, D. and May, D. (1987). *A Tutorial Introduction to occam Programming.* BSP, Oxford.

Press, W. H., Flannery, B. P., Teukolsky, S. A. and Vetterling, W. T. (1986). *Numerical Recipes. The Art of Scientific Computing.* Cambridge University Press, Cambridge.

Pritchard, D. J. and Scott, C. J. (eds) (1990). *Applications of Transputers 2.* IOS, Amsterdam.

Rice, J. R. (1975). A metalgorithm for adaptive quadrature. *J. Assoc. Comput. Mach.* **22**, pp. 61–82.

Robert, Y. (1991). Gaussian elimination on distributed memory architectures. In Spedicato (1991), pp. 253–276.

Saad, Y. (1986). Gaussian elimination on hypercubes. In Cosnard, Robert, *et al.* (1986b), pp. 5–18.

Saad, Y. (1989). Krylov subspace methods on supercomputers. *SIAM J. Sci. Statist. Comput.* **10**, pp. 1200–1232.

Sameh, A. H. (1971). On Jacobi and Jacobi-like algorithms for a parallel computer. *Math. Comput.* **25**, pp. 579–590.

Sameh, A. and Kuck, D. (1978). On stable parallel linear system solvers. *J. Assoc. Comput. Mach.* **25**, pp. 81–91.

Satyanarayanan, M. (1980). *Multiprocessors. A Comparative Study.* Prentice Hall, Englewood Cliffs, NJ.

Schnabel, R. B. (1987). Concurrent function evaluations in local and global optimization. *Comput. Meth. Appl. Mech. Engrng.* **64**, pp. 537–552.

Schnabel, R. B. (1988). *Sequential and parallel methods for unconstrained optimization.* Tech. Rept. CU-CS-414-88, Dept. of Comput. Sci., Univ. of Colorado at Boulder, CO.

Seitz, C. L. (1985). The cosmic cube. *Commun. Assoc. Comput. Mach.* **28**, pp. 22–33.

Seitz, C. L. (1990). Concurrent architectures. In Suaya and Birtwistle (1990), pp. 1–84.

Seitz, C. L. and Flaig, C. M. (1987). *VLSI mesh routing systems.* Tech. Rept. CS-TR-87-5241, California Institute of Technology, CA.

Shampine, L. F. and Watts, H. A. (1969). Block implicit one-step methods. *Math. Comput.* **23**, pp. 731–740.

Shanno, D. F. (1970). Conditioning of quasi-Newton methods for function minimization. *Math. Comput.* **24**, pp. 647 656.

Smith, G. D. (1978). *Numerical Solution of Partial Differential Equations. Finite Difference Methods* (2nd edn). Clarendon Press, Oxford.

Smith, B. T., Boyle, J. M., Dongarra, J. J., Garbow, B. S., Ikebe, Y., Klema, V. C. and Moler, C. B. (1976). *Matrix Eigensystem Routines – EISPACK Guide* (2nd edn). Lecture Notes in Computer Science **6**, Springer-Verlag, Berlin.

Spedicato, E. (1991). *Computer Algorithms for Solving Linear Algebraic Equations.* NATO ASI Series **F77**, Springer-Verlag, Berlin.

Spendley, W., Hext, G. R. and Himsworth, F. R. (1962). Sequential application of simplex designs in optimization and evolutionary design. *Technometrics* **4**, pp. 441–461.

Stone, H. S. (1990). *High Performance Computer Architecture* (2nd edn). Addison-Wesley, Reading, MA.

Straeter, T. A. (1973). *A parallel variable metric optimization algorithm.* NASA Tech. Rept. L-8986, NASA Langley Res. Cent., Hampton, VA.

Strassen, V. (1969). Gaussian elimination is not optimal. *Numer. Math.* **13**, pp. 354 356.

Suaya, R. and Birtwistle, G. (eds) (1990). *VLSI and Parallel Computation.* Morgan Kaufmann, San Mateo, CA.

Swartztrauber, P. N. (1987). Multiprocessor FFTs. *Parallel Comput.* **5**, pp. 197–210.

Sweet, R. A., Briggs, W. L., Oliveira, S., Porsche, J. L. and Turnbull, T. (1991). FFTs and three-dimensional Poisson solvers for hypercubes. *Parallel Comput.* **17**, pp. 121–131.

Tabak, D. (1990). *Multiprocessors.* Prentice Hall, Englewood Cliffs, NJ.

TopExpress (1989). *Occam Procedure Library Reference Manual.* TopExpress, Cambridge.

Torczon, V. J. (1989). *Multi-directional search: a direct search algorithm for parallel machines.* Tech. Rept. 90-7, Rice Univ., Houston, TX.

Torczon, V. J. (1991). On the convergence of the multi-directional search algorithm. *SIAM J. Optimization* **1**, pp. 123–145.

Trew, A. and Wilson, G. (eds) (1991). *Past, Present, Parallel.* Springer-Verlag, London.

Tuazon, J., Peterson, J., Pniel, M. and Liberman, D. (1985). Caltech JPL mark II hypercube concurrent processor. In DeGroot (1985), pp. 666–673.

Valiant, L. G. (1989). Optimally universal parallel computers. In Elliott and Hoare (1989), pp. 17–20.

Valiant, L. G. (1990). A bridging model for parallel computation. *Commun. Assoc. Comput. Mach.* **33**, pp. 103–111.

van de Vorst, J. G. G. (1988). The formal development of a parallel program performing LU-decomposition. *Acta Informatica* **26**, pp. 1–17.

van der Houwen, P. J. and Sommeijer, B. P. (1990). *Parallel ODE solvers.* Rept. NM-R9008, Centrum voor Wiskunde en Informatica, Amsterdam.

van Laarhoven, P. J. M. (1985). Parallel variable metric algorithms for unconstrained optimization. *Math. Prog.* **33**, pp. 68–81.

Van Loan, C. (1992). *Computational Frameworks for the Fast Fourier Transform.* SIAM, Philadelphia, PA.

van Zee, G. A. and van de Vorst, J. G. G. (eds) (1989). *Parallel Computing 1988.* Lecture Notes in Computer Science **384**, Springer-Verlag, Berlin.

Vandevender, W. H. and Haskell, K. H. (1982). The SLATEC mathematical subroutine library. *ACM SIGNUM* **17**, 3 (September), pp. 16–21.

Varga, R. S. (1962). *Matrix Iterative Analysis.* Prentice Hall, Englewood Cliffs, NJ.

von Neumann, J. (1945). *First draft of a report on the EDVAC.* Tech. Rept., Moore School of El. Engrng., Univ. of Pennsylvania, Philadelphia, PA.

von Neumann, J. (1987). First draft of a report on the EDVAC. In Aspray and Burks (1987), pp. 17–82.

Wait, R. and Mitchell, R. (1985). *Finite Element Analysis and Applications.* Wiley, Chichester.

Wallace, D. J. (1989). Scientific computation on SIMD and MIMD machines. In Elliott and Hoare (1989), pp. 125–142.

Wang, H. H. (1981). A parallel method for tridiagonal equations. *ACM Trans. Math. Software* **7**, pp. 170–183.

Weicker, R. P. (1984). Dhrystone: a synthetic systems programming benchmark. *Commun. Assoc. Comput. Mach.* **27**, pp. 1013–1030.

Welsh, J. and Bustard, D. W. (1979). Pascal Plus – another language for modular multiprogramming. *Software – Practice and Experience* **9**, pp. 947–957.

Wilkes, M. V. and Renwick, W. (1949). The EDSAC, an electronic calculating machine. *J. Sci. Instrum.* **26**, pp. 385–391.

Worland, P. B. (1976). Parallel methods for the numerical solution of ordinary differential equations. *IEEE Trans. Comput.* **C-25**, pp. 1045–1048.

Wright, K. (1991). Parallel algorithms for QR decomposition on a shared memory multiprocessor. *Parallel Comput.* **17**, pp. 779–790.

Zenios, S. A. (1989). Parallel numerical optimization: current status and an annotated bibliography. *ORSA J. Comput.* **1**, pp. 20–43.

Zmijewski, E. and Gilbert, J. R. (1988). A parallel algorithm for sparse symbolic Cholesky factorization on a multiprocessor. *Parallel Comput.* **7**, pp. 199–210.

Index

Where there are multiple entries, principal entries are in **bold face** type.